SPECIAL ALGEBRA
FOR
SPECIAL RELATIVITY

SECOND EDITION

The other book by Paul C Daiber:

ALIEN INVASION MATH STORY

SPECIAL ALGEBRA
FOR
SPECIAL RELATIVITY

Proposed Theory of Non-Finite Numbers

Paul C Daiber

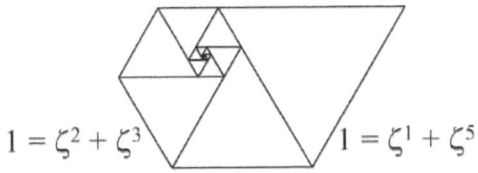

SECOND EDITION

Copyright © 2020 by Paul C Daiber

All rights reserved. This book or any portion thereof may not be reproduced or used in any manner whatsoever without the express written permission of the publisher except for the use of brief quotations in a book review, scholarly journal or other critical document.

Daiber, Paul C, 1960 –

Special Algebra for Special Relativity, Second Edition

p. cm.

Includes index.

Paperback ISBN   9798698633808

1. Special Relativity, Electricity, Waves, Algebra, Mathematics, Infinity, Math, Abstraction   I. Title.

A special thank you to amazon for making it simple to publish a book

For My Wife Sue

Table of Contents

Finite Imprecision for Numbers x

Chapter 1 – Numbers 1

1.1 Process from Descartes 1

1.2 Geometric-Vectors 2

1.3 Quaternions 4

1.4 Translation Back to Geometry 19

1.5 Singular-Label-Numbers 20

1.6 Exercises 21

Chapter 2 – Particles 35

2.1 Hypercomplex-Plane 35

2.2 Inertial Reference Frames 37

2.3 The Unspecified-Speed-Parameter 39

2.4 Compound-Label-Numbers and Components 40

2.5 Adding Hyperbolic-Angles 44

2.6 Energy, Time Dilation, Length Contraction 47

2.7 Space-Like and Time-Like Invariants 49

2.8 Electric Current Density 54

2.9 Motion Faster than Light 56

2.10 Anti-Matter 65

2.11 Distributed Material Theory 72

2.12 Exercises 83

Chapter 3 – Fields 89

3.1 Geometric-Vector Notation 89

3.2 All-Number Notation 94

3.3 Gauges and Super-Potentials 105

3.4 Lorentz Transformation 107

3.5 Biot-Savart Law 116

3.6 Electric Energy-Momentum of an Electron 118

3.7 Maxwell's Wave Equation 125

3.8 Forces Using Geometric-Vector Notation 130

3.9 Force Density Invariant 132

3.10 Area and Volume Differential Operators 142

3.11 Exercises 150

Chapter 4 – Waves 157

4.1 Differential Operator 157

4.2 Development of the Dirac Equation 160

4.3 Solutions to the Dirac Equation 164

4.4 Particle Properties 167

4.5 Two Alternative Arrangements 171

4.6 Lorentz Transformation of a Dirac Spinor 173

4.7 Exercises 178

Chapter 5 – Proposed Theory 185

5.1 Cantor's Theory of Infinite Sets 186

5.2 Idealized Real Numbers 195

5.3 Practical Real Numbers 199

5.4 Application to Special Relativity 203

5.5 Application to Dirac Spinors 215

5.6 Dirac Equation Form - Development	225
5.7 Force Density Using the Complex-Conjugate	233
5.8 Spin of a Photon	241
5.9 Exercises	244
Appendix A – Octonions and Sedonions	253
Appendix B – Spooky Action at a Distance	271
Appendix C – Discovering an Abstraction	281
The Storybook	286
Glossary	287
Index	299
Back Cover	302

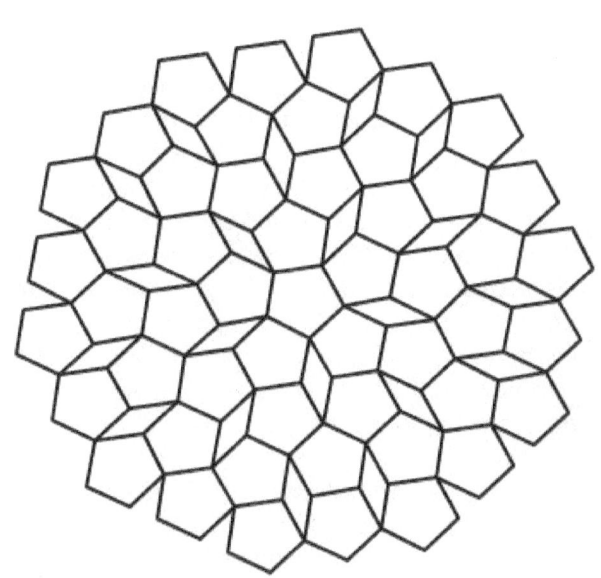

Preface

<u>Personal Note</u>. When I was a kid infinity intrigued me so much that I started digging. I wanted to know what infinity really was.

Decades passed without finding a cleanly explained set of properties for understanding and using infinity. Bookstores had, typically, a layman's explanation of Cantor's two infinities. Deeper, more professional presentations masked confounding complexity behind strange mathematical symbols. The cliché that anything well understood is easily explained applied in its contrapositive, and the cause for poor explanations appeared to be holes in logic caused by infinity's illusiveness.

Holes in logic appalled me because math was supposed to conform to logic. At first, I thought mathematicians had gone mad because they had made Cantor's infinity a belief, like a religion, by ignoring contrary evidence. But the reality why Cantor's infinity was so popular was constructive: There was no alternative to Cantor's infinity, not yet, and pure mathematics appeared to be an experimental playground, a temporary place that, eventually after modifications evolved, should transition into applied mathematics. I wanted the holes fixed so that that transition could happen, and that meant my mission transformed from researching existing theory into creating new theory.

Where to start? Proofs that "$\sqrt{2}$" and "$\log_2 3$" were irrational implicitly proved infinity was real. And legendary thinkers like Aristotle and Galileo provided speculation about what infinity was.

Einstein's motion at the speed-of-light was modelled mathematically by a division by zero, which was countered to make photon momentum finite by saying mass of a photon was zero. That calculation required zero divided by zero. I fumbled over that problem for decades, but finally, around year 2010, I found a crude property for infinity that united electromagnetic fields with matter waves, and success appeared possible. Now, a decade later, the modified math is finished. In this textbook I've written the proposed properties for infinity along with the sample application to electromagnetic fields.

Personal curiosity now satisfied, I'm hoping people read this textbook so that I can have contributed to the state-of-the-art of math by adding a step to our evolving awareness of the universe.

Returning to my obsessive wants, I'd like other people to use infinity to find actual revolutionary physics, and that's because I'd really like to read what they discover. I am a nobody, an outsider all my life, alone, with respect to math, not even in academia, and using me as an example, know that it could be you who makes the next discovery, no matter who you are. \\/\/,

Technical Summary. The first four chapters of *Special Algebra for Special Relativity* present an all-number mathematical structure for Special Relativity. The fifth chapter modifies Axiomatic Set Theory to restrict a real number to finite precision by limiting place-value digits to a maximum count before and after the decimal point. To complement finite precision, finite imprecision is formed by unspecified place-value digits beyond the maximum count.

Finite imprecision after the decimal point is finite, but is very, very small, and so is trivial. Finite imprecision to the left (before) the decimal point is not trivial and is called large-scale imprecision. In the traditional construction of real numbers, the zeros extending left of the decimal point are assumed to go to infinity and beyond, but in this book's proposed reconstruction of real numbers, there is no actual infinity and so those zeros extend only to a finite quantity.

In Special Relativity, large-scale imprecision is added to time-space hyperbolic angle "α" (alpha, that relates to speed "v" by "v = c*tanhα") using a Lorentz Transformation. Large-scale imprecision models electromagnetism after deriving Maxwell's Equations from the Dirac Equation, giving us a quantum model of photon field dynamics, something we don't currently have.

Electromagnetic field force density components are calculated using the same process by which electric current density components are calculated. Included are energy density and Poynting Vector components. Uniting those three empirically derived electromagnetic phenomena into one mathematical model is a new revelation in physics and that success suggests quantities in our geometric world actually do have finite imprecision and suggests finite imprecision real numbers should also apply to more modern theories of physics.

Axiomatic Set Theory is altered by removing the Axiom of Choice and by replacing Cantor's positive countable actual infinity (aleph null) with a time-dependent increasing largest natural number yet counted-to. A practical real number is then restricted to finite precision.

More formality is constructed by replacing Cantor's positive uncountable infinity (aleph one) with reciprocal of zero (supported by axioms) in a Modified Continuum Hypothesis, so that no positive actual infinities can be quantities of sets. An idealized real number is then given integer zero interval to the next.

Cantor's original construction of numbers had been incomplete because it had holes in logic that prevented it from transitioning from pure mathematics into applied mathematics. The proposed changes in this book remove dependency on actual infinity, and therefore remove holes in logic, and therefore complete Cantor's task of defining real numbers, to make real numbers useful.

SPECIAL ALGEBRA FOR SPECIAL RELATIVITY
Finite Imprecision for Numbers

<u>Summary</u>. Axiomatic Algebra Without Actual Infinity

This book broadens applied mathematics from Group Theory to Axiomatic Set Theory by removing actual infinity, because removing actual infinity removes paralyzing holes in logic. Transitioning Axiomatic Set Theory from temporary, experimental pure mathematics into useful applied mathematics completes the formalization of real numbers started by Cantor 150 years ago.

- <u>Modify Cantor's Continuum Hypothesis</u> by replacing countable infinity with an ever-increasing finite counting natural number and by replacing uncountable infinity with reciprocal of zero, so that positive actual infinity cannot be a quantity of a set.

- <u>Replace "Don't Divide by Zero"</u> of rational numbers with two axioms that permit $1/0=7+1/0$, $1/0=7*1/0$, and $2^{1/0}=3^{1/0}=1/0$ or 0, but not 0/0 because that leads to errors, and have $1/0$ the quantity of a set.

- <u>Remove Axiom of Choice</u> so that only bulk operations, like truncation to form an integer, apply to $1/0$ quantities. Removing the axiom of choice means there can be no individual operations to a quantity beyond the largest number yet counted-to.

- <u>Real Numbers have only a finite count</u> of knowable place-value digits both before and after the decimal point. Left of the string of zeros unknowable digits form large-scale imprecision which is akin to infinity (which is not the quantity of set) and can be analyzed using finite numbers.

- <u>Special Relativity</u>. A Lorentz Transformation adds large-scale imprecision to the time-space hyperbolic angle to derive Maxwell's Equations from the Dirac Equation. Particle properties conform to measured electromagnetic field force and energy.

- <u>That successful application</u> of the axioms to discover a quantum model for 100+ year-old electromagnetism substantiates the new axiomatic algebra is correct and is available for modern in-development theories of physics.

SYNOPSIS – FINITE IMPRECISION FOR NUMBERS

State-of-the-Art. The state-of-the-art of algebra has three axiomatic formulations, each with their own version of infinity.

- From the late 1800's, <u>Group Theory</u> pertains to finite numbers and is the only formulation applied to models of physics. The implied infinity is the unboundedness of numbers used for counting.

- More encompassing <u>Axiomatic Set Theory</u> (1908) pertains to actual infinity through the Axiom of Choice and through Cantor's Continuum Hypothesis used as a partner axiom. Applied mathematics has not utilized the Axiom of Choice or the Continuum Hypothesis and therefore has not found use for actual infinity, even though physics has identified a few division-by-zero singularities where an actual infinity seemingly should apply.

- <u>Division by zero</u> is not addressed in Axiomatic Set Theory. Division by zero is included in Group Theory only in a prohibition against division by zero in the definition of rational numbers. There is a general rule "Don't divide by zero" which is justified by errors caused by some uses of division by zero.

State-of-the-Art – Actual Infinities. Aristotle had two infinities: "potential infinity" for the unboundedness of natural numbers, and a larger "completed infinity". To visualize these infinities, Aristotle imagined dividing a line in half again and again. Galileo dropped the notion of a finite potential infinity for his two infinities. Galileo proposed the quantity of squares of natural numbers had the same quantity as the natural numbers even though squares needed bigger numbers, and for the second infinity, Galileo proposed that two line segments of different lengths had the same quantity of points.

In the early 1870's, Cantor continued the two infinities concept with his countable infinity and his uncountable infinity. The quantity of real numbers over an interval of the number-line from zero to a finite real number is Cantor's uncountable infinity "N_1", called aleph one. He proposed the uncountable infinity is calculated from the countable infinity "N_0", called aleph null, using a base of two, "$N_1 = 2^{\wedge}N_0$". The countable infinity "N_0" is the quantity of natural numbers, of integers, of rational numbers, etc. "N_0" and "N_1" are made larger than finite numbers by Cantor's Continuum Hypothesis: There is no set with a quantity between "N_0" and "N_1". Because of the equation "$N_1 = 2^{\wedge}N_0$" and because the Continuum Hypothesis requires "$N_0 < N_1$", both "N_0" and "N_1" are positive. It follows that if there are a quantity "N_1" of real numbers from zero to one, or from zero to "$\sqrt{2}$", then Cantor's real numbers have a spacing of positive "$1/N_1$", and not a spacing of integer zero.

SPECIAL ALGEBRA FOR SPECIAL RELATIVITY

Cantor provided proofs. His "diagonal proof" proved the set of rational numbers has a quantity "N_0". His long string of digits proof proved "$N_0 < N_1$". In both those proofs, only finite numbers are provided, and the person is expected to make the conceptual leap from finite to his actual infinity quantities.

Cantor had put formality to the notions of infinity that preceded him, and further formality occurred when Axiomatic Set Theory evolved out of his work on sets. Axiomatic Set Theory addresses infinity in three of its axioms:

- <u>Axiom of Infinity</u> counts nested null sets to formally introduce the unboundedness of natural numbers ("1, 2, 3, ..."). It appears null sets are used because number quantities are made devoid of connection to actual objects. The Axiom of Infinity applies to the largest finite number yet counted-to, and to all smaller natural numbers, but not to an actual infinity.

- <u>Axiom of Choice</u> allows the mathematician to assume all place-value digits for an irrational number product, for example "$\sqrt{2}*\log_2 3$", exist, even those beyond a quantity that can be counted, or beyond the countable infinity, all.

- <u>Cantor's Continuum Hypothesis</u> is used as an axiom.

In parallel to Axiomatic Set Theory, there are the axioms of Group Theory, for which an algebra for rational numbers requires:

- <u>Closed</u> and <u>inverses</u>: A sum or product of two rational numbers is a rational number, as well as negatives and reciprocals (except division by zero).

- <u>Identity Elements</u> are zero (for addition) and one (for multiplication).

- <u>Commutative</u> and <u>associative</u> properties for addition and multiplication as well as the <u>distributive</u> property of multiplication over addition.

Because there is no actual infinity in Group Theory, the proofs of irrationality of "$\sqrt{2}$" and "$\log_2 3$" only state numbers in a ratio "p/q" are larger than what can be counted-to.

- <u>Proof "$\sqrt{2}$" is irrational</u>: "$\sqrt{2} = p/q$, $2 = p^2/q^2$, $p^2 = 2*q^2$, $q^2 = 2*(p/2)^2$". Observe that p and q must both be even numbers. Also, in a ratio "p/q" both can be divided by two until one of them becomes odd, so that p or q must be able to be odd. The two observations conflict, therefore p and q are not finite.

Synopsis – Finite Imprecision for Numbers

- Proof "$\log_2 3$" is irrational: "$\log_2 3 = p/q$, $q*\log_2 3 = p$, $\log_2 3^q = p$, $3^q = 2^p$". Observe that odd number "3^q" cannot equal even number "2^p", and therefore p and q are not finite.

    Group Theory is extended to "real numbers", so that irrational numbers like "$\sqrt{2}$", "$\log_2 3$", "π", and "e" can be used in calculations.

Holes in Logic. Is "$\sqrt{2}$" included in Cantor's set of real numbers? Because countable infinity "aleph null", "N_1", is the quantity of points in the interval from "0" to "1" as well as from "0" to "$\sqrt{2}$", the ratio is "$\sqrt{2} = N_1/N_1$". Because "$N_1 = 2\wedge N_0$", it appears that "N_1" must be even and cannot be odd. Therefore, no, it appears that "$\sqrt{2}$" is not in Cantor's set of real numbers. The word "appears" applies because perhaps "N_1" was mistakenly assumed to be a finite number somewhere in this proof, perhaps because "N_1" cannot be divided by two. Similarly, proving "$\log_2 3$" is in Cantor's set of real numbers requires operational definition for "$2\wedge N_1$", but also for "$3\wedge N_1$" which does not exist per any of the axioms. There is also the open question that if a real number spans a non-integer-zero interval of "$1/N_1$" on the number-line, then does real number zero equal integer zero? And what is the value of "$\log_2 N_0$", or of "$3\wedge N_0$"?

An Algebra for Dividing by Zero. The prohibition against division by zero is formally only valid for rational numbers. But because division by zero can be used as a trick, division by zero is given the name pseudo-mathematics and shunned. Perhaps this shunning is why Dedekind and Cantor did not propose "N_1" to be "1/0" for an integer zero interval from one real number to the next.

To lighten-up on the informal prohibition against division by zero, errors are more directly targeted in a proposed Axiom of Reciprocal of Zero: No operation that includes reciprocal of integer zero can result in a non-zero finite number. Per the proposed axiom, "0/0" and "1/0 - 1/0" are not permitted operations. Also "1/0 + 1/0" is not permitted because the positive or negative feature of "0" is unspecified. Given this proposed axiom, an algebra for reciprocal of zero is possible: "1/0 + 7 = 1/0" and "(1/0)*7 = 1/0" are accepted, as well as "$2^{1/0} = 3^{1/0} = 0$ or 1/0". "1/0" is the absolute maximum magnitude of numbers and is both or either positive and negative. It was curious that our English language has the word "vertical", but mathematics had no equivalent concept.

The proposed algebra for "1/0" can be applied to the two proofs of irrational numbers. First, note that because "0/0" is not permitted, the ratio is written "$0*\sqrt{2} = 0$" and "$0*\log_2 3 = 0$", and from Group Theory we can accept that integer zero times an irrational number equals integer zero.

SPECIAL ALGEBRA FOR SPECIAL RELATIVITY

The ratio is also written "$(1/0)*\sqrt{2} = 1/0$" and "$(1/0)*\log_2 3 = 1/0$", and for these two equations propose a second axiom, the Axiom of Reciprocal of Zero being a Quantity of a Set: Reciprocal of zero is a quantity of a set such that it has no contribution after the decimal point. This axiom appears needed for "$2^{\wedge}(1/0) = 3^{\wedge}(1/0) = 0$ or $1/0$" because axioms don't define the exponent operation if there are unending random place-value digits after the decimal point in the exponent. The Axiom of Reciprocal of Zero being a Quantity of a Set holds true regardless of "$1/0 = 1/0 + \sqrt{2}$", and because this proposed axiom is not self-evident, it can be argued about. Given this second proposed axiom, "$(1/0)*\log_2 3 = 1/0$" states that "$1/0$" is the smallest number multiplied by irrational number "$\log_2 3$" so that product "$(1/0)*\log_2 3$" does not have a contribution after the decimal point. These two proposed axioms describe an irrational number with all its place-value digits, so that place-value digits extend not only unboundedly in a count, but indefinitely because of the indefiniteness of "$1/0$" through its algebra, and there is integer zero interval from one real number to the next. Because this algebra includes "$1/0$", these real numbers are called "idealized" real numbers. "Idealized" means we imagine their properties.

Referring to the two proofs of irrationality: For "$\sqrt{2}$", the number in the ratio is "$1/0$" and because of "$1/0 = 1/0 + 1$", "$1/0$" is both even and odd, and so fits the two observations without conflict. For "$\log_2 3$", "$2^{\wedge}(1/0) = 3^{\wedge}(1/0)$", and so the proof is satisfied. The close association of "$1/0$" to irrational numbers suggests "$1/0$" should be classified as an irrational number, although it is special.

Removing Uncountable Infinity. If "$(1/0)*\log_2 3$" has no contribution after the decimal point and "$1/0$" is the smallest number as a factor for which this is possible, then positive number "N_1" used as a factor for "$\log_2 3$" creates product "$N_1*\log_2 3$" that includes contribution after the decimal point, and so cannot equal "N_1". That proof of the irrelevance of "N_1" required the two axioms of "$1/0$".

Practical Mathematics. For practical, as opposed to idealized, mathematics, the Axiom of Choice is removed. Without the Axiom of Choice, the place-value digits beyond the largest number yet counted-to in the product of irrational numbers "$\sqrt{2}*\log_2 3$" cannot be known. But, more generally, there is no formal means of knowing place-value digits of any irrational number further than the largest number yet-counted-to after the decimal point. Even more generally, an integer like "0" or "1" if placed into the continuum of a number-line, is precise only down to a count of zeros after the decimal point equal to the largest number yet counted-to, which is a finite number. Rational numbers, too, have their repeat pattern ended at that finite count.

SYNOPSIS – FINITE IMPRECISION FOR NUMBERS

Replace Cantor's countable infinity "N_0" with the largest number yet counted-to, "Lmax", and replace uncountable infinity "N_1" with "1/0", to create the proposed Modified Continuum Hypothesis: There are no sets with quantity between "Lmax" and "1/0". This proposed Modified Continuum Hypothesis theorem removes the ability for there to be countable "N_0" or uncountable "N_1" because those were positive actual infinities that were quantities of sets.

Small-Scale Imprecision. Without the Axiom of Choice, a real number has a finite count of place-value digits after the decimal point, and therefore, has finite precision. The unknown and unknowable place-value digits beyond that count comprise finite imprecision, what is called "small-scale imprecision", or "local-zero" with the word "local" referring to the value of "Lmax" being local to the information represented. As a symbol, use double zero "00". In base two, the magnitude of small-scale imprecision is less than base two to the negative "Lmax" exponent, " $|\text{"00"}| < 2^{-Lmax}$ ". If finite imprecision is only positive, then the absolute value operator is not needed, and " $\text{"00"} < 2^{-Lmax}$ ".

Small-scale imprecision is written: " "00" = 0.000...bbbb...... ". Three dots "..." represent zeros extending as far as has been counted-to (to "Lmax"), and six dots "......" represent the unknown and unknowable place-value digits extending to all the larger natural numbers (to "1/0").

To visualize the six dots, create integer "1" from "1.4142......" by truncating "1.4142......" at the decimal point and discarding the unwanted ".4142......". That operation of truncation and discard was a bulk operation that did not require counting. A counting of place-value digits, that is, addressing digits one at a time for individual operations, cannot occur where there are six dots.

Symbol "b" in base two is both/either a "0" or a "1". The selection of "0" or "1" occurs when an observation is made. The observation is a reference to Schrödinger's Cat. The cat is inside a box and if a radioactive nucleus decays, the emitted particle fractures a vial of poison, and the cat dies. Alive is state "0" and dead is state "1". A radioactive decay is an event that is unique to an observing particle, and that observing particle is not us until we open the box to look. "b" is for "box" or is for "both". Each "b" symbol is converted to either a "0" or else a "1" from left to right as time progresses.

Using "b" symbols the small-scale imprecision is positive. To generalize, a symbol "d" can be used as an alternative, for which "d = b - b". The "b" boxes are still opened one at a time from left to right, but now there are two "b" boxes for each place-value digit, and the result "d" can be negative or positive. Math alone does not presently suggest which is correct, "b"s or "d"s, and perhaps a future application will settle which. (The application to electromagnetism suggests "b"s and not "d"s.)

SPECIAL ALGEBRA FOR SPECIAL RELATIVITY

<u>Large-Scale Imprecision</u>. Not only is the count after the decimal point limited to the largest number yet counted-to, "Lmax", but also the count before the decimal point. To the left are a few non-zero place-value digits and further left are all zeros. It is typically assumed the zeros extend to infinity and beyond, to "all" which is a "1/0" quantity, but in practical real numbers there is only a finite quantity. The unknown and unknowable place-value digits starting at count "Lmax + 1" and extending to "1/0" don't exist and cannot be assumed to be zero. To visualize these digits as a probability distribution, use the division reciprocal of the small-scale imprecision, but that might not be correct in models of physics. Applications to physics should eventually identify what the probability distribution is, and if it is positive only or is negative/positive.

As a symbol, use "$\Omega\Omega$". The magnitude of large-scale imprecision is greater than base two to the "Lmax" exponent, " $|\text{``}\Omega\Omega\text{''}| > 2^{Lmax}$ ". If finite imprecision is only positive, then the absolute value operator is not needed, and " "$\Omega\Omega$" > 2^{Lmax} ". As a name, large-scale imprecision can be called local-infinity, but note that it is defined in a way that means it cannot be the quantity of a set. Large-scale imprecision can be analyzed using finite number theory because in calculations we can pretend numbers greater than "Lmax" exist.

Both small-scale and large-scale imprecision are non-definite because of the use of "b"s rather than numbers for place-value digits, and aren't actual infinitesimal and infinite, respectively, but rather are something else non-finite.

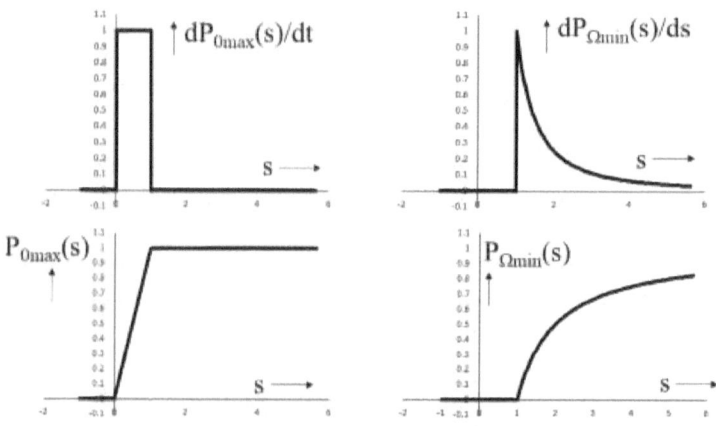

Figure A. Probability Distribution for "Lmax = 0" for small-scale imprecision " "00" = 0.bbbbb...... " left and large-scale imprecision illustrated as " "$\Omega\Omega$" = 1/"00" " right. Higher values of "Lmax" make the square thinner and taller and make the reciprocal further to the right.

SYNOPSIS – FINITE IMPRECISION FOR NUMBERS

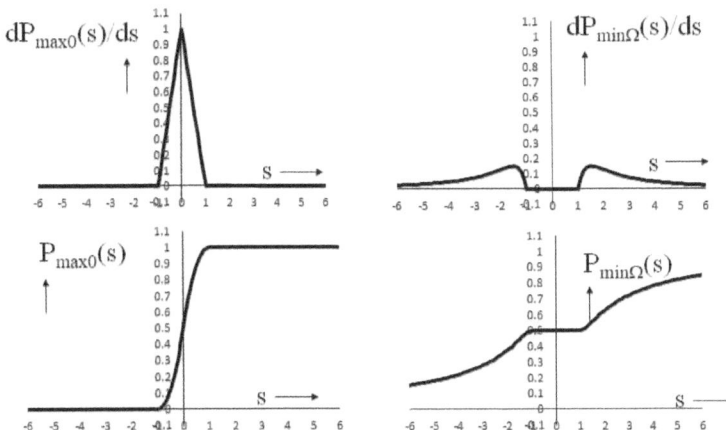

Figure B. Probability Distribution for "Lmax = 0" for small-scale imprecision " "00" = 0.dddd......" left and large-scale imprecision illustrated as " "ΩΩ" = 1/"00" " right. Higher values of "Lmax" make the spike taller and make the two arms further apart.

Truncated Number. A real number is a rational number limited to "Lmax" place-value digits before and after the decimal point, plus either large-scale or else small-scale imprecision. The rational number is given the name "truncated number". "Local-real number" is an alternative name to "practical real number". Because imprecision is best visualized in base two, truncated numbers, too, should be thought of in base two.

Summary of Axiom Changes. In summary, the axioms of Axiomatic Set Theory have been changed to remove actual infinity. This was done in two steps, the first created idealized real numbers which we can imagine, and the second created practical real numbers which apply to models of physics and have finite large-scale and small-scale imprecision.

Replacement of the countable infinity with the largest number yet counted-to ("Lmax") has coupled mathematics with physics because of the time dependency of counting.

Replacement of uncountable infinity "N_1" with "1/0" retained the indefiniteness property that the uncountable infinity had, and by careful use of axioms to not make errors, explicit use of "1/0" broke the paradigm of don't divide by zero.

SPECIAL ALGEBRA FOR SPECIAL RELATIVITY

Previously, physics was modelled only with the axioms of Group Theory because Group Theory pertained to only finite numbers, but now, with these modifications to Axiomatic Set Theory, Axiomatic Set Theory also only pertains to finite numbers (and "1/0"), and that means models of physics can use broader axiomatic mathematics. Specifically, large-scale imprecision is something new.

Application Summary. The application (that proves the validity of the proposed practical real numbers) is deriving a quantum model of photon dynamics (which includes Maxwell's Equations) after starting from the Dirac Equation. In this application, the Dirac Equation is Lorentz Transformed to an observer at the speed-of-light by adding large-scale imprecision to a hyperbolic angle "α" (which relates to speed "v" per "$v = c*\tanh(\alpha)$").

Particle properties are calculated from the Dirac Spinor. The particle property of electromagnetic field force-energy-density derived using large-scale imprecision is calculated in parallel to the particle property of current density derived using small-scale imprecision. Included in the force-energy-density is the energy density calculated as electric field squared plus magnetic field squared, and included is the photon momentum density Poynting Vector calculated as electric field cross magnetic field. The combination of these separately empirically discovered models of physics into one mathematical model is new, and that new revelation in modelling physics suggests the proposed mathematics for practical real numbers is valid and is available for more modern theories of physics.

Lorentz Transformation for a Particle. Mechanical energy-momentum components and time-space location components for an electron are given below, respectively.

$$m_B*c*\cosh\alpha_M + q*m_B*c*\sinh\alpha_M \quad ; \quad c*t_B*\cosh\alpha_M + q*c*t_B*\sinh\alpha_M$$

"m_B" is electron rest mass. "t_B" is time on a clock mounted on the electron. "c" is speed-of-light. Electron speed "v_M" relates to hyperbolic angle "α_M" by "$v_M = c*\tanh\alpha_M$". "q" is a combination of 2x2 Pauli Spin Matrices conforming to "$q^2 = 1$" and is analogous to direction in space.

Reference frame "$_B$" is stationary with respect to the particle. Reference frame "$_M$" is moving, for example, the interior of a bus in which the electron moves along the floor toward the front with speed "v_M". Reference frame "$_S$" is stationary, for example, the roadside where observer (you or me) is located. Speed of the bus relative to the roadside is "$v_{S/M} = c*\tanh\alpha_{S/M}$".

Hyperbolic angle "α_M" is a rational truncated number.

xix

SYNOPSIS – FINITE IMPRECISION FOR NUMBERS

Use Lorentz Transformation angle "$\alpha_{S/M} = (1 - q)*\Xi$". Imprecision term "$\Xi$" (xi) is either a local-infinity "$\Omega\Omega$" or else a local-zero "00". The "$1 - q$" factor keeps components finite and applies because the logarithm of rest mass is a truncated number, as is the logarithm of time measured on the electron's clock.

Lorentz Transformation "$\alpha_S = \alpha_M + \alpha_{S/M}$" has a large-scale imprecision case for "Ξ" being "$\Omega\Omega$" and a small-scale imprecision case for "Ξ" being "00". The small-scale case is trivial because "00" is nearly zero, so that "$\alpha_{S/M}$" is effectively zero such that the electron when observed in "$_S$" is the same as when observed in "$_M$". In contrast, the large-scale case has the electron observed as a photon because energy component "$E_S/c = m_B*c*\exp(\alpha_M)/2$" equals momentum component "$p_{xS} = m_B*c*\exp(\alpha_M)/2$", and time component "$c*t_S = c*t_B*\exp(\alpha_M)/2$" equals space component "$x_S = c*t_B*\exp(\alpha_M)/2$". The "1/2" factor connects the two cases with a speed-of-light signal.

$$m_B*c*\cosh\alpha_S + q*m_B*c*\sinh\alpha_S$$
$$\approx m_B*c*\exp(\alpha_M)/2 + q*m_B*c*\exp(\alpha_M)/2 \quad \text{large-scale case}$$

$$c*t_B*\cosh\alpha_S + q*c*t_B*\sinh\alpha_S$$
$$\approx c*t_B*\exp(\alpha_M)/2 + q*c*t_B*\exp(\alpha_M)/2 \quad \text{large-scale case}$$

Lorentz Transform for a Wave. Electron dynamics are modeled by the Dirac Equation. The Dirac Equation is developed below with a large-scale and a small-scale case. Begin with a relationship of momentum components per the Pythagorean Theorem, and then split it into two equations to form a matrix equation.

$$(m_B*c*(1 + \cosh\alpha_S)*PP_S)*(m_B*c*(-1 + \cosh\alpha_S)*-QQ_S)$$
$$= (q*m_B*c*\sinh\alpha_S*-QQ_S)*(q*m_B*c*\sinh\alpha_S*PP_S)$$

$$\begin{pmatrix} m_B*c*(1 + \cosh\alpha_S) & q*m_B*c*\sinh\alpha_S \\ q*m_B*c*\sinh\alpha_S & m_B*c*(-1 + \cosh\alpha_S) \end{pmatrix} * \begin{pmatrix} PP_S \\ QQ_S \end{pmatrix} = \begin{pmatrix} 0 \\ 0 \end{pmatrix}$$

For the large-scale case, rest mass "m_B" terms are zero because the "q" in "$\alpha_{S/M} = (1 - q)*\Xi$" creates zero factor "$\exp(-\text{"}\Omega\Omega\text{"})$". And infinity factor "$\cosh(\text{"}\Omega\Omega\text{"}) + q*\sinh(\text{"}\Omega\Omega\text{"})$" cancels that zero factor so that "$E_S/c = m_B*c*\cosh\alpha_S = m_B*c*\exp(\alpha_M)/2$" and "$p_{xS} = m_B*c*\sinh\alpha_S = m_B*c*\exp(\alpha_M)/2$". To account for electric charge, the right-side is given non-zero value "a".

SPECIAL ALGEBRA FOR SPECIAL RELATIVITY

$$\begin{pmatrix} (m_B*c*\exp(\alpha_M)/2) & (q*m_B*c*\exp(\alpha_M)/2) \\ (q*m_B*c*\exp(\alpha_M)/2) & (m_B*c*\exp(\alpha_M)/2) \end{pmatrix} * \begin{pmatrix} PP_S \\ QQ_S \end{pmatrix} = \begin{pmatrix} a \\ q*a \end{pmatrix}$$

Mechanical equals total minus electrical.

$$m_B*c*\cosh(\alpha_S) = i*\hbar*\nabla_{tS} - Q_B*V_{tS}$$

$$q*m_B*c*\sinh(\alpha_S) = q_x*(-i*\hbar*\nabla_{xS} - Q_B*V_{xS}) \\ + q_y*(-i*\hbar*\nabla_{yS} - Q_B*V_{yS}) + q_z*(-i*\hbar*\nabla_{zS} - Q_B*V_{zS})$$

"$\nabla_{tS} = \partial/\partial ct_S$", "$\nabla_{xS} = \partial/\partial x_S$", "$\nabla_{yS} = \partial/\partial y_S$", and "$\nabla_{zS} = \partial/\partial z_S$". "V" components form the external voltage invariant. "$\hbar$" is Planck's constant.

For the large-scale case, zero electron rest mass "m_B" coincides to zero electron electric charge "Q_B", as can be derived from the classic radius of an electron calculation, and the above operators reduce to:

$$m_B*c*\exp(\alpha_M)/2 = i*\hbar*\nabla_{tS}$$

$$q*m_B*c*\exp(\alpha_M)/2 = q_x*(-i*\hbar*\nabla_{xS}) + q_y*(-i*\hbar*\nabla_{yS}) + q_z*(-i*\hbar*\nabla_{zS})$$

Pauli Spin Matrices substitute for "q_x", "$-q_y$", and "q_z". Divide by "$i*\hbar$".

$$q_x \Rightarrow \begin{pmatrix} 0 & 1 \\ 1 & 0 \end{pmatrix} \quad q_y \Rightarrow \begin{pmatrix} 0 & i \\ -i & 0 \end{pmatrix} \quad q_z \Rightarrow \begin{pmatrix} 1 & 0 \\ 0 & -1 \end{pmatrix} \quad 1 \Rightarrow \begin{pmatrix} 1 & 0 \\ 0 & 1 \end{pmatrix}$$

$$\partial/\partial ct \begin{pmatrix} 1 & 0 & 0 & 0 \\ 0 & 1 & 0 & 0 \\ 0 & 0 & 1 & 0 \\ 0 & 0 & 0 & 1 \end{pmatrix} + -\partial/\partial x \begin{pmatrix} 0 & 0 & 0 & 1 \\ 0 & 0 & 1 & 0 \\ 0 & 1 & 0 & 0 \\ 1 & 0 & 0 & 0 \end{pmatrix} + -\partial/\partial y \begin{pmatrix} 0 & 0 & 0 & i \\ 0 & 0 & -i & 0 \\ 0 & i & 0 & 0 \\ -i & 0 & 0 & 0 \end{pmatrix}$$

$$+ -\partial/\partial z \begin{pmatrix} 0 & 0 & 1 & 0 \\ 0 & 0 & 0 & -1 \\ 1 & 0 & 0 & 0 \\ 0 & -1 & 0 & 0 \end{pmatrix} * \begin{pmatrix} \Psi_{1\_LargeCase} \\ \Psi_{2\_LargeCase} \\ \Psi_{3\_LargeCase} \\ \Psi_{4\_LargeCase} \end{pmatrix} = \begin{pmatrix} \Phi_{1\_LargeCase} \\ \Phi_{2\_LargeCase} \\ \Phi_{3\_LargeCase} \\ \Phi_{4\_LargeCase} \end{pmatrix}$$

"PP_S" and "$QQ_S = q*PP_S$" were replaced by the Dirac Spinor (Ψ) (psi). Similarly, "a" and "$q*a$" were replaced by spinor (Φ) (phi).

SYNOPSIS – FINITE IMPRECISION FOR NUMBERS

Electric field components are "E". Magnetic field "B". Electric current density components are "J", with adjusted measurement units.

$$\Psi_{1\_LargeCase} = \Psi_{3\_LargeCase} = E_z + i*c*B_z$$
$$\Psi_{2\_LargeCase} = \Psi_{4\_LargeCase} = (E_x + i*c*B_x) - i*(E_y + i*c*B_y)$$
$$\Phi_{1\_LargeCase} = \Phi_{3\_LargeCase} = -J_t - J_z$$
$$\Phi_{1\_LargeCase} = \Phi_{3\_LargeCase} = -J_x + i*J_y$$

First Portion:

$$\partial/\partial ct \begin{vmatrix} 1&0&0&0 \\ 0&1&0&0 \\ 0&0&1&0 \\ 0&0&0&1 \end{vmatrix} + -\partial/\partial x \begin{vmatrix} 0&0&0&1 \\ 0&0&1&0 \\ 0&1&0&0 \\ 1&0&0&0 \end{vmatrix} + -\partial/\partial y \begin{vmatrix} 0&0&0&i \\ 0&0&-i&0 \\ 0&i&0&0 \\ -i&0&0&0 \end{vmatrix}$$

$$+ -\partial/\partial z \begin{vmatrix} 0&0&1&0 \\ 0&0&0&-1 \\ 1&0&0&0 \\ 0&-1&0&0 \end{vmatrix} * \begin{vmatrix} (E_z + i*c*B_z) \\ (E_x + i*c*B_x) - i*(E_y + i*c*B_y) \\ (E_z + i*c*B_z) \\ (E_x + i*c*B_x) - i*(E_y + i*c*B_y) \end{vmatrix} = \begin{vmatrix} (-J_t - J_z) \\ (-J_x + i*J_y) \\ (-J_t - J_z) \\ (-J_x + i*J_y) \end{vmatrix}$$

Rotate the three Pauli Spin Matrices, each through "q_x", "q_y" and "q_z".

Second Portion:

$$\partial/\partial ct \begin{vmatrix} 1&0&0&0 \\ 0&1&0&0 \\ 0&0&1&0 \\ 0&0&0&1 \end{vmatrix} + -\partial/\partial y \begin{vmatrix} 0&0&0&1 \\ 0&0&1&0 \\ 0&1&0&0 \\ 1&0&0&0 \end{vmatrix} + -\partial/\partial z \begin{vmatrix} 0&0&0&i \\ 0&0&-i&0 \\ 0&i&0&0 \\ -i&0&0&0 \end{vmatrix}$$

$$+ -\partial/\partial x \begin{vmatrix} 0&0&1&0 \\ 0&0&0&-1 \\ 1&0&0&0 \\ 0&-1&0&0 \end{vmatrix} * \begin{vmatrix} (E_x + i*c*B_x) \\ (E_y + i*c*B_y) - i*(E_z + i*c*B_z) \\ (E_x + i*c*B_x) \\ (E_y + i*c*B_y) - i*(E_z + i*c*B_z) \end{vmatrix} = \begin{vmatrix} (-J_t - J_x) \\ (-J_y + i*J_z) \\ (-J_t - J_x) \\ (-J_y + i*J_z) \end{vmatrix}$$

xxii

SPECIAL ALGEBRA FOR SPECIAL RELATIVITY

Third Portion:

$$\partial/\partial ct * \begin{pmatrix} 1&0&0&0 \\ 0&1&0&0 \\ 0&0&1&0 \\ 0&0&0&1 \end{pmatrix} + -\partial/\partial z * \begin{pmatrix} 0&0&0&1 \\ 0&0&1&0 \\ 0&1&0&0 \\ 1&0&0&0 \end{pmatrix} + -\partial/\partial x * \begin{pmatrix} 0&0&0&i \\ 0&0&-i&0 \\ 0&i&0&0 \\ -i&0&0&0 \end{pmatrix}$$

$$+ -\partial/\partial y * \begin{pmatrix} 0&0&1&0 \\ 0&0&0&-1 \\ 1&0&0&0 \\ 0&-1&0&0 \end{pmatrix} * \begin{pmatrix} (E_y + i*c*B_y) \\ (E_z + i*c*B_z) - i*(E_x + i*c*B_x) \\ (E_y + i*c*B_y) \\ (E_z + i*c*B_z) - i*(E_x + i*c*B_x) \end{pmatrix} = \begin{pmatrix} (-J_t - J_y) \\ (-J_z + i*J_x) \\ (-J_t - J_y) \\ (-J_z + i*J_x) \end{pmatrix}$$

The three portions contain all eight component equations of Maxwell's Equations.

Small-scale case Dirac Equation:

$$m_B*c* \begin{pmatrix} 1&0&0&0 \\ 0&1&0&0 \\ 0&0&-1&0 \\ 0&0&0&-1 \end{pmatrix} + (i*\hbar*\partial/\partial ct - Q_B*V_t)* \begin{pmatrix} 1&0&0&0 \\ 0&1&0&0 \\ 0&0&1&0 \\ 0&0&0&1 \end{pmatrix} + (-i*\hbar*\partial/\partial x - Q_B*V_x)* \begin{pmatrix} 0&0&0&1 \\ 0&0&1&0 \\ 0&1&0&0 \\ 1&0&0&0 \end{pmatrix}$$

$$+ (-i*\hbar*\partial/\partial y - Q_B*V_y)* \begin{pmatrix} 0&0&0&i \\ 0&0&-i&0 \\ 0&i&0&0 \\ -i&0&0&0 \end{pmatrix} + (-i*\hbar*\partial/\partial z - Q_B*V_z)* \begin{pmatrix} 0&0&1&0 \\ 0&0&0&-1 \\ 1&0&0&0 \\ 0&-1&0&0 \end{pmatrix} * \begin{pmatrix} \Psi_1 \\ \Psi_2 \\ \Psi_3 \\ \Psi_4 \end{pmatrix} = \begin{pmatrix} 0 \\ 0 \\ 0 \\ 0 \end{pmatrix}$$

Dirac Spinor is "Ψ_1", "Ψ_2", "Ψ_3", "Ψ_4". Electric charge is "Q_B". Voltage invariant components are "V". Electron rest mass is "m_B".

Large-scale case and small-scale case, together, as one mathematical model, pertain to a combined photon/electron particle and combined electromagnetic field and electron matter-wave. It appears that as time progresses, "α_M" becomes more precise to change observations of the projected photon.

To conceptually justify this new model, refer to the classic radius of the electron, for which matter mass "m_B" is equated to electric field energy "E_B/c^2".

SYNOPSIS – FINITE IMPRECISION FOR NUMBERS

Conclusion. The Dirac Spinor is post-processed into components of an electric current density invariant. By the same equations, the large-scale case is post-processed into components of an electromagnetic field force density invariant, and it includes the empirically discovered energy density and Poynting Vector components. Because the large-scale case Dirac Equation unites separately discovered empirical models, the large-scale case Dirac Equation is fundamental to electromagnetic field theory. The discovery of something more fundamental justifies a claim that the modified Axiomatic Set Theory for finite imprecision real numbers is valid in applied mathematics and is ready for use in more modern theories.

Also note that one particle is at two places, to violate a preconceived notion for geometric space that that isn't possible. And note material (fermion electrons) is one in the same as force (boson photons), to unite opposites. Take this radical notion further by supposing perceived reality results from rational numbers inside exponential functions, alone from objects, interacting by becoming more precise with respect to each other, to form patterns we see as the Dirac Equation and other mathematical models of physics. Justify this notion through the observation that there is no geometric space in which Dirac spinors exist. Per this radical notion, the universe is fundamentally rational numbers and is not physical stuff.

For further information, please read *Special Algebra for Special Relativity*.

SPECIAL ALGEBRA FOR SPECIAL RELATIVITY

CHAPTER 1 - NUMBERS

Chapter 1 – Numbers

1.1 Process from Descartes

Formal mathematics in the written record began about twenty-five hundred years ago in ancient Greece, when Pythagoras presented his proof for the Pythagorean Theorem. Along with that proof came the fresh idea that numbers could be separated from items being counted.

About four hundred years ago in *La Géométrie*, Descartes solved geometric problems using an all-number algebra and plotted solutions on a (Des)Cartesian grid. He had separated algebra from geometry.

The three steps that form the "Process from Descartes" are:

1) Mathematically model a physically real phenomenon with geometry (by utilizing geometric vectors)

2) Translate the geometric mathematical model into more abstract (and theory development friendly) all-number algebra to do the analysis

3) Translate component values from the all-number algebraic finished-calculation back into geometry as a final-result

The Process from Descartes, what he called "analytic geometry", explicitly removed analysis from geometry and measurements but did so temporarily. The Process from Descartes is the basis of applied mathematics in this book.

This book will show that physical material and force results from rational numbers becoming more precise relative to each other inside exponential functions. And that means step 3 is only preparation for a measurement and is not a reversion away from numbers. To make sense of the math, we can speculate that numbers are fundamental, and not material (electrons/fermions) or force (photons/bosons) or even time-space.

Pythagoras said, "All is number," and yes, I think he got it right.

SPECIAL ALGEBRA FOR SPECIAL RELATIVITY

1.2 Geometric-Vectors

<u>Geometric-Unit-Vectors and Components</u>. Newton's Second Law is our example: Force "**F**" equals mass "m" times acceleration "**a**". "**F**" and "**a**" are geometric-vectors, as indicated by **bold** font.

$$\mathbf{F} = m*\mathbf{a} \quad ; \quad F_x*\mathbf{i}_x + F_y*\mathbf{i}_y + F_z*\mathbf{i}_z = m*a_x*\mathbf{i}_x + m*a_y*\mathbf{i}_y + m*a_z*\mathbf{i}_z$$

$$F_x = m*a_x \quad ; \quad F_y = m*a_y \quad ; \quad F_z = m*a_z \quad \text{(Component equations)}$$

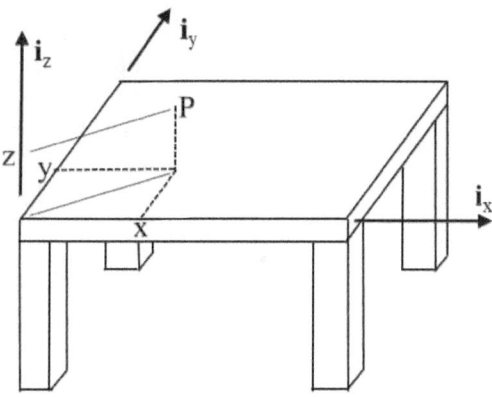

Figure 1. Tabletop Coordinate System, Location "P" is "$_3\mathbf{r} = x*\mathbf{i}_x + y*\mathbf{i}_y + z*\mathbf{i}_z$"

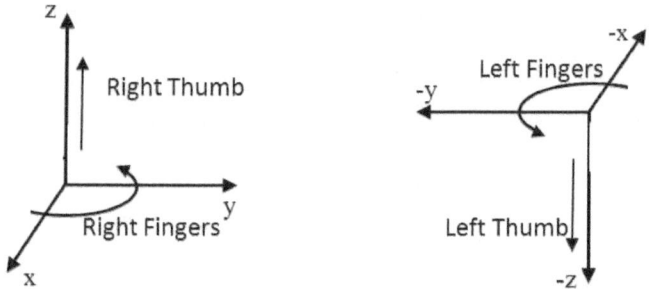

Figure 2. Right-hand coordinate system, and Left-hand system

CHAPTER 1 - NUMBERS

Three perpendicular geometric-unit-vectors "i_x", "i_y", and "i_z" identify directions along which length components "x", "y" and "z" are measured by counting measurement units (inches or centimeters) on measuring tapes that have their zero points at the origin.

Per tradition, a right-hand coordinate system for "x, y, z" is always selected: Right-hand fingers curl through positive "i_x" and then positive "i_y", and the thumb points to positive "i_z".

Push slippery ice of mass "m" with force "F_x", and the ice moves with acceleration "a_x" per "$F_x = m*a_x$". The push was parallel with the "x" direction and not necessarily along the front edge of the table.

Dot-Product Operations:
$i_x \bullet i_x = 1$; $i_x \bullet i_y = 0$; $i_x \bullet i_z = 0$
$i_y \bullet i_x = 0$; $i_y \bullet i_y = 1$; $i_y \bullet i_z = 0$
$i_z \bullet i_x = 0$; $i_z \bullet i_y = 0$; $i_z \bullet i_z = 1$

Cross-Product Operations:
$i_x \mathbf{x} i_x = 0$; $i_x \mathbf{x} i_y = i_z$; $i_x \mathbf{x} i_z = -i_y$
$i_y \mathbf{x} i_x = -i_z$; $i_y \mathbf{x} i_y = 0$; $i_y \mathbf{x} i_z = i_x$
$i_z \mathbf{x} i_x = i_y$; $i_z \mathbf{x} i_y = -i_x$; $i_z \mathbf{x} i_z = 0$

Time Geometric-Unit-Vector. The time component in time-space is "c*t". Time "t" is measured by a clock in seconds relative to a time origin by counting a repeated ticking internal to the clock mechanism.

"c ≈ 300,000,000" meters/second is the speed-of-light and is a measurement unit conversion factor from time units to length units.

"i_t" is the time geometric-unit-vector. Location "P" in time-space:

$$_4\mathbf{r} = c*t*i_t + x*i_x + y*i_y + z*i_z$$

"$_4$" pre-subscript indicates the quantity of terms in the expression. Each term consists of a component factor ("c*t", "x", "y", or "z") multiplied by a geometric direction factor ("i_t", "i_x", "i_y", or "i_z").

We could speculate "$i_x \mathbf{x} i_t$" should include an "i" factor ("$i^2 = -1$") because, perhaps, "$i_t \bullet i_t = -1$", but we don't go there because geometric-vectors are only used to set up a problem.

Engineering-Calculation-Algebra. Lack of division inverse to dot- and cross-products illustrates that physical properties of geometric vectors took priority over mathematical properties. Priority to physical properties is appropriate in engineering because an engineer applies known theory in a simple and efficient calculation. In the time of Maxwell (late 1800's) and Minkowski (early 1900's) (slightly over a hundred years ago) geometric vectors had just been introduced by Heaviside and was, perhaps, the best algebra available.

1.3 Quaternions

An all-number algebra using quaternions fits Special Relativity so well it appears quaternions are inherently the correct algebra. An all-number algebra has only mathematical properties, no physical properties. It might seem counter-intuitive, but algebra separated away from the physical, geometric world better models subtle symmetries.

Theory-Development-Algebra is the name for the new intention of the all-number algebra, which is to expand known theories into new theories, as opposed to using known theories in engineering.
    The difference between engineering-calculation-algebra and theory-development-algebra evolved (in the writing of this book) from the difference in Pythagorean mathematics between logistica (routine calculations) and arithmetica (theory).

Quaternions and Label-Numbers. Three quaternions "j_x", "j_y", and "j_z" (with the number "1" as the fourth) are in Hamilton's paper dated 1843. Each quaternion squares to negative one.
    Quaternions "j_x", "j_y", and "j_z", complex number factor "i", integer one "1", and unit magnitude constructions of these special numbers, are given the categorical name (in this book) "label-numbers".

Translation to all-number. Place label-numbers in front of components. "$_4\mathbf{r} = c*t*\mathbf{i}_t + x*\mathbf{i}_x + y*\mathbf{i}_y + z*\mathbf{i}_z$" becomes

$$i*_4r = i*c*t + j_x*x + j_y*y + j_z*z \quad \text{or} \quad _4r = 1*c*t + q_x*x + q_y*y + q_z*z$$

$$q_x = j_x/i \; ; \; q_y = j_y/i \; ; \; q_z = j_z/i \; ; \qquad q_x^2 = q_y^2 = q_z^2 = +1$$

Exponential Function with Quaternions. Argument "s" can have a label-number included as a factor, for example "$s = q_x*\alpha$" (α is alpha). In contrast, a geometric-unit-vector cannot be in the argument of an exponential function because there is no algebra by which to calculate.

$$\exp(s) = 1 + s + s^2/2 + s^3/6 + s^4/24 + s^5/120 + s^6/6! + \ldots$$

CHAPTER 1 - NUMBERS

$$\begin{aligned}\exp(q_x*\alpha) &= 1 + q_x*\alpha + (q_x*\alpha)^2/2 + (q_x*\alpha)^3/6 + (q_x*\alpha)^4/24 + \ldots \\ &= 1 + q_x*\alpha + \alpha^2/2 + q_x*\alpha^3/6 + \alpha^4/24 + \ldots \\ &= (1 + \alpha^2/2 + \alpha^4/24 + \ldots) + q_x*(\alpha + \alpha^3/6 + \ldots) \\ &= \cosh(\alpha) + q_x*\sinh(\alpha)\end{aligned}$$

<u>Complex-conjugate</u> operator "^{*i}" swaps "i" for "$-i$" and reverses factors.

$$i^{*i} = -i \qquad\qquad\qquad 1^{*i} = 1$$

<u>Quaternion hypercomplex-conjugate</u> operator "^{*j}" swaps "j_x", "j_y", "j_z" for "$-j_x$", "$-j_y$", "$-j_z$", respectively, and reverses factors.

$$i^{*j} = i \qquad\qquad ; \qquad 1^{*j} = 1$$

$$j_x^{*j} = -j_x \;\; ; \;\; j_y^{*j} = -j_y \;\; ; \;\; j_z^{*j} = -j_z \;\; ; \;\; q_x^{*j} = -q_x \;\; ; \;\; q_y^{*j} = -q_y \;\; ; \;\; q_z^{*j} = -q_z$$

$$-q_x = (q_x)^{*j} = (j_y*q_z)^{*j} = (q_z)^{*j}*(j_y)^{*j} = (-q_z)*(-j_y) = q_z*j_y = -q_x$$

Both an all-number expression and its conjugate relate to the same geometric-vector expression. The "^{*j}" symbol is removed after "$_4r^{*j}$" is used as a factor in a multiplication operation, and not before.

$$_4r^{*j} = (1*c*t + q_x*x + q_y*y + q_z*z)^{*j} = c*t*1^{*j} + x*q_x^{*j} + y*q_y^{*j} + z*q_z^{*j}$$

Division reciprocal can substitute for negative.

$$i^{*i} = -i = 1/i \;\; ; \qquad j_x^{*j} = -j_x = 1/j_x \;\; ; \;\; j_y^{*j} = -j_y = 1/j_y \;\; ; \;\; j_z^{*j} = -j_z = 1/j_z$$

<u>Matrix Isomorphs of Quaternions</u>. An "isomorph" is a functional equal with respect to an algebra. Products of label-numbers are calculated using matrix isomorphs because matrix isomorphs provide internal mathematical structure for the calculation. A traditional association of matrix isomorphs to quaternion label-numbers is given below.

$$j_x \Rightarrow \begin{matrix} 0 & i \\ i & 0 \end{matrix} \qquad j_y \Rightarrow \begin{matrix} 0 & -1 \\ 1 & 0 \end{matrix} \qquad j_z \Rightarrow \begin{matrix} i & 0 \\ 0 & -i \end{matrix} \qquad i \Rightarrow \begin{matrix} i & 0 \\ 0 & i \end{matrix}$$

SPECIAL ALGEBRA FOR SPECIAL RELATIVITY

$$q_x \Rightarrow \begin{pmatrix} 0 & 1 \\ 1 & 0 \end{pmatrix} \quad q_y \Rightarrow \begin{pmatrix} 0 & i \\ -i & 0 \end{pmatrix} \quad q_z \Rightarrow \begin{pmatrix} 1 & 0 \\ 0 & -1 \end{pmatrix} \quad 1 \Rightarrow \begin{pmatrix} 1 & 0 \\ 0 & 1 \end{pmatrix}$$

("q_y" is the negative of its corresponding Pauli Spin Matrix.)
The "1" and "i" terms in the matrices also have isomorphs.

$$1 \Rightarrow \begin{pmatrix} 1 & 0 \\ 0 & 1 \end{pmatrix} \quad i \Rightarrow \begin{pmatrix} 0 & -1 \\ 1 & 0 \end{pmatrix}$$

"j_x", "j_y", "j_z" can be rotated to "j_y", "j_z", "j_x" or to "j_z", "j_x", "j_y", but with matrices not rotated.

Multiply matrices per the below equation. It applies only if matrix terms ("a, b, c, d, e, f, g, h") commute (and so cannot be quaternions).

$$\begin{pmatrix} a & b \\ c & d \end{pmatrix} * \begin{pmatrix} e & f \\ g & h \end{pmatrix} = \begin{pmatrix} a*e + b*g & a*f + b*h \\ c*e + d*g & c*f + d*h \end{pmatrix}$$

Label-numbers have the following product equations.

$j_x*j_y = -j_y*j_x = j_z$; $\quad i*j_x = j_x*i$
$j_y*j_z = -j_z*j_y = j_x$; $\quad i*j_y = j_y*i$
$j_z*j_x = -j_x*j_z = j_y$; $\quad i*j_z = j_z*i$

$j_x^2 = j_y^2 = j_z^2 = i^2 = -1$; $\quad q_x^2 = q_y^2 = q_z^2 = 1^2 = +1$

$q_x = j_x/i$; $q_y = j_y/i$; $q_z = j_z/i$

$q_x*j_y = -j_y*q_x = -q_y*j_x = j_x*q_y = q_z$; $\quad q_x*q_y = -q_y*q_x = -i*q_z = -j_z$
$q_y*j_z = -j_z*q_y = -q_z*j_y = j_y*q_z = q_x$; $\quad q_y*q_z = -q_z*q_y = -i*q_x = -j_x$
$q_z*j_x = -j_x*q_z = -q_x*j_z = j_z*q_x = q_y$; $\quad q_z*q_x = -q_x*q_z = -i*q_y = -j_y$

Quaternion Division Operation. To divide one matrix by another, alter the numerator to create a matrix divided by itself. All possible factorings and regroupings end with the same result. Alternatively, use a reciprocal.

$q_y/j_z = (j_z*q_x)/j_z = -(q_x*j_z)/j_z = -q_x*(j_z/j_z) = -q_x$; $\quad q_y/j_z = q_y*(-j_z) = -q_x$

CHAPTER 1 - NUMBERS

<u>Comparing Quaternions to Geometric-Unit-Vectors</u>. Geometric-unit-vectors had no "i" factor in a cross-product result ("$\mathbf{i}_x \mathbf{x} \mathbf{i}_y = \mathbf{i}_z$"). In contrast, quaternions did have an "i" factor ("$q_x * q_y = -i * q_z$").

An engineer does not want an "i" factor in a calculation that predicts a measurement because every measurement is a real number. In contrast, the "i" is wanted for theory development.

<u>Group Theory - An Algebra that Includes Quaternions</u>. Product equations above must be organized into a mathematical group. The group's criteria are used in a proof so that the proof is not ambiguous or illogical. Criteria for a group:

- Closure
- Identity element
- Commutative property
- Associative property
- Inverse operation

Two groups combine to form an algebra field with:

- Distributive property

<u>Number Sets</u>. Every natural number, integer, and rational number is finite. Finite means the natural number can be counted to. Natural numbers "N" used for counting start with one and have one added repeatedly and unboundedly. Include zero and negatives to form integers "Z".

$$N = \{1, 2, 3, ...\} \quad ; \quad Z = \{..., -3, -2, -1, 0, 1, 2, 3, ...\}$$

Rational numbers "Q" are each a ratio of two integers: "$q \in Q$" if "$q = m/n$" with "m, n $\in Z$" and "n $\neq$ 0". ("$\in$" is "element of".) A rational number "q" has multiple selections of "m" and "n", for example "$1.5 = 3/2 = -6/-4$". Because natural numbers are finite, a rational number is finite, not only in magnitude but also in that its place-value digits have a repeat pattern that begins a finite count left and right of the decimal point.

Integers are a subset of rational numbers: "$Z \subset Q$" because the denominator "n" may equal "1". An integer lacks a decimal point.

Descartes assigned the name "real" to real numbers to distinguish them from complex or imaginary numbers. Today, real numbers "R" are rational numbers united with numbers proven to be irrational. "$Q \subset R$"

SPECIAL ALGEBRA FOR SPECIAL RELATIVITY

Algebra Field for Real Numbers. Integers "Z", rational numbers "Q", and real numbers "R" each form a group with respect to addition. The example group "$\{R, +\}$" uses real numbers "R". Criteria are:

- .1. Addition Closure Property. Each element "a" formed by addition is an element of "R": "a = b + c". "a, b, c $\in R$"

- .2. Addition Identity Property. An addition identity element "e" is an element of the set "R", for which "a = a + e". "a, e $\in R$" ("e" is integer zero "0")

- .3. Addition Commutative Property. The order of elements in an addition operation has no effect on the result of the operation, so that "a + b = b + a". "a, b $\in R$"

- .4. Addition Associative Property. The order of addition operations has no effect on the result of the operations, so that "(a + b) + c = a + (b + c)". "a, b, c $\in R$"

- .5. Addition Inverse Property. Each element "a" formed by the addition inverse operation, "a = -b" (such that "a + b = e"), is an element of "R". "a, b, e $\in R$"

Rational numbers "Q" and real numbers "R" each form a group with respect to multiplication. "$\{R, *\}$" Criteria:

- .6. Multiplication Closure Property. Each element "a" formed by multiplication "a = b*c" is an element of "R". "a, b, c $\in R$"

- .7. Multiplication Identity Property. A multiplication identity element "f" is an element of the set "R", for which "a = a*f".
  "a, f $\in R$" ("f" is integer one "1")

- .8. Multiplication Commutative Property. The order of elements in a multiplication operation has no effect on the result of the operation, so that "a*b = b*a". "a, b $\in R$"

CHAPTER 1 - NUMBERS

- .9. Multiplication Associative Property. The order of multiplication operations has no effect on the result of the operations, so that "(a*b)*c = a*(b*c)". "a, b, c ∈ R"

- .10. Multiplication Inverse Property. Each element "a" formed by the multiplication inverse operation, "a = 1/b" (such that "a*b = f"), is an element of "R". "a, b, f ∈ R" with the exception "b ≠ 0"

The two groups above are brought together to form an algebra field "{R, +, *}" (or "{Q, +, *}") by use of the distributive property.

- .11. Distributive Property of Multiplication over Addition. "a*(b + c) = a*b + a*c" and "(a + b)*c = a*c + b*c". "a, b, c ∈ R"

Use algebra field criteria to prove properties.

- .12. Property of One Addition Identity Element "e": If "e_1 + a = a" and if also there is "e_2 + a = a" then from "a = a" we have "e_1 + a = e_2 + a", and after subtracting "a" we have "e = e_1 = e_2".

- .13. Property of One Multiplication Identity Element "f": If "f_1*a = a" and if "f_2*a = a", then from "a = a" we have "f_1*a = f_2*a", and, after dividing out "a" (for "a ≠ 0"), we have "f = f_1 = f_2".

- .14. Property of "0 = -0": The addition identity element "0" is its own addition inverse, so that "0 = -0".

- .15. Property of "1 = 1/1" and "-1 = 1/(-1)": The multiplication identity element "1" is its own multiplication inverse, "1 = 1/1". And, the addition inverse of the multiplication identity element is its own multiplication inverse, "-1 = 1/(-1)".

- .16. Property of "b*0 = 0": Any number "b" multiplied by the additive identity element zero "0" equals the additive identity element zero "0" as the product. If "a = a + 0", then "0 = a - a" and "b*0 = b*a - b*a = 0".

- .17. Inverse Anti-Commutative Properties: The addition inverse anti-commutative property is "-(a - b) = b - a". The multiplication inverse anti-commutative property is "1/(a/b) = b/a".

SPECIAL ALGEBRA FOR SPECIAL RELATIVITY

Exponent operation properties for rational numbers and integers are listed below. The exponent operation "^" is repeated multiplication (for example, "4^5 = 4^5 = 4*4*4*4*4"). Similarly, the multiplication operation "*" is repeated addition "+" (for example, "4*5 = 4 + 4 + 4 + 4 + 4").

- .18. Exponent Closure Property: Each element "a" formed by the exponent operation "a = b^c" is an element of "Q". "a, b ∈ Q", "c ∈ Z", and "b ≠ 0" if "c < 0".

- .19. Exponent Identity Property: An exponent identity element "g" is an element of set "Z" for which "a = a^g". "a ∈ Q", "g ∈ Z" ("g" is integer one "1")

- .20. No Exponent Inverse Property: Element "c" formed by the root inverse exponent operation, "c = a^(1/b)", is not an element of "Q" if "b ≠ 1" or "b ≠ -1". "a ∈ Q", "b ∈ Z"

- .21. Base Inverse Property: Element "b" formed by the logarithm inverse exponent operation, "$\log_a c$ = b", is an element of "Z" only if "c = a^b". "a, c ∈ Q"

- .22. No Exponent Associative Property: The order of exponent operations cannot be altered, so that "(a^b)^c" does not necessarily equal "a^(b^c)". "a ∈ Q", "b, c ∈ Z"

- .23. No Exponent Commutative Property: The order of elements in an exponent operation has an effect on the result of the operation so that "a^b ≠ b^a". "a, b ∈ Z", "a ≠ b"

- .24. Distributive Property for Exponent Operation. "a^(b + c) = (a^b)*(a^c)". "a ∈ Q", "b, c ∈ Z"

<u>Complex Numbers</u>. "{1, i}" in a group with multiplication, "{{1, i}, *}":

- Closure: No holes in the multiplication table
- Identity: Identity element is integer one.
- Commutative Property: Applies without exception
- Associative Property: Applies without exception
- Inverse: The ratio of any two numbers is in the set of numbers

CHAPTER 1 - NUMBERS

$1/1 = 1$; $i/1 = i$; $1/i = 1*(-i) = -i$; $i/i = 1$

"$\{1, i\}$" combines with real numbers "R" to form "complex numbers", "C", that conform to the criteria of an algebra with the same properties as for real numbers "R", "$\{C, +, *\}$".

Conjugate operation "$*^i$" does not affect "$\{C, +, *\}$".

Through an algebraic manipulation "i" is placed in the numerator.

$$1/(5 + i*7) = (5 - i*7)/((5 - i*7)*(5 + i*7))$$
$$= (5 - i*7)/(25 + 49)$$
$$= 5/74 - i*7/74$$

"$a + i*b$" with "a, b $\in$ R" is called the "summation-form".
2x2 matrix isomorphs apply.

$$1 \Rightarrow \begin{matrix} 1 & 0 \\ 0 & 1 \end{matrix} \quad ; \quad i \Rightarrow \begin{matrix} 0 & -1 \\ 1 & 0 \end{matrix} \quad ; \quad a + i*b \Rightarrow \begin{matrix} a & -b \\ b & a \end{matrix}$$

*	+1	-1	+i	-i
+1	$(+1)*(+1) = +1$	$(+1)*(-1) = -1$	$(+1)*(+i) = +i$	$(+1)*(-i) = -i$
-1	$(-1)*(+1) = -1$	$(-1)*(-1) = +1$	$(-1)*(+i) = -i$	$(-1)*(-i) = +i$
-i	$(-i)*(+1) = -i$	$(-i)*(-1) = +i$	$(-i)*(+i) = +1$	$(-i)*(-i) = -1$
+i	$(+i)*(+1) = +i$	$(+i)*(-1) = -i$	$(+i)*(+i) = -1$	$(+i)*(-i) = +1$

Table 1. Multiplication Table for Complex Number Factor. "+1" is along the major diagonal by using a conjugate in the left column (first factor).

<u>Quaternion-Hypercomplex Numbers</u>. "$\{\{1, j_x, j_y, j_z\}, *\}$" criteria:

- Closure: No holes in the multiplication table
- Identity: Identity element is integer one
- Commutative Property: Applies with "1" or "-1" as a factor or as the product. The anti-commutative property applies when two different (non-"1") quaternions are factors.
- Associative Property: Applies without exception
- Inverse: The ratio of any two numbers is in the set of numbers

SPECIAL ALGEBRA FOR SPECIAL RELATIVITY

$j_x/j_y = (j_y{}^*j_z)/j_y = -(j_z{}^*j_y)/j_y = -j_z{}^*(j_y/j_y) = -j_z{}^*1 = -j_z$; $j_x/j_y = j_x{}^*(-j_y) = -j_z$

"$\{1, j_x, j_y, j_z\}$" with "R" forms "quaternion-hypercomplex numbers", "QH", "$\{QH, +, *(\text{anti-commute})\}$".

Conjugate operation "$^*/$" does not affect the algebra field.

*	+1	-1	+jx	-jx
+1	(+1)*(+1) = +1	(+1)*(-1) = -1	(+1)*(+jx) = +jx	(+1)*(-jx) = -jx
-1	(-1)*(+1) = -1	(-1)*(-1) = +1	(-1)*(+jx) = -jx	(-1)*(-jx) = +jx
-jx	(-jx)*(+1) = -jx	(-jx)*(-1) = +jx	(-jx)*(+jx) = +1	(-jx)*(-jx) = -1
+jx	(+jx)*(+1) = +jx	(+jx)*(-1) = -jx	(+jx)*(+jx) = -1	(+jx)*(-jx) = +1
-jy	(-jy)*(+1) = -jy	(-jy)*(-1) = +jy	(-jy)*(+jx) = +jz	(-jy)*(-jx) = -jz
+jy	(+jy)*(+1) = +jy	(+jy)*(-1) = -jy	(+jy)*(+jx) = -jz	(+jy)*(-jx) = +jz
-jz	(-jz)*(+1) = -jz	(-jz)*(-1) = +jz	(-jz)*(+jx) = -jy	(-jz)*(-jx) = +jy
+jz	(+jz)*(+1) = +jz	(+jz)*(-1) = -jz	(+jz)*(+jx) = +jy	(+jz)*(-jx) = -jy

*	+jy	-jy	+jz	-jz
+1	(+1)*(+jy) = +jy	(+1)*(-jy) = -jy	(+1)*(+jz) = +jz	(+1)*(-jz) = -jz
-1	(-1)*(+jy) = -jy	(-1)*(-jy) = +jy	(-1)*(+jz) = -jz	(-1)*(-jz) = +jz
-jx	(-jx)*(+jy) = -jz	(-jx)*(-jy) = +jz	(-jx)*(+jz) = +jy	(-jx)*(-jz) = -jy
+jx	(+jx)*(+jy) = +jz	(+jx)*(-jy) = -jz	(+jx)*(+jz) = -jy	(+jx)*(-jz) = +jy
-jy	(-jy)*(+jy) = +1	(-jy)*(-jy) = -1	(-jy)*(+jz) = -jx	(-jy)*(-jz) = +jx
+jy	(+jy)*(+jy) = -1	(+jy)*(-jy) = +1	(+jy)*(+jz) = +jx	(+jy)*(-jz) = -jx
-jz	(-jz)*(+jy) = +jx	(-jz)*(-jy) = -jx	(-jz)*(+jz) = +1	(-jz)*(-jz) = -1
+jz	(+jz)*(+jy) = -jx	(+jz)*(-jy) = +jx	(+jz)*(+jz) = -1	(+jz)*(-jz) = +1

Table 2. Multiplication Table for Quaternions. "+1" is along the major diagonal using the conjugate in the left column.

CHAPTER 1 - NUMBERS

Quaternions can be placed exclusively in the numerator.

$$1/(3 + j_x*5 + j_y*7) = (3 - j_x*5 - j_y*7)/((3 - j_x*5 - j_y*7)*(3 + j_x*5 + j_y*7))$$
$$= (3 - j_x*5 - j_y*7)/(9 + 25 + 49 - 35*(j_x*j_y + j_y*j_x))$$
$$= (3 - j_x*5 - j_y*7)/83$$
$$= 3/83 - j_x*5/83 - j_y*7/83$$

Any quaternion-hypercomplex number can be written in summation-form "a + j_x*b + j_y*c + j_z*d" with "a, b, c, d ∈ R". Each of three rotation options for 2x2 matrix isomorph substitutions apply. The determinant equals "$a^2 + b^2 + c^2 + d^2$". Example using one rotation option:

$$a + j_x*b + j_y*c + j_z*d \Rightarrow \begin{matrix}(a + i*d) & -(c + i*b)^{*i} \\ (c + i*b) & (a + i*d)^{*i}\end{matrix}$$

<u>Quaternion-Complex-Hypercomplex Numbers</u>. "{{1, i, j_x, q_x}, *}" criteria:

- Closure: No holes in the multiplication table

- Identity: The identity element is integer one

- Commutative Property: Applies without exception

- Associative Property: Applies without exception

- Inverse: The ratio of any two numbers is in the set of numbers

"{1, i, j_x, q_x}" combines with "{1, j_x, j_y, j_z}" to form a 16x16 multiplication table for "{1, $i, j_x, j_y, j_z, q_x, q_y, q_z$}". Inverse example:

$$q_x/j_y = (j_x/i)/j_y = -i*(j_x/j_y) = -i*(-j_z) = -j_z/i = -q_z$$

"{1, $i, j_x, j_y, j_z, q_x, q_y, q_z$}" combines with "R" to form "quaternion-complex-hypercomplex numbers", "{QCH, +, *(anti-commute)}". Conjugates apply.
Label-numbers can be placed exclusively in the numerator.

$$1/(3 + i*5 + q_y*7) = (3 + i*5 - q_y*7)/((3 + i*5 - q_y*7)*(3 + i*5 + q_y*7))$$
$$= (3 + i*5 - q_y*7)/(-16 + i*30 - 49) = (3 + i*5 - q_y*7)/(-65 + i*30)$$
$$= (3 + i*5 - q_y*7)*(-65 - i*30)/((-65 - i*30)*(-65 + i*30))$$

SPECIAL ALGEBRA FOR SPECIAL RELATIVITY

$$= (-45 - i*415 + q_y*455 + j_y*210)/5125$$

Summation-form "$a_r + i*a_i + q_x*b_r + j_x*b_i + q_y*c_r + j_y*c_i + q_z*d_r + j_z*d_i$" with "$a_r, a_i, b_r, b_i, c_r, c_i, d_r, d_i \in R$" has three 2x2 matrix isomorphs.

$$((a_r + d_r) + i*(a_i + d_i)) \quad -((c_i - b_r) + i*(c_r + b_i))^{*i}$$

$$((c_i + b_r) + i*(-c_r + b_i)) \quad ((a_r - d_r) + i*(-a_i + d_i))^{*i}$$

"$det = (a_r^2 - d_r^2 - a_i^2 + d_i^2 + c_i^2 - b_r^2 - c_r^2 + b_i^2) + i*2*(a_r*a_i - d_r*d_i - c_r*c_i - b_r*b_i)$".

*	+1	-1	+i	-i
+1	(+1)*(+1) = +1	(+1)*(-1) = -1	(+1)*(+i) = +i	(+1)*(-i) = -i
-1	(-1)*(+1) = -1	(-1)*(-1) = +1	(-1)*(+i) = -i	(-1)*(-i) = +i
-i	(-i)*(+1) = -i	(-i)*(-1) = +i	(-i)*(+i) = +1	(-i)*(-i) = -1
+i	(+i)*(+1) = +i	(+i)*(-1) = -i	(+i)*(+i) = -1	(+i)*(-i) = +1
-jx	(-jx)*(+1) = -jx	(-jx)*(-1) = +jx	(-jx)*(+i) = +qx	(-jx)*(-i) = -qx
+jx	(+jx)*(+1) = +jx	(+jx)*(-1) = -jx	(+jx)*(+i) = -qx	(+jx)*(-i) = +qx
+qx	(+qx)*(1) = +qx	(+qx)*(-1) = -qx	(+qx)*(+i) = +jx	(+qx)*(-i) = -jx
-qx	(-qx)*(+1) = -qx	(-qx)*(-1) = +qx	(-qx)*(+i) = -jx	(-qx)*(-i) = +jx

*	+jx	-jx	+qx	-qx
+1	(+1)*(+jx) = +jx	(+1)*(-jx) = -jx	(1)*(+qx) = +qx	(+1)*(-qx) = -qx
-1	(-1)*(+jx) = -jx	(-1)*(-jx) = +jx	(-1)*(+qx) = -qx	(-1)*(-qx) = +qx
-i	(-i)*(+jx) = +qx	(-i)*(-jx) = -qx	(-i)*(+qx) = -jx	(-i)*(-qx) = +jx
+i	(+i)*(+jx) = -qx	(+i)*(-jx) = +qx	(+i)*(+qx) = +jx	(+i)*(-qx) = -jx
-jx	(-jx)*(+jx) = +1	(-jx)*(-jx) = -1	(-jx)*(+qx) = -i	(-jx)*(-qx) = +i
+jx	(+jx)*(+jx) = -1	(+jx)*(-jx) = +1	(+jx)*(+qx) = +i	(+jx)*(-qx) = -i
+qx	(+qx)*(+jx) = +i	(+qx)*(-jx) = -i	(+qx)*(+qx) = 1	(+qx)*(-qx) = -1
-qx	(-qx)*(+jx) = -i	(-qx)*(-jx) = +i	(-qx)*(+qx) = -1	(-qx)*(-qx) = +1

Table 3. Multiplication Table for a Quaternion and the Complex Number Factor. Conjugates for the first factor place "+1" on the major diagonal.

CHAPTER 1 - NUMBERS

Dot-Product and Cross-Product. Analogous to geometric vectors, all-number expressions, for example "$j_z*q_z = i$" and "$j_y*q_z = q_x$", can be separated using dot-product "•" and cross-product "**x**".

$$j_z*q_z = j_z \bullet q_z + j_z \mathbf{x} q_z = i + 0 = i \quad ; \quad j_y*q_z = j_y \bullet q_z + j_y \mathbf{x} q_z = 0 + q_x = q_x$$

$$\begin{array}{llll}
1 \bullet 1 = 1 & 1 \bullet j_x = 0 & 1 \bullet j_y = 0 & 1 \bullet j_z = 0 \\
j_x \bullet 1 = 0 & j_x \bullet j_x = -1 & j_x \bullet j_y = 0 & j_x \bullet j_z = 0 \\
j_y \bullet 1 = 0 & j_y \bullet j_x = 0 & j_y \bullet j_y = -1 & j_y \bullet j_z = 0 \\
j_z \bullet 1 = 0 & j_z \bullet j_x = 0 & j_z \bullet j_y = 0 & j_z \bullet j_z = -1 \\
\end{array}$$

$$\begin{array}{llll}
1\mathbf{x}1 = 0 & 1\mathbf{x}j_x = j_x & 1\mathbf{x}j_y = j_y & 1\mathbf{x}j_z = j_z \\
j_x\mathbf{x}1 = j_x & j_x\mathbf{x}j_x = 0 & j_x\mathbf{x}j_y = +j_z & j_x\mathbf{x}j_z = -j_y \\
j_y\mathbf{x}1 = j_y & j_y\mathbf{x}j_x = -j_z & j_y\mathbf{x}j_y = 0 & j_y\mathbf{x}j_z = +j_x \\
j_z\mathbf{x}1 = j_z & j_z\mathbf{x}j_x = +j_y & j_z\mathbf{x}j_y = -j_x & j_z\mathbf{x}j_z = 0 \\
\end{array}$$

"$_4r_C$" and "$_4r_D$" are factors in "$_6A$". "$_3A$" is "area".

$$_4r_C = {_1r_C} + {_3r_C} \quad ; \quad {_1r_C} = 1*c*t_C \quad ; \quad {_3r_C} = q_x*x_C + q_y*y_C + q_z*z_C$$

$$_1r_C{}^{*j} = c*t_C*1^{*j} \quad ; \quad {_3r_C}{}^{*j} = x_C*q_x{}^{*j} + y_C*q_y{}^{*j} + z_C*q_z{}^{*j}$$

$$_4r_C{}^{*j}*{_4r_D} = {_4r_C}{}^{*j} \bullet {_4r_D} + {_4r_C}{}^{*j} \mathbf{x} {_4r_D}$$

$$_1r_C{}^{*j} \bullet {_1r_D} = c*t_C*(1^{*j}*1)*c*t_D = c*t_C*c*t_D$$

$$_3r_C{}^{*j} \bullet {_3r_D} = x_C*q_x{}^{*j}*q_x*x_D + y_C*q_y{}^{*j}*q_y*y_D + z_C*q_z{}^{*j}*q_z*z_D$$
$$= -(x_C*x_D + y_C*y_D + z_C*z_D)$$

$$_4r_C{}^{*j} \bullet {_4r_D} = {_1r_C}{}^{*j} \bullet {_1r_D} + {_3r_C}{}^{*j} \bullet {_3r_D} = c*t_C*c*t_D - (x_C*x_D + y_C*y_D + z_C*z_D)$$

$$_6A = {_4r_C}{}^{*j} \mathbf{x} {_4r_D} = {_1r_C}{}^{*j}*{_3r_D} + {_3r_C}{}^{*j}*{_1r_D} + {_3r_C}{}^{*j} \mathbf{x} {_3r_D}$$

$$_3B = {_1r_C}{}^{*j}*{_3r_D} + {_3r_C}{}^{*j}*{_1r_D}$$

$$= (c*t_C*q_x*x_D + c*t_C*q_y*y_D + c*t_C*q_z*z_D)$$
$$+ (q_x{}^{*j}*x_C*c*t_D + q_y{}^{*j}*y_C*c*t_D + q_z{}^{*j}*z_C*c*t_D)$$

$$= (c*t_C*x_D - c*t_D*x_C)*q_x + (c*t_C*y_D - c*t_D*y_C)*q_y + (c*t_C*z_D - c*t_D*z_C)*q_z$$
$$= B_x*q_x + B_y*q_y + B_z*q_z$$

$$_3A = {_3r_C}^{*j}\mathbf{x}{_3r_D}$$

$$= x_C*q_x^{*j}*q_y*y_D + x_C*q_x^{*j}*q_z*z_D$$
$$+ y_C*q_y^{*j}*q_x*x_D + y_C*q_y^{*j}*q_z*z_D$$
$$+ z_C*q_z^{*j}*q_x*x_D + z_C*q_z^{*j}*q_y*y_D$$

$$= (y_C*z_D - z_C*y_D)*j_x + (z_C*x_D - x_C*z_D)*j_y + (x_C*y_D - y_C*x_D)*j_z$$
$$= A_x*j_x + A_y*j_y + A_z*j_z$$

$$_6A = {_3B} + {_3A} = B_x*q_x + B_y*q_y + B_z*q_z + A_x*j_x + A_y*j_y + A_z*j_z$$
$$= (B_x + i*A_x)*q_x + (B_y + i*A_y)*q_y + (B_z + i*A_z)*q_z$$

Note from component placement that "$({_4r_C}^{*j}\mathbf{x}{_4r_D}) = -({_4r_D}^{*j}\mathbf{x}{_4r_C})$", which is the opposite of "$({_4r_C}^{*j}\bullet{_4r_D}) = ({_4r_D}^{*j}\bullet{_4r_C})$".

Division operation as an inverse applies only to a complete multiplication operation "*" and not to a dot-product or cross-product (or triple-vector-product or remnant-product) alone. Quite often, though, in practice either the dot-product or cross-product equals zero due to an identity.

Triple-Vector-Product and Remnant-Product. A four-term summation-form number "$_4r_B$" multiplied by a six-term summation-form number "$_6A$" may be split between a triple-vector-product "■" and a remnant-product "♦". The triple-vector-product "■" has a mathematically imaginary result. The remnant-product "♦" has a mathematically real result (as given by the "$_r$" (real) and "$_i$" (imaginary) subscripts in the summation-form in quaternion-complex-hypercomplex algebra).

$$_4r_B*{_6A} = {_4r_B}*({_4r_C}^{*j}\mathbf{x}{_4r_D}) = {_4r_B}\blacksquare({_4r_C}^{*j}\mathbf{x}{_4r_D}) + {_4r_B}\blacklozenge({_4r_C}^{*j}\mathbf{x}{_4r_D})$$

Triple-Vector-Product. "■"

$$_4r_B\blacksquare{_6A} = {_1r_B}*{_3A} + {_3r_B}\mathbf{x}{_3B} + {_3r_B}\bullet{_3A}$$

$$_4V_{BCD} = {_4r_B}\blacksquare({_4r_C}^{*j}\mathbf{x}{_4r_D}) = {_1r_B}*({_3r_C}^{*j}\mathbf{x}{_3r_D}) + {_3r_B}\mathbf{x}({_1r_C}^{*j}*{_3r_D})$$
$$+ {_3r_B}\mathbf{x}({_3r_C}^{*j}*{_1r_D}) + {_3r_B}\bullet({_3r_C}^{*j}\mathbf{x}{_3r_D})$$

CHAPTER 1 - NUMBERS

$$= c*t_B*((y_C*z_D - z_C*y_D)*j_x + (z_C*x_D - x_C*z_D)*j_y + (x_C*y_D - y_C*x_D)*j_z)$$
$$- ((y_B*B_z - z_B*B_y)*j_x + (z_B*B_x - x_B*B_z)*j_y + (x_B*B_y - y_B*B_x)*j_z)$$
$$+ x_B*(y_C*z_D - z_C*y_D)*q_x*j_x + y_B*(z_C*x_D - x_C*z_D)*q_y*j_y$$
$$+ z_B*(x_C*y_D - y_C*x_D)*q_z*j_z$$

$$= c*t_B*y_C*z_D*j_x - c*t_B*z_C*y_D*j_x + c*t_B*z_C*x_D*j_y$$
$$- c*t_B*x_C*z_D*j_y + c*t_B*x_C*y_D*j_z - c*t_B*y_C*x_D*j_z$$
$$- ((y_B*c*t_C*z_D - y_B*c*t_D*z_C - z_B*c*t_C*y_D + z_B*c*t_D*y_C)*j_x$$
$$+ (z_B*c*t_C*x_D - z_B*c*t_D*x_C - x_B*c*t_C*z_D + x_B*c*t_D*z_C)*j_y$$
$$+ (x_B*c*t_C*y_D - x_B*c*t_D*y_C - y_B*c*t_C*x_D - y_B*c*t_D*x_C)*j_z)$$
$$+ x_B*y_C*z_D*i - x_B*z_C*y_D*i + y_B*z_C*x_D*i - y_B*x_C*z_D*i$$
$$+ z_B*x_C*y_D*i - z_B*y_C*x_D*i$$

$$= i*(x_B*y_C*z_D + y_B*z_C*x_D + z_B*x_C*y_D$$
$$- z_B*y_C*x_D - x_B*z_C*y_D - y_B*x_C*z_D)$$
$$+ j_x*(c*t_B*y_C*z_D + z_B*c*t_C*y_D + y_B*z_C*c*t_D$$
$$- c*t_B*z_C*y_D - y_B*c*t_C*z_D - z_B*y_C*c*t_D)$$
$$+ j_y*(c*t_B*z_C*x_D + x_B*c*t_C*z_D + z_B*x_C*c*t_D$$
$$- c*t_B*x_C*z_D - z_B*c*t_C*x_D - x_B*z_C*c*t_D)$$
$$+ j_z*(c*t_B*x_C*y_D + y_B*c*t_C*x_D + x_B*y_C*c*t_D$$
$$- c*t_B*y_C*x_D - x_B*c*t_C*y_D - y_B*x_C*c*t_D)$$

Notice there is no term with two "x" dimension factors, etc. Also note that a reversal of any two terms causes the result to be negative, "$_4V = {_4r_B}\blacksquare({_4r_C}^{*j}\mathbf{X}{_4r_D})$ $= -{_4r_C}\blacksquare({_4r_B}^{*j}\mathbf{X}{_4r_D}) = {_4r_C}\blacksquare({_4r_D}^{*j}\mathbf{X}{_4r_B})$".

Notice "$i*{_4r_B}\blacksquare{_6A} \neq {_4r_B}\blacksquare(i*{_6A})$".

World-Volume.

$$_1W = {_4r_A}^{*j}\bullet({_4r_B}\blacksquare({_4r_C}^{*j}\mathbf{X}{_4r_D}))$$
$$= (c*t_A*1^{*j} + x_A*q_x^{*j} + y_A*q_y^{*j} + z_A*q_z^{*j})\bullet({_4r_B}\blacksquare({_4r_C}^{*j}\mathbf{X}{_4r_D}))$$

$$= i*(c*t_A*x_B*y_C*z_D + c*t_A*y_B*z_C*x_D + c*t_A*z_B*x_C*y_D$$
$$- c*t_A*z_B*y_C*x_D - c*t_A*x_B*z_C*y_D - c*t_A*y_B*x_C*z_D)$$
$$- i*(x_A*c*t_B*y_C*z_D + x_A*z_B*c*t_C*y_D + x_A*y_B*z_C*c*t_D$$
$$- x_A*c*t_B*z_C*y_D - x_A*y_B*c*t_C*z_D - x_A*z_B*y_C*c*t_D)$$
$$- i*(y_A*c*t_B*z_C*x_D + y_A*x_B*c*t_C*z_D + y_A*z_B*x_C*c*t_D$$
$$- y_A*c*t_B*x_C*z_D - y_A*z_B*c*t_C*x_D - y_A*x_B*z_C*c*t_D)$$
$$- i*(z_A*c*t_B*x_C*y_D + z_A*y_B*c*t_C*x_D + z_A*x_B*y_C*c*t_D$$
$$- z_A*c*t_B*y_C*x_D - z_A*x_B*c*t_C*y_D - z_A*y_B*x_C*c*t_D)$$

SPECIAL ALGEBRA FOR SPECIAL RELATIVITY

$$_1w = {_4r_A}^{*j} \bullet ({_4r_B} \blacksquare ({_4r_C}^{*j} \mathbf{x} {_4r_D})) = -{_4r_D}^{*j} \bullet ({_4r_A} \blacksquare ({_4r_B}^{*j} \mathbf{x} {_4r_C}))$$
$$= {_4r_C}^{*j} \bullet ({_4r_D} \blacksquare ({_4r_A}^{*j} \mathbf{x} {_4r_B})) = -{_4r_B}^{*j} \bullet ({_4r_C} \blacksquare ({_4r_D}^{*j} \mathbf{x} {_4r_A}))$$

$$_1w = {_4r_A}^{*j} \bullet ({_4r_B} \blacksquare ({_4r_C}^{*j} \mathbf{x} {_4r_D})) = -{_4r_B}^{*j} \bullet ({_4r_A} \blacksquare ({_4r_C}^{*j} \mathbf{x} {_4r_D}))$$

Contravariant Vectors. In flat space a basis vector set can be non-orthogonal (which is unlike General Relativity for which the basis vectors are orthogonal). If four non-parallel time-space location vectors are the covariant basis vectors, then the four corresponding contravariant basis vectors will have units of per length and abide by specific dot-product rules.

Contravariant vector "$_4k_A$" is perpendicular (orthogonal or normal) to each covariant vector "$_4r_B$", "$_4r_C$", and "$_4r_D$". "World volume" "$_1w = ({_4r_A}^{*j} \bullet ({_4r_B} \blacksquare ({_4r_C}^{*j} \mathbf{x} {_4r_D})))$" is in the denominator so that "$_4k_A$" is real. See an exercise problem for an example.

$$_4k_A = {_4r_B} \blacksquare ({_4r_C}^{*j} \mathbf{x} {_4r_D}) / ({_4r_A}^{*j} \bullet ({_4r_B} \blacksquare ({_4r_C}^{*j} \mathbf{x} {_4r_D})))$$

$$_4k_A{}^{*j} \bullet {_4r_B} = {_4k_A}^{*j} \bullet {_4r_C} = {_4k_A}^{*j} \bullet {_4r_D} = 0 \; ; \; {_4k_A}^{*j} \bullet {_4r_A} = 1$$

$$_4r_A = {_4k_B} \blacksquare ({_4k_C}^{*j} \mathbf{x} {_4k_D}) / ({_4k_A}^{*j} \bullet ({_4k_B} \blacksquare ({_4k_C}^{*j} \mathbf{x} {_4k_D})))$$

Remnant-product "♦" is the sum of terms not in the triple-vector-product.

$$_4r_B \blacklozenge ({_4r_C}^{*j} \mathbf{x} {_4r_D}) = {_4r_B}^{*} ({_4r_C}^{*j} \mathbf{x} {_4r_D}) - {_4r_B} \blacksquare ({_4r_C}^{*j} \mathbf{x} {_4r_D})$$

$$_4r_B \blacklozenge {_6A} = {_1r_B}^{*}{_3B} + {_3r_B} \mathbf{x} {_3A} + {_3r_B} \bullet {_3B}$$

$$_4r_B \blacklozenge ({_3r_C}^{*j} \mathbf{x} {_3r_D}) = {_1r_B}^{*}({_1r_C}^{*j*}{_3r_D}) + {_1r_B}^{*}({_3r_C}^{*j*}{_1r_D}) + {_3r_B} \mathbf{x} ({_3r_C}^{*j} \mathbf{x} {_3r_D})$$
$$+ {_3r_B} \bullet ({_1r_C}^{*j*}{_3r_D}) + {_3r_B} \bullet ({_3r_C}^{*j*}{_1r_D})$$

$$= c^*t_B{}^*(c^*t_C{}^*x_D - c^*t_D{}^*x_C)^*q_x + c^*t_B{}^*(c^*t_C{}^*y_D - c^*t_D{}^*y_C)^*q_y$$
$$+ c^*t_B{}^*(c^*t_C{}^*z_D - c^*t_D{}^*z_C)^*q_z$$
$$+ (y_B{}^*A_z - z_B{}^*A_y)^*q_x + (z_B{}^*A_x - x_B{}^*A_z)^*q_y + (x_B{}^*A_y - y_B{}^*A_x)^*q_z$$
$$+ x_B{}^*(c^*t_C{}^*x_D - c^*t_D{}^*x_C)^*q_x{}^*q_x + y_B{}^*(c^*t_C{}^*y_D - c^*t_D{}^*y_C)^*q_y{}^*q_y$$
$$+ z_B{}^*(c^*t_C{}^*z_D - c^*t_D{}^*z_C)^*q_z{}^*q_z$$

CHAPTER 1 - NUMBERS

$$= (c*t_B*c*t_C*x_D - c*t_B*c*t_D*x_C)*q_x$$
$$+ (c*t_B*c*t_C*y_D - c*t_B*c*t_D*y_C)*q_y$$
$$+ (c*t_B*c*t_C*z_D - c*t_B*c*t_D*z_C)*q_z$$
$$+ (y_B*(x_C*y_D - y_C*x_D) - z_B*(z_C*x_D - x_C*z_D))*q_x$$
$$+ (z_B*(y_C*z_D - z_C*y_D) - x_B*(x_C*y_D - y_C*x_D))*q_y$$
$$+ (x_B*(z_C*x_D - x_C*z_D) - y_B*(y_C*z_D - z_C*y_D))*q_z$$
$$+ x_B*c*t_C*x_D - x_B*c*t_D*x_C + y_B*c*t_C*y_D - y_B*c*t_D*y_C$$
$$+ z_B*c*t_C*z_D - z_B*c*t_D*z_C$$

$$= 1*(x_B*c*t_C*x_D + y_B*c*t_C*y_D + z_B*c*t_C*z_D$$
$$\quad - x_B*c*t_D*x_C - y_B*c*t_D*y_C - z_B*c*t_D*z_C)$$
$$+ q_x*(c*t_B*c*t_C*x_D + y_B*x_C*y_D + z_B*x_C*z_D$$
$$\quad - c*t_B*c*t_D*x_C - y_B*y_C*x_D - z_B*z_C*x_D)$$
$$+ q_y*(c*t_B*c*t_C*y_D + z_B*y_C*z_D + x_B*y_C*x_D$$
$$\quad - c*t_B*c*t_D*y_C - z_B*z_C*y_D - x_B*x_C*y_D)$$
$$+ q_z*(c*t_B*c*t_C*z_D + x_B*z_C*x_D + y_B*z_C*y_D$$
$$\quad - c*t_B*c*t_D*z_C - x_B*x_C*z_D - y_B*y_C*z_D)$$

Triple-vector-product "■" and remnant-product "♦" each result in a four-term summation-form number, one imaginary (which is the triple-vector product "■") and the other real (remnant product "♦").

There is a pattern to the count of terms in the products: One (for a real number), four (for time-space location), six (for area), four (for volume), and one (for world volume). This pattern of numbers – 1,4,6,4,1 – is a row of Pascal's Triangle.

1.4 Translation Back to Geometry

The translation for step three has the following equivalences.

"1" and "i" become "i_t"   ;   "q_y" and "j_y" become "i_y"
"q_x" and "j_x" become "i_x"   ;   "q_z" and "j_z" become "i_z"

<u>Axial Vectors</u>. "Polar vectors" have "1", "q_x", "q_y", and/or "q_z" with real components. Examples are time-space location, contravariant location, energy-momentum, frequency-wavenumber, and electric field.

"Axial vectors" have "j_x", "j_y", and/or "j_z" such that their components are imaginary. Examples are area (only the imaginary portion), torque, and magnetic field. There are other imaginary all-number expressions, but we do

SPECIAL ALGEBRA FOR SPECIAL RELATIVITY

not consider these to be axial vectors: volume four-term summation-form, and world volume.

Axial geometric-vectors are called "pseudo-vectors" because direction of an axial geometric-vector changes if the coordinate system is left-hand rather than right-hand.

In all-number algebra, "q_x", "q_y", and "q_z" may be thought of as direction indicated by the thumb and "j_x", "j_y", and "j_z" may be thought of as direction of the curl of fingers in a plane, with no thumb applied. Step three translation forces the thumb to be applied.

1.5 Singular-Label-Numbers

$\alpha_1 = 1 + q_1$ alpha (includes "$i - j_x$")
$\beta_1 = q_2 + i*q_3 = q_2*\alpha_1$ beta
$\gamma_1 = \alpha_1 + \beta_1 = \alpha_2*\alpha_1$ gamma

(1, 2, 3) are substituted by (x, y, z), (-x, z, y), (z, x, y), etc. with handedness retained. Therefore, (x, z, y) and (x, -y, z) are not valid.

$$\alpha_x => \begin{array}{cc} 1 & 1 \\ 1 & 1 \end{array} \quad ; \quad \beta_x => \begin{array}{cc} i & i \\ -i & -i \end{array} \quad ; \quad \gamma_x => \begin{array}{cc} 1+i & 1+i \\ 1-i & 1-i \end{array}$$

The above 2x2 matrix isomorphs are singular matrices because the determinant is zero. For terms that commute, the determinant is calculated by multiplying upper left by lower right and subtracting from it upper right by lower left: "$|\gamma_x| = (1+i)*(1-i) - (1-i)*(1+i) = 0$".

A singular-label-number multiplied by its quaternion hypercomplex-conjugate equals zero:

$\alpha_1^{*j}*\alpha_1 = (1 - q_1)*(1 + q_1) = 0 \quad ; \quad \beta_1^{*j}*\beta_1 = (-q_2 - i*q_3)*(q_2 + i*q_3) = 0$

$\gamma_1^{*j}*\gamma_1 = (\alpha_1^{*j} + \beta_1^{*j})*(\alpha_1 + \beta_1)$
$= \alpha_1^{*j}*\alpha_1 + \alpha_1^{*j}*\beta_1 + \beta_1^{*j}*\alpha_1 + \beta_1^{*j}*\beta_1$
$= 0 + (1 - q_1)*((q_2)*(1 + q_1)) + ((1 - q_1)*(-q_2))*(1 + q_1) + 0 = 0$

Squares differ between the three varieties.

CHAPTER 1 - NUMBERS

$\alpha_1{}^2 = (1 + q_1)*(1 + q_1)$; $\beta_1{}^2 = (q_2 + i*q_3)*(q_2 + i*q_3)$
  $= 1 + 2*q_1 + q_1*q_1$ $= q_2*q_2 + q_2*i*q_3 + i*q_3*q_2 + i*q_3*i*q_3$
  $= 2*\alpha_1$ $= 0$

$\gamma_1{}^2 = (\alpha_1 + \beta_1)*(\alpha_1 + \beta_1)$
  $= \alpha_1{}^2 + \alpha_1*\beta_1 + \beta_1*\alpha_1 + \beta_1{}^2$
  $= 2*\alpha_1 + q_2*\alpha_1{}^{*j}*\alpha_1 + q_2*\alpha_1*\alpha_1 + 0$
  $= 2*\alpha_1 + q_2*0 + 2*\beta_1 + 0$
  $= 2*\gamma_1$

A singular-label-number in the denominator is a division by zero and is not allowed, and that gives us "singularity theorems". Examples:

$$((1 + q_x)/2)*((3 - q_x)/2) = (1 + q_x)/2$$

$$((1 \pm q_x)/2)*\exp(q_x*\alpha) = ((1 \pm q_x)/2)*\exp(\pm\alpha)$$

An exponential function with a singular label number argument has a result of unit magnitude, for example, "$\exp((i + j_x)*\pi/2) = -q_x$".

Singular-label-numbers should be thought of as part of a group rather than as individual anomalies. Singular-label-numbers are used for phenomena at the speed-of-light, specifically, in this book, electromagnetic waves.

1.6 Exercises

1) Write geometric-unit-vector representations for time-space "$_4\mathbf{r}$" and energy-momentum "$_4\mathbf{p}$": Use "E/c" for the energy component (time component), and "p_x", "p_y", and "p_z" for the three momentum components (three space components). Write all-number representations for time-space location "$_4r$" and "$_4p$". Write hypercomplex-conjugates of "$_4r$" and "$_4p$".

2) Write matrix multiplication operations for the below.

$$j_x{}^2 = j_y{}^2 = j_z{}^2 = i^2 = -1 \quad ; \quad q_x{}^2 = q_y{}^2 = q_z{}^2 = 1^2 = +1$$

$j_x*j_y = -j_y*j_x = j_z$; $i*j_x = j_x*i$ $q_x = j_x/i$
$j_y*j_z = -j_z*j_y = j_x$; $i*j_y = j_y*i$ $q_y = j_y/i$
$j_z*j_x = -j_x*j_z = j_y$; $i*j_z = j_z*i$ $q_z = j_z/i$

SPECIAL ALGEBRA FOR SPECIAL RELATIVITY

$$q_x{}^*j_y = -j_y{}^*q_x = -q_y{}^*j_x = j_x{}^*q_y = q_z \quad ; \quad q_x{}^*q_y = -q_y{}^*q_x = -i{}^*q_z = -j_z$$
$$q_y{}^*j_z = -j_z{}^*q_y = -q_z{}^*j_y = j_y{}^*q_z = q_x \quad ; \quad q_y{}^*q_z = -q_z{}^*q_y = -i{}^*q_x = -j_x$$
$$q_z{}^*j_x = -j_x{}^*q_z = -q_x{}^*j_z = j_z{}^*q_x = q_y \quad ; \quad q_z{}^*q_x = -q_x{}^*q_z = -i{}^*q_y = -j_y$$

$$\alpha_x{}^{*j}{}^*\alpha_x = (1 - q_x)^*(1 + q_x) = 0 \quad ; \quad \alpha_x{}^2 = (1 + q_x)^*(1 + q_x) = 2^*\alpha_x$$
$$\alpha_y{}^{*j}{}^*\alpha_y = (1 - q_y)^*(1 + q_y) = 0 \quad ; \quad \alpha_y{}^2 = (1 + q_y)^*(1 + q_y) = 2^*\alpha_y$$
$$\alpha_z{}^{*j}{}^*\alpha_z = (1 - q_z)^*(1 + q_z) = 0 \quad ; \quad \alpha_z{}^2 = (1 + q_z)^*(1 + q_z) = 2^*\alpha_z$$

$$\beta_x{}^{*j}{}^*\beta_x = (-q_y - i^*q_z)^*(q_y + i^*q_z) = 0 \quad ; \quad \beta_x{}^2 = (q_y + i^*q_z)^*(q_y + i^*q_z) = 0$$
$$\beta_y{}^{*j}{}^*\beta_y = (-q_z - i^*q_x)^*(q_z + i^*q_x) = 0 \quad ; \quad \beta_y{}^2 = (q_z + i^*q_x)^*(q_z + i^*q_x) = 0$$
$$\beta_z{}^{*j}{}^*\beta_z = (-q_x - i^*q_y)^*(q_x + i^*q_y) = 0 \quad ; \quad \beta_z{}^2 = (q_x + i^*q_y)^*(q_x + i^*q_y) = 0$$

$$\gamma_x{}^{*j}{}^*\gamma_x = (\alpha_x{}^{*j} + \beta_x{}^{*j})^*(\alpha_x + \beta_x) \quad ; \quad \gamma_x{}^2 = (\alpha_x + \beta_x)^*(\alpha_x + \beta_x) = 2^*\gamma_x$$
$$\gamma_y{}^{*j}{}^*\gamma_y = (\alpha_y{}^{*j} + \beta_y{}^{*j})^*(\alpha_y + \beta_y) \quad ; \quad \gamma_y{}^2 = (\alpha_y + \beta_y)^*(\alpha_y + \beta_y) = 2^*\gamma_y$$
$$\gamma_z{}^{*j}{}^*\gamma_z = (\alpha_z{}^{*j} + \beta_z{}^{*j})^*(\alpha_z + \beta_z) \quad ; \quad \gamma_z{}^2 = (\alpha_z + \beta_z)^*(\alpha_z + \beta_z) = 2^*\gamma_z$$

3) Write dot product and cross product multiplication tables using "q"'s rather than "j"'s. And, again for "q^{*j}" on the left.

4) Translate the below summation-form all-number expressions into summation-form expressions that have geometric-unit-vectors as factors. Explain in words what the geometric-unit-vector translations represent in our physical world.

$$_4r_C = c^*t_C + q_x{}^*x_C + q_y{}^*y_C + q_z{}^*z_C$$
$$_3r_C{}^{*j} = c^*t_C + x_C{}^*q_x{}^{*j} + y_C{}^*q_y{}^{*j} + z_C{}^*q_z{}^{*j}$$

$$_6A = B_x{}^*q_x + B_y{}^*q_y + B_z{}^*q_z + A_x{}^*j_x + A_y{}^*j_y + A_z{}^*j_z$$
$$_4V_A = i^*V_{At} + j_x{}^*V_{Ax} + j_y{}^*V_{Ay} + j_z{}^*V_{Az}$$

World Volume = i^*w

5) For the below set of time-space location geometric-vectors "$_4r_A$", "$_4r_B$", "$_4r_C$" and "$_4r_D$", find the corresponding set of contravariant geometric-vectors "$_4k_A$", "$_4k_B$", "$_4k_C$" and "$_4k_D$". Check with the dot product. And, then, as another check, repeat the process but use the derived contravariant geometric-vectors "$_4k_A$", "$_4k_B$", "$_4k_C$" and "$_4k_D$" substituting for the original (covariant) geometric-vectors, to find "$_4r_A$", "$_4r_B$", "$_4r_C$" and "$_4r_D$".

CHAPTER 1 - NUMBERS

First Exercise:

$_4r_A = 1*i_x$ $_4r_B = 1*i_y$ $_4r_C = 3*i_t + 5*i_z$ $_4r_D = 2*i_t + 7*i_z$

Second Exercise:

$_4r_A = 1*i_x$ $_4r_B = 1*i_y$ $_4r_C = 2*i_t + 5*i_z$ $_4r_D = 3*i_t + 7*i_z$

Third Exercise:

$_4r_A = 2*i_t + 1*i_x$ $_4r_B = 3*i_t + 1*i_y$ $_4r_C = 5*i_z$ $_4r_D = 7*i_z$

6) Prove the two time-space location geometric-vectors "$_4r_C$" and "$_4r_D$" are each perpendicular to the cross-product of "$_4r_C$" and "$_4r_D$". Use the triple-vector-product set equal to zero.

7) A row in Pascal's Triangle is 1 3 3 1. Using three-dimensional geometric-unit-vectors, relate the first "1" to a real number, the first "3" to the polar vector for location, the second "3" to the axial vector for area formed by the cross-product, and the second "1" to the volume formed by the dot-product. Next, for the row 1 2 1, model space using complex numbers: For two locations on the complex-plane modeled by "$_2r_A$" and "$_2r_B$", take the complex-conjugate of one of them, and then multiply them using the dot product, and relate the result to the 1 2 1 model of terms in vector expressions for the complex-plane.

8) .A. Prove "$a*b = ((a^2 + b^2) - (a^2 - b^2))/4$" for rational numbers by stating which criteria are used in each step of the proof. .B. Prove "$-(a - b) = b - a$" and "$1/(a/b) = b/a$" for rational numbers, also by referencing criteria. .C. Prove quadratic equation solution "$x = -b/(2*a) \pm (\sqrt{(b*b - 4*a*c)})/(2*a)$" is the solution for "$a*x^2 + b*x + c = 0$". For the quadratic equation solution, what limitations must be placed onto "a", "b", "c", and "x" for integers, rational numbers, real numbers, and for complex numbers to apply. .D. Prove the distributive property of the exponential operation over multiplication with reference to an algebra and criteria. Explain why this distributive property does not apply to "$\exp(q_x*x + q_y*y)$".

9) Prove the below singularity theorem. Why cannot "$1 \pm q_x$" be divided by both sides to equate "$\exp(q_x*\alpha)$" to "$\exp(\pm\alpha)$"?

$$(1 \pm q_x)*\exp(q_x*\alpha) = (1 \pm q_x)*\exp(\pm\alpha)$$

SPECIAL ALGEBRA FOR SPECIAL RELATIVITY

Select Solutions

1) $_4\mathbf{r} = c*t*\mathbf{i}_t + x*\mathbf{i}_x + y*\mathbf{i}_y + z*\mathbf{i}_z$; $_4r = 1*c*t + q_x*x + q_y*y + q_z*z$
$_4r^{*j} = c*t*1^{*j} + x*q_x^{*j} + y*q_y^{*j} + z*q_z^{*j}$
$= c*t*1 - x*q_x - y*q_y - z*q_z$

$_4\mathbf{p} = (E/c)*\mathbf{i}_t + p_x*\mathbf{i}_x + p_y*\mathbf{i}_y + p_z*\mathbf{i}_z$; $_4p = 1*(E/c) + q_x*p_x + q_y*p_y + q_z*p_z$
$_4p^{*j} = (E/c)*1^{*j} + p_x*q_x^{*j} + p_y*q_y^{*j} + p_z*q_z^{*j}$
$= (E/c)*1 - p_x*q_x - p_y*q_y - p_z*q_z$

2)

$j_x^2 = -1 \Rightarrow$ $\begin{matrix} 0 & i \\ i & 0 \end{matrix} * \begin{matrix} 0 & i \\ i & 0 \end{matrix} = \begin{matrix} 0*0+i*i & 0*i+i*0 \\ i*0+0*i & i*i+0*0 \end{matrix} = \begin{matrix} -1 & 0 \\ 0 & -1 \end{matrix}$

$j_x*j_y = j_z \Rightarrow$ $\begin{matrix} 0 & i \\ i & 0 \end{matrix} * \begin{matrix} 0 & -1 \\ 1 & 0 \end{matrix} = \begin{matrix} 0*0+i*1 & 0*-1+i*0 \\ i*0+0*1 & i*-1+0*0 \end{matrix} = \begin{matrix} i & 0 \\ 0 & -i \end{matrix}$

$i*j_x = j_x*i$

$\begin{matrix} i & 0 \\ 0 & i \end{matrix} * \begin{matrix} 0 & i \\ i & 0 \end{matrix} = \begin{matrix} i*0+0*i & i*i+0*0 \\ 0*0+i*i & 0*i+i*0 \end{matrix} = \begin{matrix} 0*i+i*0 & 0*0+i*i \\ i*i+0*0 & i*0+0*i \end{matrix} = \begin{matrix} 0 & i \\ i & 0 \end{matrix} * \begin{matrix} i & 0 \\ 0 & i \end{matrix}$

$q_x = j_x/i$ $j_x/i = -(j_x*i)*(i/i) = -(j_x*i) = q_x$

$\begin{matrix} 0 & i \\ i & 0 \end{matrix} / \begin{matrix} i & 0 \\ 0 & i \end{matrix} = -(\begin{matrix} 0 & i \\ i & 0 \end{matrix} * \begin{matrix} i & 0 \\ 0 & i \end{matrix})*(\begin{matrix} i & 0 \\ 0 & i \end{matrix} / \begin{matrix} i & 0 \\ 0 & i \end{matrix}) = -\begin{matrix} 0 & i \\ i & 0 \end{matrix} * \begin{matrix} i & 0 \\ 0 & i \end{matrix} = -\begin{matrix} 0*i+i*0 & 0*0+i*i \\ i*i+0*0 & i*0+0*i \end{matrix} = \begin{matrix} 0 & 1 \\ 1 & 0 \end{matrix}$

$q_x^2 = +1 \Rightarrow$ $\begin{matrix} 0 & 1 \\ 1 & 0 \end{matrix} * \begin{matrix} 0 & 1 \\ 1 & 0 \end{matrix} = \begin{matrix} 0*0+1*1 & 0*1+1*0 \\ 1*0+0*1 & 1*1+0*0 \end{matrix} = \begin{matrix} 1 & 0 \\ 0 & 1 \end{matrix}$

$q_x*j_y = q_z \Rightarrow$ $\begin{matrix} 0 & 1 \\ 1 & 0 \end{matrix} * \begin{matrix} 0 & -1 \\ 1 & 0 \end{matrix} = \begin{matrix} 0*0+1*1 & 0*-1+1*0 \\ 1*0+0*1 & 1*-1+0*0 \end{matrix} = \begin{matrix} 1 & 0 \\ 0 & -1 \end{matrix}$

CHAPTER 1 - NUMBERS

$q_x * q_y = -j_z \Rightarrow$ $\begin{matrix} 0 & 1 \\ 1 & 0 \end{matrix}$ * $\begin{matrix} 0 & i \\ -i & 0 \end{matrix}$ = $\begin{matrix} 0*0+1*-i & 0*i+1*0 \\ 1*0+0*-i & 1*i+0*0 \end{matrix}$ = $\begin{matrix} -i & 0 \\ 0 & i \end{matrix}$

$\alpha_x^{*j} * \alpha_x = (1 - q_x)*(1 + q_x) = 0$

$\begin{matrix} 1 & -1 \\ -1 & 1 \end{matrix}$ * $\begin{matrix} 1 & 1 \\ 1 & 1 \end{matrix}$ = $\begin{matrix} 1*1+-1*1 & 1*1+-1*1 \\ -1*1+1*1 & -1*1+1*1 \end{matrix}$ = $\begin{matrix} 0 & 0 \\ 0 & 0 \end{matrix}$

$\beta_x^{*j} * \beta_x = (-q_y - i*q_z)*(q_y + i*q_z) = 0$

$\begin{matrix} -i*1 & -i \\ -i & -i*-1 \end{matrix}$ * $\begin{matrix} i*1 & i \\ -i & i*-1 \end{matrix}$ $\begin{matrix} -i & -i \\ i & i \end{matrix}$ * $\begin{matrix} i & i \\ -i & -i \end{matrix}$ = $\begin{matrix} -i*i+-i*-i & -i*i+-i*-i \\ i*i+i*-i & i*i+i*-i \end{matrix}$ = $\begin{matrix} 0 & 0 \\ 0 & 0 \end{matrix}$

$\gamma_x^{*j} * \gamma_x = (\alpha_x^{*j} + \beta_x^{*j})*(\alpha_x + \beta_x) = 0$

$\begin{matrix} 1-i & -1-i \\ -1+i & 1+i \end{matrix}$ * $\begin{matrix} 1+i & 1+i \\ 1-i & 1-i \end{matrix}$ = $\begin{matrix} (1^2-i^2)-(1^2-i^2) & (1^2-i^2)-(1^2-i^2) \\ -(1^2-i^2)+(1^2-i^2) & -(1^2-i^2)+(1^2-i^2) \end{matrix}$ = $\begin{matrix} 0 & 0 \\ 0 & 0 \end{matrix}$

$\alpha_x^2 = (1 + q_x)*(1 + q_x) = 2*\alpha_x$

$\begin{matrix} 1 & 1 \\ 1 & 1 \end{matrix}$ * $\begin{matrix} 1 & 1 \\ 1 & 1 \end{matrix}$ = $\begin{matrix} 1*1+1*1 & 1*1+1*1 \\ 1*1+1*1 & 1*1+1*1 \end{matrix}$ = $\begin{matrix} 2 & 2 \\ 2 & 2 \end{matrix}$

$\beta_x^2 = (q_y + i*q_z)*(q_y + i*q_z) = 0$

$\begin{matrix} i & i \\ -i & -i \end{matrix}$ * $\begin{matrix} i & i \\ -i & -i \end{matrix}$ = $\begin{matrix} i^2-i^2 & i^2-i^2 \\ -(i^2-i^2) & -(i^2-i^2) \end{matrix}$ = $\begin{matrix} 0 & 0 \\ 0 & 0 \end{matrix}$

$\gamma_x^2 = (\alpha_x + \beta_x)*(\alpha_x + \beta_x) = 2*\gamma_x$

$\begin{matrix} 1+i & 1+i \\ 1-i & 1-i \end{matrix}$ * $\begin{matrix} 1+i & 1+i \\ 1-i & 1-i \end{matrix}$ = $\begin{matrix} (1^2+2*i+i^2)+(1^2-i^2) & (1^2+2*i+i^2)+(1^2-i^2) \\ (1^2-i^2)+(1^2-2*i+i^2) & (1^2-i^2)+(1^2-2*i+i^2) \end{matrix}$ = $\begin{matrix} 2+2*i & 2+2*i \\ 2-2*i & 2+2*i \end{matrix}$

SPECIAL ALGEBRA FOR SPECIAL RELATIVITY

3) Solution:

$1 \bullet 1 = 1$ $1 \bullet q_x = 0$ $1 \bullet q_y = 0$ $1 \bullet q_z = 0$
$q_x \bullet 1 = 0$ $q_x \bullet q_x = +1$ $q_x \bullet q_y = 0$ $q_x \bullet q_z = 0$
$q_y \bullet 1 = 0$ $q_y \bullet q_x = 0$ $q_y \bullet q_y = +1$ $q_y \bullet q_z = 0$
$q_z \bullet 1 = 0$ $q_z \bullet q_x = 0$ $q_z \bullet q_y = 0$ $q_z \bullet q_z = +1$

$1 \mathbf{x} 1 = 0$ $1 \mathbf{x} q_x = q_x$ $1 \mathbf{x} q_y = q_y$ $1 \mathbf{x} q_z = q_z$
$q_x \mathbf{x} 1 = q_x$ $q_x \mathbf{x} q_x = 0$ $q_x \mathbf{x} q_y = -j_z$ $q_x \mathbf{x} q_z = +j_y$
$q_y \mathbf{x} 1 = q_y$ $q_y \mathbf{x} q_x = +j_z$ $q_y \mathbf{x} q_y = 0$ $q_y \mathbf{x} q_z = -j_x$
$q_z \mathbf{x} 1 = q_z$ $q_z \mathbf{x} q_x = -j_y$ $q_z \mathbf{x} q_y = +j_x$ $q_z \mathbf{x} q_z = 0$

$1^{*j} \bullet 1 = 1$ $1^{*j} \bullet q_x = 0$ $1^{*j} \bullet q_y = 0$ $1^{*j} \bullet q_z = 0$
$q_x^{*j} \bullet 1 = 0$ $q_x^{*j} \bullet q_x = -1$ $q_x^{*j} \bullet q_y = 0$ $q_x^{*j} \bullet q_z = 0$
$q_y^{*j} \bullet 1 = 0$ $q_y^{*j} \bullet q_x = 0$ $q_y^{*j} \bullet q_y = -1$ $q_y^{*j} \bullet q_z = 0$
$q_z^{*j} \bullet 1 = 0$ $q_z^{*j} \bullet q_x = 0$ $q_z^{*j} \bullet q_y = 0$ $q_z^{*j} \bullet q_z = -1$

$1^{*j} \mathbf{x} 1 = 0$ $1^{*j} \mathbf{x} q_x = q_x$ $1^{*j} \mathbf{x} q_y = q_y$ $1^{*j} \mathbf{x} q_z = q_z$
$q_x^{*j} \mathbf{x} 1 = -q_x$ $q_x^{*j} \mathbf{x} q_x = 0$ $q_x^{*j} \mathbf{x} q_y = +j_z$ $q_x^{*j} \mathbf{x} q_z = -j_y$
$q_y^{*j} \mathbf{x} 1 = -q_y$ $q_y^{*j} \mathbf{x} q_x = -j_z$ $q_y^{*j} \mathbf{x} q_y = 0$ $q_y^{*j} \mathbf{x} q_z = +j_x$
$q_z^{*j} \mathbf{x} 1 = -q_z$ $q_z^{*j} \mathbf{x} q_x = +j_y$ $q_z^{*j} \mathbf{x} q_y = -j_x$ $q_z^{*j} \mathbf{x} q_z = 0$

4) Solutions:

$_4\mathbf{r}_C = c*t_C*\mathbf{i}_t + x_C*\mathbf{i}_x + y_C*\mathbf{i}_y + z_C*\mathbf{i}_z$; $_3\mathbf{B} = B_x*\mathbf{i}_x + B_y*\mathbf{i}_y + B_z*\mathbf{i}_z$
$_3\mathbf{A} = A_x*\mathbf{i}_x + A_y*\mathbf{i}_y + A_z*\mathbf{i}_z$

$_4\mathbf{V}_A = V_{At}*\mathbf{i}_t + V_{Ax}*\mathbf{i}_x + V_{Ay}*\mathbf{i}_y + V_{Az}*\mathbf{i}_z$; World Volume = w

5) Select Solution, First Exercise:

$_4\mathbf{r}_A = 1*\mathbf{i}_x$ $_4\mathbf{r}_B = 1*\mathbf{i}_y$ $_4\mathbf{r}_C = 3*\mathbf{i}_t + 5*\mathbf{i}_z$ $_4\mathbf{r}_D = 2*\mathbf{i}_t + 7*\mathbf{i}_z$

$_4\mathbf{r}_A = q_x*1$ $_4\mathbf{r}_B = q_y*1$ $_4\mathbf{r}_C = 3 + q_z*5$ $_4\mathbf{r}_D = 2 + q_z*7$

$_4\mathbf{r}_A^{*j} = 1*q_x^{*j}$ $_4\mathbf{r}_B^{*j} = 1*q_y^{*j}$ $_4\mathbf{r}_C^{*j} = 3*1^{*j} + 5*q_z^{*j}$ $_4\mathbf{r}_D^{*j} = 2*1^{*j} + 7*q_z^{*j}$

$x_A = 1$, $y_B = 1$, $c*t_C = 3$, $z_C = 5$, $c*t_D = 2$, $z_D = 7$

CHAPTER 1 - NUMBERS

$_3r_D{}^{*j}\mathbf{x}_3r_A = (7*q_z{}^{*j})\mathbf{x}(q_x*1) = 7*q_z{}^{*j}*q_x*1 = j_y*7$
$_1r_D{}^{*j}*_3r_A = (2*1^{*j})\mathbf{x}(q_x*1) = 2*1^{*j}*q_x*1 = q_x*2$
$_3r_D{}^{*j}*_1r_A = (7*q_z{}^{*j})\mathbf{x}(0) = 0$
$_4r_D{}^{*j}\mathbf{x}_4r_A = (2*1^{*j} + 7*q_z{}^{*j})\mathbf{x}(q_x*1) = 2*1^{*j}*q_x*1 + 7*q_z{}^{*j}*q_x*1$
$\qquad\qquad = q_x*2 + j_y*7$

$_3r_A{}^{*j}\mathbf{x}_3r_B = (1*q_x{}^{*j})\mathbf{x}(q_y*1) = 1*q_x{}^{*j}*q_y = j_z*1$
$_1r_A{}^{*j}*_3r_B = (0)\mathbf{x}(q_y*1) = 0$
$_3r_A{}^{*j}*_1r_B = (1*q_x{}^{*j})\mathbf{x}(0) = 0$
$_4r_A{}^{*j}\mathbf{x}_4r_B = (1*q_x{}^{*j})\mathbf{x}(q_y*1) = 1*q_x{}^{*j}*q_y = j_z*1$

$_3r_B{}^{*j}\mathbf{x}_3r_C = (1*q_y{}^{*j})\mathbf{x}(q_z*5) = 1*q_y{}^{*j}*q_z*5 = j_x*5$
$_1r_B{}^{*j}*_3r_C = (0)\mathbf{x}(q_z*5) = 0$
$_3r_B{}^{*j}*_1r_C = (1*q_y{}^{*j})\mathbf{x}(3) = 1*q_y{}^{*j}*3 = q_y*(-3)$
$_4r_B{}^{*j}\mathbf{x}_4r_C = (1*q_y{}^{*j})\mathbf{x}(3 + q_z*5) = 1*q_y{}^{*j}*3 + 1*q_y{}^{*j}*q_z*5$
$\qquad\qquad = q_y*(-3) + j_x*5$

$_3r_C{}^{*j}\mathbf{x}_3r_D = (5*q_z{}^{*j})\mathbf{x}(q_z*7) = 0$
$_1r_C{}^{*j}*_3r_D = (3*1^{*j})\mathbf{x}(q_z*7) = 3*1^{*j}*q_z*7 = q_z*21$
$_3r_C{}^{*j}*_1r_D = (5*q_z{}^{*j})\mathbf{x}(2) = 5*q_z{}^{*j}*2 = q_z*(-10)$
$_4r_C{}^{*j}\mathbf{x}_4r_D = (3*1^{*j} + 5*q_z{}^{*j})\mathbf{x}(2 + q_z*7) = 3*1^{*j}*q_z*7 + 5*q_z{}^{*j}*2$
$\qquad\qquad = q_z*(21 - 10) = q_z*11$

$_4V_{CDA} = {}_4r_C\blacksquare({}_4r_D{}^{*j}\mathbf{x}_4r_A)$
$= {}_1r_C*({}_3r_D{}^{*j}\mathbf{x}_3r_A) + {}_3r_C\mathbf{x}({}_1r_D{}^{*j}*_3r_A) + {}_3r_C\mathbf{x}({}_3r_D{}^{*j}*_1r_A) + {}_3r_C\bullet({}_3r_D{}^{*j}\mathbf{x}_3r_A)$
$= (3)*(j_y*7) + (q_z*5)\mathbf{x}(q_x*2) + (q_z*5)\mathbf{x}(0) + (q_z*5)\bullet(j_y*7)$
$= j_y*21 + j_y*(-10) + 0 + 0$
$= j_y*11$
$\qquad\qquad$ Check:
$_4V_{CDA} = {}_4r_C\blacksquare({}_4r_D{}^{*j}\mathbf{x}_4r_A) = {}_1r_C*({}_3r_D{}^{*j}\mathbf{x}_3r_A) + {}_3r_C\mathbf{x}({}_1r_D{}^{*j}*_3r_A)$
$\qquad\qquad + {}_3r_C\mathbf{x}({}_3r_D{}^{*j}*_1r_A) + {}_3r_C\bullet({}_3r_D{}^{*j}\mathbf{x}_3r_A)$

$= i*(x_C*y_D*z_A + y_C*z_D*x_A + z_C*x_D*y_A$
$\quad - z_C*y_D*x_A - x_C*z_D*y_A - y_C*x_D*z_A)$
$+ j_x*(c*t_C*y_D*z_A + z_C*c*t_D*y_A + y_C*z_D*c*t_A$
$\quad - c*t_C*z_D*y_A - y_C*c*t_D*z_A - z_C*y_D*c*t_A)$
$+ j_y*(c*t_C*z_D*x_A + x_C*c*t_D*z_A + z_C*x_D*c*t_A$
$\quad - c*t_C*x_D*z_A - z_C*c*t_D*x_A - x_C*z_D*c*t_A)$
$+ j_z*(c*t_C*x_D*y_A + y_C*c*t_D*x_A + x_C*y_D*c*t_A$
$\quad - c*t_C*y_D*x_A - x_C*c*t_D*y_A - y_C*x_D*c*t_A)$

SPECIAL ALGEBRA FOR SPECIAL RELATIVITY

$$= i*(0) + j_x*(0) + j_y*(c*t_C*z_D*x_A - z_C*c*t_D*x_A) + j_z*(0)$$
$$= j_y*(3*7*1 - 5*2*1) = j_y*(21 - 10) = j_y*(11) \quad \text{Good}$$

$${}_4V_{DAB} = {}_4r_D\blacksquare({}_4r_A{}^{*j}\mathbf{X}{}_4r_B)$$
$$= {}_1r_D*({}_3r_A{}^{*j}\mathbf{X}{}_3r_B) + {}_3r_D\mathbf{X}({}_1r_A{}^{*j}*{}_3r_B) + {}_3r_D\mathbf{X}({}_3r_A{}^{*j}*{}_1r_B) + {}_3r_D\bullet({}_3r_A{}^{*j}\mathbf{X}{}_3r_B)$$
$$= (2)*(j_z*1) + (q_z*7)\mathbf{X}(0) + (q_z*7)\mathbf{X}(0) + (q_z*7)\bullet(j_z*1)$$
$$= j_z*2 + i*7$$
Check:
$${}_4V_{DAB} = {}_4r_D\blacksquare({}_4r_A{}^{*j}\mathbf{X}{}_4r_B) = {}_1r_D*({}_3r_A{}^{*j}\mathbf{X}{}_3r_B) + {}_3r_D\mathbf{X}({}_1r_A{}^{*j}*{}_3r_B)$$
$$+ {}_3r_D\mathbf{X}({}_3r_A{}^{*j}*{}_1r_B) + {}_3r_D\bullet({}_3r_A{}^{*j}\mathbf{X}{}_3r_B)$$

$$= i*(x_D*y_A*z_B + y_D*z_A*x_B + z_D*x_A*y_B$$
$$- z_D*y_A*x_B - x_D*z_A*y_B - y_D*x_A*z_B)$$
$$+ j_x*(c*t_D*y_A*z_B + z_D*c*t_A*y_B + y_D*z_A*c*t_B$$
$$- c*t_D*z_A*y_B - y_D*c*t_A*z_B - z_D*y_A*c*t_B)$$
$$+ j_y*(c*t_D*z_A*x_B + x_D*c*t_A*z_B + z_D*x_A*c*t_B$$
$$- c*t_D*x_A*z_B - z_D*c*t_A*x_B - x_D*z_A*c*t_B)$$
$$+ j_z*(c*t_D*x_A*y_B + y_D*c*t_A*x_B + x_D*y_A*c*t_B$$
$$- c*t_D*y_A*x_B - x_D*c*t_A*y_B - y_D*x_A*c*t_B)$$

$$= i*(z_D*x_A*y_B) + j_x*(0) + j_y*(0) + j_z*(c*t_D*x_A*y_B)$$
$$= i*(7*1*1) + j_x*(0) + j_y*(0) + j_z*(2*1*1)$$
$$= i*(7) + j_z*(2) \quad \text{Good}$$

$${}_4V_{ABC} = {}_4r_A\blacksquare({}_4r_B{}^{*j}\mathbf{X}{}_4r_C)$$
$$= {}_1r_A*({}_3r_B{}^{*j}\mathbf{X}{}_3r_C) + {}_3r_A\mathbf{X}({}_1r_B{}^{*j}*{}_3r_C) + {}_3r_A\mathbf{X}({}_3r_B{}^{*j}*{}_1r_C) + {}_3r_A\bullet({}_3r_B{}^{*j}\mathbf{X}{}_3r_C)$$
$$= (0)*(j_x*5) + (q_x*1)\mathbf{X}(0) + (q_x*1)\mathbf{X}(q_y*(-3)) + (q_x*1)\bullet(j_x*5)$$
$$= j_z*3 + i*5$$
Check:
$${}_4V_{ABC} = {}_4r_A\blacksquare({}_4r_B{}^{*j}\mathbf{X}{}_4r_C) = {}_1r_A*({}_3r_B{}^{*j}\mathbf{X}{}_3r_C) + {}_3r_A\mathbf{X}({}_1r_B{}^{*j}*{}_3r_C)$$
$$+ {}_3r_A\mathbf{X}({}_3r_B{}^{*j}*{}_1r_C) + {}_3r_A\bullet({}_3r_B{}^{*j}\mathbf{X}{}_3r_C)$$

$$= i*(x_A*y_B*z_C + y_A*z_B*x_C + z_A*x_B*y_C$$
$$- z_A*y_B*x_C - x_A*z_B*y_C - y_A*x_B*z_C)$$
$$+ j_x*(c*t_A*y_B*z_C + z_A*c*t_B*y_C + y_A*z_B*c*t_C$$
$$- c*t_A*z_B*y_C - y_A*c*t_B*z_C - z_A*y_B*c*t_C)$$
$$+ j_y*(c*t_A*z_B*x_C + x_A*c*t_B*z_C + z_A*x_B*c*t_C$$

CHAPTER 1 - NUMBERS

$\qquad - c*t_A*x_B*z_C - z_A*c*t_B*x_C - x_A*z_B*c*t_C)$
$\qquad + j_z*(c*t_A*x_B*y_C + y_A*c*t_B*x_C + x_A*y_B*c*t_C$
$\qquad - c*t_A*y_B*x_C - x_A*c*t_B*y_C - y_A*x_B*c*t_C)$

$= i*(x_A*y_B*z_C) + j_x*(0) + j_y*(0) + j_z*(x_A*y_B*c*t_C)$
$= i*(1*1*5) + j_z*(1*1*3)$
$= i*(5) + j_z*(3)$ Good

$\qquad _4V_{BCD} = {_4r_B}\blacksquare({_4r_C}^{*j}\mathbf{x}{_4r_D})$
$\qquad\qquad = {_1r_B}*({_3r_C}^{*j}\mathbf{x}{_3r_D}) + {_3r_B}\mathbf{x}({_1r_C}^{*j}*{_3r_D}) + {_3r_B}\mathbf{x}({_3r_C}^{*j}*{_1r_D}) + {_3r_B}\bullet({_3r_C}^{*j}\mathbf{x}{_3r_D})$
$\qquad\qquad = (0)*(0) + (q_y*1)\mathbf{x}(q_z*21) + (q_y*1)\mathbf{x}(q_z*(-10)) + (q_y*1)\bullet(0)$
$\qquad\qquad = j_x*(-21+10) = j_x*(-11)$
$\qquad\qquad\qquad\qquad\qquad$ Check:
$\qquad _4V_{BCD} = {_4r_B}\blacksquare({_4r_C}^{*j}\mathbf{x}{_4r_D}) = {_1r_B}*({_3r_C}^{*j}\mathbf{x}{_3r_D}) + {_3r_B}\mathbf{x}({_1r_C}^{*j}*{_3r_D})$
$\qquad\qquad\qquad\qquad + {_3r_B}\mathbf{x}({_3r_C}^{*j}*{_1r_D}) + {_3r_B}\bullet({_3r_C}^{*j}\mathbf{x}{_3r_D})$

$= i*(x_B*y_C*z_D + y_B*z_C*x_D + z_B*x_C*y_D$
$\qquad - z_B*y_C*x_D - x_B*z_C*y_D - y_B*x_C*z_D)$
$+ j_x*(c*t_B*y_C*z_D + z_B*c*t_C*y_D + y_B*z_C*c*t_D$
$\qquad - c*t_B*z_C*y_D - y_B*c*t_C*z_D - z_B*y_C*c*t_D)$
$+ j_y*(c*t_B*z_C*x_D + x_B*c*t_C*z_D + z_B*x_C*c*t_D$
$\qquad - c*t_B*x_C*z_D - z_B*c*t_C*x_D - x_B*z_C*c*t_D)$
$+ j_z*(c*t_B*x_C*y_D + y_B*c*t_C*x_D + x_B*y_C*c*t_D$
$\qquad - c*t_B*y_C*x_D - x_B*c*t_C*y_D - y_B*x_C*c*t_D)$

$= i*(0) + j_x*(y_B*z_C*c*t_D - y_B*c*t_C*z_D) + j_y*(0) + j_z*(0)$
$= j_x*(1*5*2 - 1*3*7) = j_x*(10 - 21)$
$= j_x*(-11)$ Good

$\qquad _1W = ({_4r_A}^{*j}\bullet({_4r_B}\blacksquare({_4r_C}^{*j}\mathbf{x}{_4r_D}))) = (1*q_x^{*j})\bullet(j_x*(-11)) = i*11$
$\qquad _1W = -({_4r_D}^{*j}\bullet({_4r_A}\blacksquare({_4r_B}^{*j}\mathbf{x}{_4r_C}))) = -(2*1^{*j} + 7*q_z^{*j})\bullet(j_z*3 + i*5)$
$\qquad\qquad\qquad\qquad = -(-i*21 + i*10) = i*11$
$\qquad _1W = ({_4r_C}^{*j}\bullet({_4r_D}\blacksquare({_4r_A}^{*j}\mathbf{x}{_4r_B}))) = (3*1^{*j} + 5*q_z^{*j})\bullet(j_z*2 + i*7)$
$\qquad\qquad\qquad\qquad = i*21 - i*10 = i*11$
$\qquad _1W = -({_4r_B}^{*j}\bullet({_4r_C}\blacksquare({_4r_D}^{*j}\mathbf{x}{_4r_A}))) = -(1*q_y^{*j})\bullet(j_y*11) = i*(11)$

Checks: $x_A = 1$, $y_B = 1$, $c*t_C = 3$, $z_C = 5$, $c*t_D = 2$, $z_D = 7$
$\qquad _1W = {_4r_A}^{*j}\bullet({_4r_B}\blacksquare({_4r_C}^{*j}\mathbf{x}{_4r_D}))$
$\qquad\qquad = (c*t_A*1^{*j} + x_A*q_x^{*j} + y_A*q_y^{*j} + z_A*q_z^{*j})\bullet({_4r_B}\blacksquare({_4r_C}^{*j}\mathbf{x}{_4r_D}))$

SPECIAL ALGEBRA FOR SPECIAL RELATIVITY

$= i*(c*t_A*x_B*y_C*z_D + c*t_A*y_B*z_C*x_D + c*t_A*z_B*x_C*y_D$
$\quad - c*t_A*z_B*y_C*x_D - c*t_A*x_B*z_C*y_D - c*t_A*y_B*x_C*z_D)$
$- i*(x_A*c*t_B*y_C*z_D + x_A*z_B*c*t_C*y_D + x_A*y_B*z_C*c*t_D$
$\quad - x_A*c*t_B*z_C*y_D - x_A*y_B*c*t_C*z_D - x_A*z_B*y_C*c*t_D)$
$- i*(y_A*c*t_B*z_C*x_D + y_A*x_B*c*t_C*z_D + y_A*z_B*x_C*c*t_D$
$\quad - y_A*c*t_B*x_C*z_D - y_A*z_B*c*t_C*x_D - y_A*x_B*z_C*c*t_D)$
$- i*(z_A*c*t_B*x_C*y_D + z_A*y_B*c*t_C*x_D + z_A*x_B*y_C*c*t_D$
$\quad - z_A*c*t_B*y_C*x_D - z_A*x_B*c*t_C*y_D - -z_A*y_B*x_C*c*t_D)$

$= i*(0) - i*(x_A*y_B*z_C*c*t_D - x_A*y_B*c*t_C*z_D) - i*(0) - i*(0)$
$= -i*(1*1*5*2 - 1*1*3*7) = -i*(10 - 21)$
$= i*(11)$ Good

$_1W = -(_4r_B{}^{*j} \bullet (_4r_C \blacksquare (_4r_D{}^{*j} \mathbf{X}_4 r_A))) = -(1*q_y{}^{*j}) \bullet (j_y * 11) = i*(11)$

$_1W = -_4r_D{}^{*j} \bullet (_4r_A \blacksquare (_4r_B{}^{*j} \mathbf{X}_4 r_C))$
$= (c*t_D*1^{*j} + x_D*q_x{}^{*j} + y_D*q_y{}^{*j} + z_D*q_z{}^{*j}) \bullet (_4r_A \blacksquare (_4r_B{}^{*j} \mathbf{X}_4 r_C))$

$= -(i*(c*t_D*x_A*y_B*z_C + c*t_D*y_A*z_B*x_C + c*t_D*z_A*x_B*y_C$
$\quad - c*t_D*z_A*y_B*x_C - c*t_D*x_A*z_B*y_C - c*t_D*y_A*x_B*z_C)$
$- i*(x_D*c*t_A*y_B*z_C + x_D*z_A*c*t_B*y_C + x_D*y_A*z_B*c*t_C$
$\quad - x_D*c*t_A*z_B*y_C - x_D*y_A*c*t_B*z_C - x_D*z_A*y_B*c*t_C)$
$- i*(y_D*c*t_A*z_B*x_C + y_D*x_A*c*t_B*z_C + y_D*z_A*x_B*c*t_C$
$\quad - y_D*c*t_A*x_B*z_C - y_D*z_A*c*t_B*x_C - y_D*x_A*z_B*c*t_C)$
$- i*(z_D*c*t_A*x_B*y_C + z_D*y_A*c*t_B*x_C + z_D*x_A*y_B*c*t_C$
$\quad - z_D*c*t_A*y_B*x_C - z_D*x_A*c*t_B*y_C - z_D*y_A*x_B*c*t_C))$

$= -(i*(c*t_D*x_A*y_B*z_C) - i*(0) - i*(0) - i*(z_D*x_A*y_B*c*t_C))$
$= -(i*(2*1*1*5) - i*(0) - i*(0) - i*(7*1*1*3)) = -(i*(10) - i*(21))$
$= i*(11)$ Good

$_1W = {}_4r_C{}^{*j} \bullet (_4r_D \blacksquare (_4r_A{}^{*j} \mathbf{X}_4 r_B))$
$= (c*t_C*1^{*j} + x_C*q_x{}^{*j} + y_C*q_y{}^{*j} + z_C*q_z{}^{*j}) \bullet (_4r_D \blacksquare (_4r_A{}^{*j} \mathbf{X}_4 r_B))$

$= i*(c*t_C*x_D*y_A*z_B + c*t_C*y_D*z_A*x_B + c*t_C*z_D*x_A*y_B$
$\quad - c*t_C*z_D*y_A*x_B - c*t_C*x_D*z_A*y_B - c*t_C*y_D*x_A*z_B)$
$- i*(x_C*c*t_D*y_A*z_B + x_C*z_D*c*t_A*y_B + x_C*y_D*z_A*c*t_B$
$\quad - x_C*c*t_D*z_A*y_B - x_C*y_D*c*t_A*z_B - x_C*z_D*y_A*c*t_B)$
$- i*(y_C*c*t_D*z_A*x_B + y_C*x_D*c*t_A*z_B + y_C*z_D*x_A*c*t_B$
$\quad - y_C*c*t_D*x_A*z_B - y_C*z_D*c*t_A*x_B - y_C*x_D*z_A*c*t_B)$

CHAPTER 1 - NUMBERS

$$- i*(z_C*c*t_D*x_A*y_B + z_C*y_D*c*t_A*x_B + z_C*x_D*y_A*c*t_B$$
$$- z_C*c*t_D*y_A*x_B - z_C*x_D*c*t_A*y_B - z_C*y_D*x_A*c*t_B)$$

$$= i*(c*t_C*z_D*x_A*y_B) - i*(0) - i*(0) - i*(z_C*c*t_D*x_A*y_B)$$
$$= i*(3*7*1*1) - i*(5*2*1*1) = i*(21) - i*(10)$$
$$= i*(11) \quad \text{Good}$$

$$_1w = -_4r_B{}^{*j}\bullet(_4r_C\blacksquare(_4r_D{}^{*j}\mathbf{X}_4r_A))$$
$$= (c*t_B*1^{*j} + x_B*q_x{}^{*j} + y_B*q_y{}^{*j} + z_B*q_z{}^{*j})\bullet(_4r_C\blacksquare(_4r_D{}^{*j}\mathbf{X}_4r_A))$$

$$= -(i*(c*t_B*x_C*y_D*z_A + c*t_B*y_C*z_D*x_A + c*t_B*z_C*x_D*y_A$$
$$- c*t_B*z_C*y_D*x_A - c*t_B*x_C*z_D*y_A - c*t_B*y_C*x_D*z_A)$$
$$- i*(x_B*c*t_C*y_D*z_A + x_B*z_C*c*t_D*y_A + x_B*y_C*z_D*c*t_A$$
$$- x_B*c*t_C*z_D*y_A - x_B*y_C*c*t_D*z_A - x_B*z_C*y_D*c*t_A)$$
$$- i*(y_B*c*t_C*z_D*x_A + y_B*x_C*c*t_D*z_A + y_B*z_C*x_D*c*t_A$$
$$- y_B*c*t_C*x_D*z_A - y_B*z_C*c*t_D*x_A - y_B*x_C*z_D*c*t_A)$$
$$- i*(z_B*c*t_C*x_D*y_A + z_B*y_C*c*t_D*x_A + z_B*x_C*y_D*c*t_A$$
$$- z_B*c*t_C*y_D*x_A - z_B*x_C*c*t_D*y_A - z_B*y_C*x_D*c*t_A))$$

$$= -(i*(0) - i*(0) - i*(y_B*c*t_C*z_D*x_A - y_B*z_C*c*t_D*x_A) - i*(0))$$
$$= -(-i*(1*3*7*1 - 1*5*2*1)) = i*(21 - 10)$$
$$= i*(11) \quad \text{Good}$$

$$_4k_B = {}_4r_C\blacksquare({}_4r_D{}^{*j}\mathbf{X}_4r_A)/(-_1w) = (j_y*11)/(-i*11) = q_y*(-1)$$
$$_4k_C = {}_4r_D\blacksquare({}_4r_A{}^{*j}\mathbf{X}_4r_B)/_1w = (j_z*2 + i*7)/(i*11) = (7/11) + q_z*(2/11)$$
$$_4k_D = {}_4r_A\blacksquare({}_4r_B{}^{*j}\mathbf{X}_4r_C)/(-_1w) = (j_z*3 + i*5)/(-i*11) = (-5/11) + q_z*(-3/11)$$
$$_4k_A = {}_4r_B\blacksquare({}_4r_C{}^{*j}\mathbf{X}_4r_D)/_1w = (j_x*(-11))/(i*11) = q_x*(-1)$$

$_4k_B{}^{*j}\bullet_4r_A = 0$	$_4k_B{}^{*j}\bullet_4r_C = 0$	$_4k_B{}^{*j}\bullet_4r_D = 0$
$_4k_C{}^{*j}\bullet_4r_A = 0$	$_4k_C{}^{*j}\bullet_4r_B = 0$	$_4k_C{}^{*j}\bullet_4r_D = (7/11)*2 - (2/11)*7 = 0$
$_4k_D{}^{*j}\bullet_4r_A = 0$	$_4k_D{}^{*j}\bullet_4r_B = 0$	$_4k_D{}^{*j}\bullet_4r_C = (5/11)*3 - (3/11)*5 = 0$
$_4k_A{}^{*j}\bullet_4r_B = 0$	$_4k_A{}^{*j}\bullet_4r_C = 0$	$_4k_A{}^{*j}\bullet_4r_D = 0$

$$_4k_A{}^{*j}\bullet_4r_A = 1 \quad _4k_B{}^{*j}\bullet_4r_B = 1 \quad _4k_C{}^{*j}\bullet_4r_C = 1 \quad _4k_D{}^{*j}\bullet_4r_D = 1$$
$$_4r_A{}^{*j}\bullet_4k_A = 1 \quad _4r_B{}^{*j}\bullet_4k_B = 1 \quad _4r_C{}^{*j}\bullet_4k_C = 1 \quad _4r_D{}^{*j}\bullet_4k_D = 1$$

$$_4r_A = q_x*1 \quad _4r_B = q_y*1 \quad _4r_C = 3 + q_z*5 \quad _4r_D = 2 + q_z*7$$

$$_4r_A{}^{*j} = 1*q_x{}^{*j} \quad _4r_B{}^{*j} = 1*q_y{}^{*j} \quad _4r_C{}^{*j} = 3*1^{*j} + 5*q_z{}^{*j} \quad _4r_D{}^{*j} = 2*1^{*j} + 7*q_z{}^{*j}$$

SPECIAL ALGEBRA FOR SPECIAL RELATIVITY

6) The triple-vector-product being equal to zero indicates the geometric-vector is perpendicular to the cross-product of that geometric-vector with another geometric-vector.

$$_4r_C \blacksquare (_4r_C{}^{*j} \mathbf{X} _4r_D)$$
$$= {}_1r_C{}^{*}({}_3r_C{}^{*j}\mathbf{X}{}_3r_D) + {}_3r_C\mathbf{X}({}_1r_C{}^{*j}{}^{*}{}_3r_D) + {}_3r_C\mathbf{X}({}_3r_C{}^{*j}{}^{*}{}_1r_D) + {}_3r_C\bullet({}_3r_C{}^{*j}\mathbf{X}{}_3r_D)$$
$$= i^{*}(x_C{}^{*}y_C{}^{*}z_D + y_C{}^{*}z_C{}^{*}x_D + z_C{}^{*}x_C{}^{*}y_D$$
$$\quad - z_C{}^{*}y_C{}^{*}x_D - x_C{}^{*}z_C{}^{*}y_D - y_C{}^{*}x_C{}^{*}z_D)$$
$$+ j_x{}^{*}(c^{*}t_C{}^{*}y_C{}^{*}z_D + z_C{}^{*}c^{*}t_C{}^{*}y_D + y_C{}^{*}z_C{}^{*}c^{*}t_D$$
$$\quad - c^{*}t_C{}^{*}z_C{}^{*}y_D - y_C{}^{*}c^{*}t_C{}^{*}z_D - z_C{}^{*}y_C{}^{*}c^{*}t_D)$$
$$+ j_y{}^{*}(c^{*}t_C{}^{*}z_C{}^{*}x_D + x_C{}^{*}c^{*}t_C{}^{*}z_D + z_C{}^{*}x_C{}^{*}c^{*}t_D$$
$$\quad - c^{*}t_C{}^{*}x_C{}^{*}z_D - z_C{}^{*}c^{*}t_C{}^{*}x_D - x_C{}^{*}z_C{}^{*}c^{*}t_D)$$
$$+ j_z{}^{*}(c^{*}t_C{}^{*}x_C{}^{*}y_D + y_C{}^{*}c^{*}t_C{}^{*}x_D + x_C{}^{*}y_C{}^{*}c^{*}t_D$$
$$\quad - c^{*}t_C{}^{*}y_C{}^{*}x_D - x_C{}^{*}c^{*}t_C{}^{*}y_D - y_C{}^{*}x_C{}^{*}c^{*}t_D)$$
$$= i^{*}(0) + j_x{}^{*}(0) + j_y{}^{*}(0) + j_z{}^{*}(0)$$
$$= 0$$

7) Solutions:

The first "1" in "1 3 3 1" pertains to real numbers.
The first "3" in "1 3 3 1" pertains to polar vectors.

$$_3r_B = x_B{}^{*}i_x + y_B{}^{*}i_y + z_B{}^{*}i_z$$
$$_3r_C = x_C{}^{*}i_x + y_C{}^{*}i_y + z_D{}^{*}i_z$$
$$_3r_D = x_D{}^{*}i_x + y_D{}^{*}i_y + z_D{}^{*}i_z$$

The second "3" in "1 3 3 1" pertains to axial vectors.

$$_3\mathbf{A} = {}_3r_C\mathbf{X}{}_3r_D = (x_C{}^{*}i_x + y_C{}^{*}i_y + z_C{}^{*}i_z)\mathbf{X}(x_D{}^{*}i_x + y_D{}^{*}i_y + z_D{}^{*}i_z)$$
$$= ((y_C{}^{*}z_D{}^{*} - z_C{}^{*}y_D){}^{*}i_x + (z_C{}^{*}x_D{}^{*} - x_C{}^{*}z_D){}^{*}i_y + (x_C{}^{*}y_D{}^{*} - y_C{}^{*}x_D){}^{*}i_z)$$
$$= A_x{}^{*}i_x + A_y{}^{*}i_y + A_z{}^{*}i_z$$

The second "1" in "1 3 3 1" pertains to the dot-product of a cross-product to form a scalar (non-vector) result.

$$V = {}_3r_B\bullet({}_3r_C\mathbf{X}{}_3r_D) = (x_B{}^{*}i_x + y_B{}^{*}i_y + z_B{}^{*}i_z)\bullet(A_x{}^{*}i_x + A_y{}^{*}i_y + A_z{}^{*}i_z)$$
$$= x_B{}^{*}A_x + y_B{}^{*}A_y + z_B{}^{*}A_z$$

The first "1" in "1 2 1" pertains to real numbers.

CHAPTER 1 - NUMBERS

The "2" in "1 2 1" pertains to a complex number, including one complex number times another.

$_2r_A = x_A + i*y_A$; $C = {_2r_A}^{*j} * {_2r_B} = (x_A + y_A*i^{*i})*(x_B + i*y_B)$
$_2r_A = x_A + y_A*i^{*i}$ $\quad\quad\quad = (x_A*x_B + y_A*y_B*(i^{*i}*i)) + (i*x_A*y_B + i^{*i}*y_A*x_B)$
$_2r_B = x_B + i*y_B$ $\quad\quad\quad\quad = (x_A*x_B + y_A*y_B) + i*(x_A*y_B - y_A*x_B)$
$\quad\quad\quad\quad\quad\quad\quad\quad\quad = A + i*B$

The second "1" in "1 2 1" pertains to a scalar number that is the result of a dot-product operation.

$_2r_A = x_A + i*y_A$ $\quad\quad ;\quad\quad A = {_2r_A}^{*j} \bullet {_2r_B}$
$_2r_A = x_A + y_A*i^{*i}$ $\quad\quad\quad\quad\quad = (x_A*x_B + y_A*y_B*(i^{*i}*i))$
$_2r_B = x_B + i*y_B$ $\quad\quad\quad\quad\quad\quad = (x_A*x_B + y_A*y_B)$

8) Solution not given.

9) Solution: $\quad (1 \pm q_x)*\exp(q_x*\alpha)$
$\quad\quad\quad\quad\quad = (1 \pm q_x)*(\cosh(\alpha) + q_x*\sinh(\alpha))$
$\quad\quad\quad\quad\quad = (1 \pm q_x)*\cosh(\alpha) + q_x*(1 \pm q_x)*\sinh(\alpha)$
$\quad\quad\quad\quad\quad = (1 \pm q_x)*\cosh(\pm\alpha) \pm q_x*(1 \pm q_x)*\sinh(\pm\alpha)$
$\quad\quad\quad\quad\quad = (1 \pm q_x)*\cosh(\pm\alpha) + (1 \pm q_x)*\sinh(\pm\alpha)$
$\quad\quad\quad\quad\quad = (1 \pm q_x)*(\cosh(\pm\alpha) + \sinh(\pm\alpha))$
$\quad\quad\quad\quad\quad = (1 \pm q_x)*\exp(\pm\alpha)$

Division by "$1 \pm q_x$" is prohibited because a singular-label-number cannot be in the denominator, to avoid a division by zero that would lead to the non-sense result of equating "$\exp(q_x*\alpha)$" to "$\exp(\pm\alpha)$".

SPECIAL ALGEBRA FOR SPECIAL RELATIVITY

Further Thought.

1) By what criteria are some theories of pure mathematics segregated away from applied mathematics?

2) Label-numbers "i", "j", and "k" in "$\{1, i, j, k, i{*}j, j{*}k, k{*}i, i{*}j{*}k\}$" square to negative one and commute. With negatives, prove this set forms a division algebra by reviewing each of the five criteria.

3) Write analogous equations to those below, using Pauli Spin Matrices ("$PSM_x = j_x/i$", "$PSM_y = j_y{*}i$" and "$PSM_z = j_z/i$").

$$q_x = j_x/i \;;\; q_y = j_y/i \;;\; q_z = j_z/i \;; \qquad q_x^2 = q_y^2 = q_z^2 = 1^2 = +1$$

$$q_x{*}j_y = -j_y{*}q_x = -q_y{*}j_x = j_x{*}q_y = q_z \;;\; q_x{*}q_y = -q_y{*}q_x = -i{*}q_z = -j_z$$
$$q_y{*}j_z = -j_z{*}q_y = -q_z{*}j_y = j_y{*}q_z = q_x \;;\; q_y{*}q_z = -q_z{*}q_y = -i{*}q_x = -j_x$$
$$q_z{*}j_x = -j_x{*}q_z = -q_x{*}j_z = j_z{*}q_x = q_y \;;\; q_z{*}q_x = -q_x{*}q_z = -i{*}q_y = -j_y$$

4) 2x2 matrix isomorphs for quaternions did not include intermediate combinations of the three 2x2 matrix isomorphs. Assign "$a = \exp(i{*}\theta)$" and "$b = \exp(i{*}\phi)$" in the below 2x2 matrix. Select "θ" (theta), "ϕ" (phi) and "δ" (delta) to create "1", "j_x", "j_y", and "j_z". Try to select other values for "θ", "ϕ" and "δ" retaining "$j_x = j_y{*}j_z = -j_z{*}j_y$", "$j_y = j_z{*}j_x = -j_x{*}j_z$", and "$j_z = j_x{*}j_y = -j_y{*}j_x$". Try a different format for the 2x2 matrix, too.

$$\begin{pmatrix} a{*}\sin(\delta) & -b^{*i}{*}\cos(\delta) \\ b{*}\cos(\delta) & a^{*i}{*}\sin(\delta) \end{pmatrix} = \begin{pmatrix} \exp(i{*}\theta){*}\sin(\delta) & -\exp(-i{*}\phi){*}\cos(\delta) \\ \exp(i{*}\phi){*}\cos(\delta) & \exp(-i{*}\theta){*}\sin(\delta) \end{pmatrix}$$

5) If organized as a group, singular-label-numbers would have each of the five criteria of a mathematical group addressed. Try to define an algebra for singular-label-numbers.

6) For a better understanding, push math to extremes, and further. Quaternions are pushed to an extreme in octonions, sedonions, and beyond. Read the Appendix on octonion-sedonion algebra.

Chapter 2 – Particles

Einstein's Special Theory of Relativity provides the method for transforming mathematical descriptions of physical entities, called invariants, from one constant speed observer vantage to a different constant speed observer vantage. Two examples of invariants are time-space location and energy-momentum.

A person seated in a moving bus has a speed vantage from which they measure time, space-location, energy, and momentum component values for a baseball thrown forward from the back of the bus. Because the bus is moving, a person standing on the roadside has a different vantage, and so measures different component values for the same baseball. The change in component values is quantified through the mathematics of Special Relativity by use of the Lorentz Transformation.

A Lorentz Transformation is addition of time-space hyperbolic angles.

2.1 Hypercomplex-Plane

In this chapter, Special Relativity is presented with the "x" component the only space component. Time-space location "$_2r = 1*c*t + q_x*x$" is visualized by plotting components "c*t" and "x" on a Cartesian grid called a time-space "hypercomplex-plane". The hypercomplex-plane is an analogy to the "complex-plane" used to illustrate a complex number.

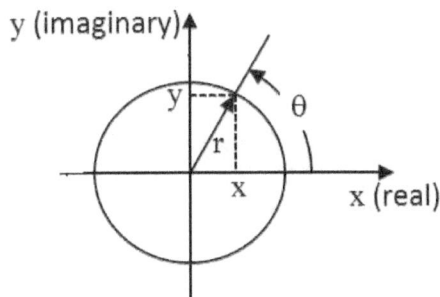

Figure 3. Complex-Plane and Radial Coordinate System.

A complex-plane is drawn on a sheet of paper using two space dimensions "x" and "y". "θ" is theta.

$_2z = x + i*y = r*\cos\theta + i*r*\sin\theta = r*\exp(i*\theta)$
$1 = \sqrt{(\cos^2\theta + \sin^2\theta)}$; $r = \sqrt{(x^2 + y^2)}$

SPECIAL ALGEBRA FOR SPECIAL RELATIVITY

Rotate the complex-plane coordinate system by rotating the sheet of paper.

For the hypercomplex-plane, substitute "j_x*x" for "$i*y$" and "$i*c*t$" for "x" and find "$i*_2r = i*c*t + j_x*x$".

The right-side of "$i*_2r = i*c*t + j_x*x$" has two label-numbers that square to negative one, "$i^2 = -1$" and "$j_x^2 = -1$". That makes it different from the right side of "$_2z = x + i*y$". Because of that difference, the hypercomplex-plane cannot be rotated by rotating the sheet of paper.

Divide "$i*_2r = i*c*t + j_x*x$" by "i" to get "$_2r = c*t + q_x*x$", in which "$q_x^2 = +1$". "$r_{hyperbolic}$" is the hyperbolic-radius. "α" (alpha) is the time-space hyperbolic-angle.

$$_2r = 1*c*t + q_x*x = 1*r_{hyperbolic}*\cosh\alpha + q_x*r_{hyperbolic}*\sinh\alpha$$
$$= r_{hyperbolic}*\exp(q_x*\alpha)$$

$1 = \sqrt{(\cosh^2\alpha - \sinh^2\alpha)}$; $r_{hyperbolic} = \sqrt{((c*t)^2 - x^2)}$

$\cosh\alpha = c*t/r_{hyperbolic}$; $\sinh\alpha = x/r_{hyperbolic}$; $\tanh\alpha = x/(c*t)$

$\exp(q_x*\alpha) = \cosh\alpha + q_x*\sinh\alpha$

To illustrate a time-space hypercomplex-plane on a space-space sheet of paper, replace "q_x" with "i" and call it "$q_{x\text{-illustrated}}$".

$$_2r_{illustrated} = 1*c*t + i*x = 1*r_{illustrated}*\cos\alpha_{illustrated} + i*r_{illustrated}*\sin\alpha_{illustrated}$$
$$= r_{illustrated}*\exp(i*\alpha_{illustrated})$$

$1 = \sqrt{(\cos^2\alpha_{illustrated} + \sin^2\alpha_{illustrated})}$; $r_{illustrated} = \sqrt{((c*t)^2 + x^2)}$

$\cos\alpha_{illustrated} = c*t/r_{illustrated}$; $\sin\alpha_{illustrated} = x/r_{illustrated}$; $\tan\alpha_{illustrated} = x/(c*t)$

$\exp(i*\alpha_{illustrated}) = \cos\alpha_{illustrated} + i*\sin\alpha_{illustrated}$

The difference between circular trigonometric functions used in the "$_2r_{illustrated}$" illustration and hyperbolic trigonometric functions used in the "$_2r$" time-space location expression causes distortion. Vertical and horizontal displacements are illustrated correctly. Lines at an angle are distorted by appearing longer than in time-space reality (because "$r_{illustrated} = \sqrt{((c*t)^2 + x^2)} > r_{hyperbolic} = \sqrt{((c*t)^2 - x^2)}$"). Diagonal lines (for which "$(c*t)^2 = x^2$") are completely distorted because any point illustrated on a diagonal has a zero hyperbolic-radius in time-space reality.

CHAPTER 2 — PARTICLES

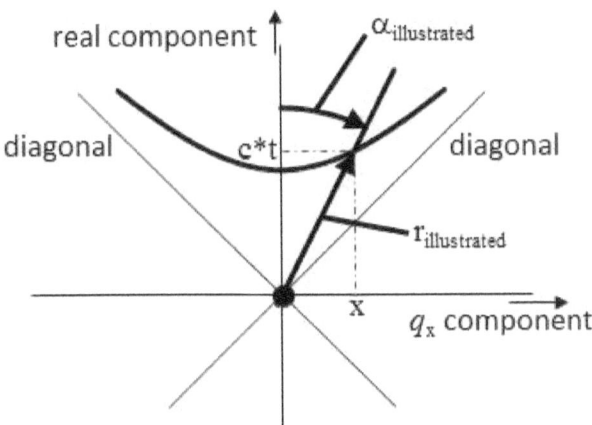

Figure 4. The Illustrated Hypercomplex-Plane with the Hyperbolic Radial Coordinate System.

A light source at the origin emits one photon that travels along the positive "x" axis at the speed-of-light. The plot of points "c*t" and "x" is called the world-line which, for the photon, is the diagonal from the origin to the upper-right.

The world-line of the light source is straight up the "c*t" axis, because "x = 0", and the hyperbolic-radius equals "c*t".

In contrast to the light source, the hyperbolic radius for the photon remains zero because "c*t = x". Because "x = $r_{hyperbolic}$*sinhα", "α" is large (to the ultimate "1/0" if we allow division by zero).

2.2 Inertial Reference Frames

Selected origins for "c*t" and "x" are valid for both a moving and a stationary inertial (constant speed) reference frame. The visualization is a big rectangular prism bus moving down a road toward the right.

Inertial reference frame "M" is the interior of the moving bus. A person seated on the moving bus takes measurements of component values. These are written with subscript "$_M$".

Inertial reference frame "S" is the stationary roadside. A person standing on the roadside also takes measurements. Those component values are written with subscript "$_S$".

SPECIAL ALGEBRA FOR SPECIAL RELATIVITY

"$(c*t_M, x_M) = (c*t_S, x_S) = (0, 0)$": When "$c*t_M = c*t_S = 0$", the back of the bus ("$x_M = 0$") and the speed limit sign ("$x_S = 0$") identify the same point in space. "x_M" is measured using measurement tape along the floor of the bus. "x_S" is measured using measurement tape along the roadside. "t_M" is measured using a clock mounted on the bus wall. "t_S" is measured using a clock mounted on the roadside.

Define a physical system (of components and label-numbers) inside bus "M", and then apply a "Lorentz Transformation" using the speed of the bus "S/M" to describe the same physical system observed from roadside "S".

A Lorentz Transformation is a hyperbolic-angle rotation.

Rest Frame "B" of the object in motion is a third inertial reference frame. The baseball's own rest frame "B" has the ball stopped. Therefore, in "B", there is advancement of time "t_B", but no change in space location ("$x_B = 0$"). And there is rest mass energy per "$E_B = m_B*c^2$", but no momentum, "$p_{xB} = 0$".

Speed "v_M" could be written "$v_{M/B}$" because "$v_B = 0$".

The "Relative" in "Special Relativity". "Relativity" is the ancient concept that all inertial frames have the same preference.

In the movie *Agora* watch a re-enactment of Hypatia of Alexandria (who lived around 390 AD / 390 CE) measure locations on a ship's deck of weights dropped from the mast to verify there is no preferred reference frame. Galileo (around 1600 AD / 1600 CE) described in one of his books the experiment of dropping weights from a ship's mast. And Galileo created a visualization for the lack of a preferred reference frame by describing what a person would observe when isolated inside a moving ship.

Einstein (in year 1905) added to relativity the requirement that the speed-of-light be the same regardless of reference frame and that meant time and space measurements were specific to a reference frame.

The change in space is easy to imagine. Place a measuring tape along the floor inside the bus and another measuring tape on the ground along the roadside and see the two coincide only when time equals zero (at a non-relativistic speed). The change in time is not so easy to imagine, and that's the trick.

Geometric Problem Definition. On bus "M" we have measured or have calculated component values for time-space location "$_2r$" and energy-momentum "$_2p$" for a moving baseball.

$$_2r = c*t_M*\mathbf{i}_{tM} + x_M*\mathbf{i}_{xM} \quad ; \quad _2p = (E_M/c)*\mathbf{i}_{tM} + p_{xM}*\mathbf{i}_{xM}$$

CHAPTER 2 – PARTICLES

Baseball "B" of rest mass "m_B" was thrown at speed "v_M" in the positive "$+x_M$" direction inside the bus at "$t_M = 0$". "x_M" can be calculated from "$x_M = v_M * t_M$".

Total energy "E_M" equals rest mass energy "$E_B = m_B * c^2$" added to kinetic energy. Mechanical momentum "p_{xM}" equals relativistic mass "E_M/c^2" times speed "v_M", "$p_{xM} = v_M * (E_M/c^2)$".

Geometric-unit-vectors "$\mathbf{i}_{tM}$" and "$\mathbf{i}_{xM}$" are the same in energy-momentum "$_2\mathbf{p}$" as they are in time-space "$_2\mathbf{r}$".

The bus moves at speed "$v_{S/M}$". Our objective is to use "M" components and bus speed "$v_{S/M}$" to calculate the same four components relative to roadside "S".

$$_2\mathbf{r} = c*t_S*\mathbf{i}_{tS} + x_S*\mathbf{i}_{xS} \quad ; \quad _2\mathbf{p} = (E_S/c)*\mathbf{i}_{tS} + p_{xS}*\mathbf{i}_{xS}$$

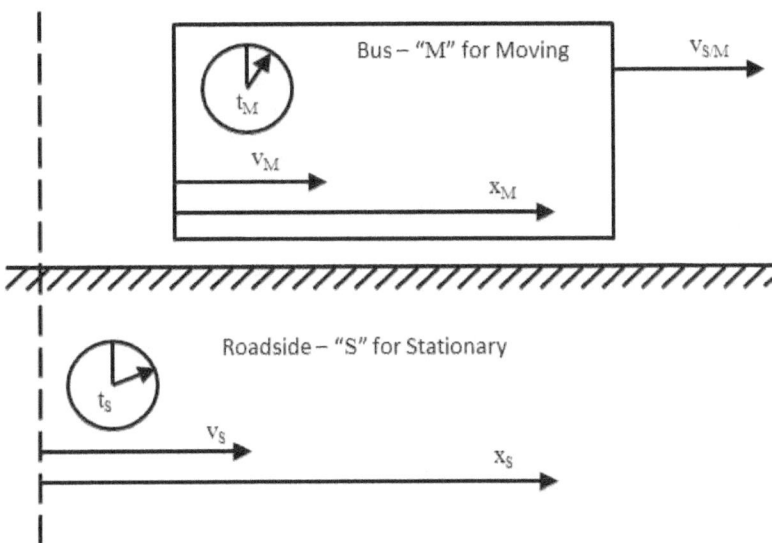

Figure 5. Moving and Stationary Reference Frames.

2.3 The Unspecified-Speed-Parameter

Begin the second step by replacing:

i_{tM} with $\exp(-q_x*\varsigma)$; i_{xM} with $q_x*\exp(-q_x*\varsigma)$

If "i_{tM}" had been replaced with "1" and "i_{xM}" with "q_x", then the replacement would have implied preference for moving coordinate system "M". To avoid that preference, "$\exp(-q_x*\varsigma)$" uses "unspecified-speed-parameter", "ς" (sigma, or, in English, esse).

If "ς" is included in a label-number, then we call that label-number a "compound-label-number". A compound-label-number contrasts with a "simple-label-number", such as "1" or "q_x".

"ς" is unknown and unknowable. It is different from independent variable "x" because "x" is intended to be substituted with a selected single valued real number. In other words, although "x" is not known, "x" is knowable. In contrast, "ς" is unknowable.

We are less likely to mistakenly assume a preferred reference frame if continuously reminded about the unspecified-speed-parameter "ς" by using compound-label-numbers. (Explicit use of "ς" ensures we do not violate "gauge invariance".)

$$1_M = \exp(-q_x*\varsigma) = \cosh\varsigma - q_x*\sinh\varsigma$$

$$q_{xM} = q_x*1_M = q_x*\exp(-q_x*\varsigma) = -\sinh\varsigma + q_x*\cosh\varsigma = \exp(-q_x*\varsigma)*q_x = 1_M*q_x$$

2.4 Compound-Label-Numbers and Components

Retain component values in the translation from a geometric representation to the all-number (more abstract) representation.

$$_2\mathbf{r} = c*t_M*i_{tM} + x_M*i_{xM}$$

$$\begin{aligned}
_2r &= 1_M*c*t_M + q_{xM}*x_M \\
&= \exp(-q_x*\varsigma)*(1*c*t_M + q_x*x_M) \\
&= \exp(-q_x*\varsigma)*(1*c*t_B*\cosh(\alpha_M) + q_x*c*t_B*\sinh(\alpha_M)) \\
&= \exp(-q_x*\varsigma)*c*t_B*\exp(q_x*\alpha_M)
\end{aligned}$$

$$_2\mathbf{p} = (E_M/c)*i_{tM} + p_{xM}*i_{xM}$$

$$_2p = 1_M*(E_M/c) + q_{xM}*p_{xM}$$

CHAPTER 2 – PARTICLES

$= \exp(-q_x*\varsigma)*(1*E_M/c + q_x*p_{xM})$
$= \exp(-q_x*\varsigma)*(1*m_B*c*\cosh(\alpha_M) + q_x*m_B*c*\sinh(\alpha_M))$
$= \exp(-q_x*\varsigma)*m_B*c*\exp(q_x*\alpha_M)$

An advantage of all-number expressions is use of "exp".

$c*t_B*\exp(q_x*\alpha_M) = c*t_B*\cosh\alpha_M + q_x*c*t_B*\sinh\alpha_M = c*t_M + q_x*x_M$

$c*t_M = c*t_B*\cosh\alpha_M$; $x_M = c*t_B*\sinh\alpha_M$

$E_M/c = m_B*c*\cosh\alpha_M$; $p_{xM} = m_B*c*\sinh\alpha_M$

"α_M" (alpha) relates to speed "v_M" of the baseball on the bus.

$(v_M/c) = \tanh\alpha_M$ $\alpha_M = \text{atanh}(v_M/c)$

For the above equations, a baseball was thrown forward from the back of the bus at time "$t_M = 0$", and the baseball has constant speed "v_M" in the positive "x_M" direction. Speed-parameter "α_M" is the hyperbolic-angle for motion of the baseball inside the bus.

"$c*t_B$" and "m_B*c" are each a hyperbolic-radius in their respective "$c*t_M$"/"x_M" or "E_M/c"/"p_{xM}" hypercomplex-planes and are calculated from the Pythagorean Theorem.

$1^2 = \cosh^2\alpha_M - \sinh^2\alpha_M$; $(c*t_B)^2 = (c*t_M)^2 - x_M^2$; $(m_B*c)^2 = (E_M/c)^2 - p_{xM}^2$

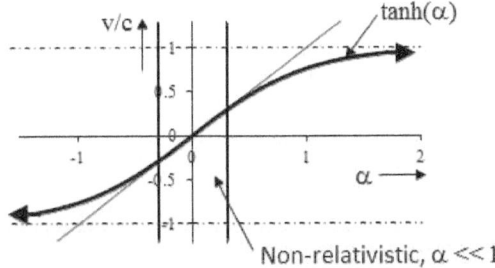

Figure 6. Hyperbolic tangent relationship between speed "v" and hyperbolic-angle "α".

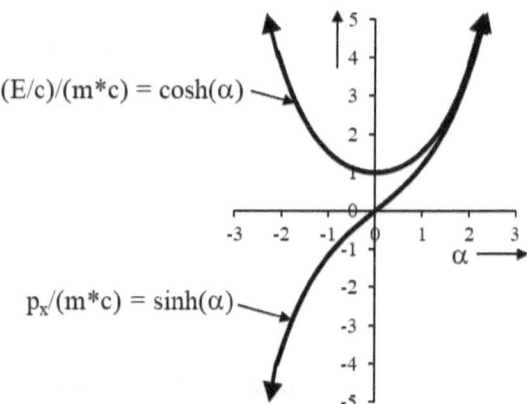

Figure 7. Energy "E" and momentum "p_x" approach infinity as speed "v" approaches speed-of-light "c" per "cosh" and "sinh" functions.

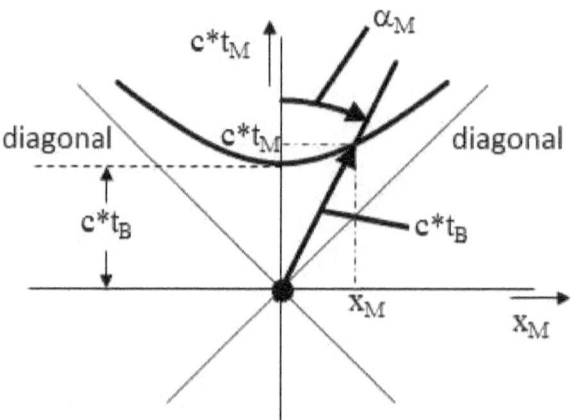

Figure 8. Time and space in "M" for an object "B" at time "t_M".

CHAPTER 2 – PARTICLES

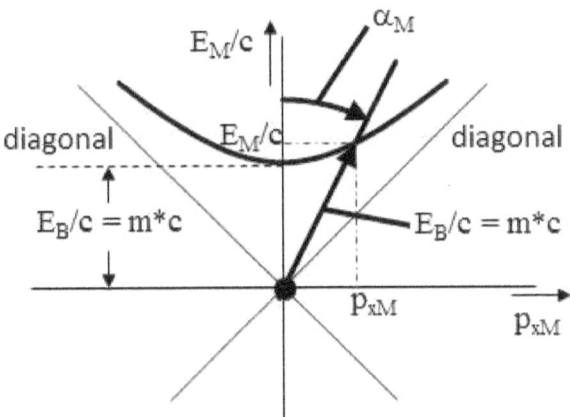

Figure 9. Energy and momentum for an object "B" of mass "m" moving in "M".

"$x_M = v_M * t_M$" and "$p_{xM} = v_M * E_M / c^2$" are proven valid:

$x_M = c*t_B*\sinh\alpha_M = c*t_B*\cosh\alpha_M*\tanh\alpha_M = c*t_M*\tanh\alpha_M = c*t_M*v_M/c = v_M*t_M$

$p_{xM} = m_B*c*\sinh\alpha_M = m_B*c*\cosh\alpha_M*\tanh\alpha_M = (E_M/c)*\tanh\alpha_M = (E_M/c)*v_M/c$
$= v_M*E_M/c^2$

Non-Relativistic Approximation "$\alpha_M < 0.3$" means "tanh" and "sinh" functions are not noticeable (unless extreme precision is needed).

$\tanh\alpha_M \approx \sinh\alpha_M \approx v_M/c \approx \alpha_M \qquad \alpha_M \ll 1$

$\cosh\alpha_M \approx 1 + (v_M/c)^2/2 \qquad \alpha_M \ll 1$

For "$\alpha_M \ll 1$", momentum "p_{xM}" is rest mass times speed. And "$(v_M/c)^2/2$" is not noticeable in "$t_M \approx t_B$" but is noticeable as kinetic energy "$m_B*v_M^2/2$". Add non-relativistic speeds "$v_S \approx v_M + v_{S/M}$" and "$x_S \approx x_M + v_{S/M}*t_M$".

Very Large Hyperbolic-Angles. Per "$(v_M/c) = \tanh\alpha_M$", if "α_M" approaches positive infinity, then speed "v_M" nearly equals speed-of-light "c", and "α_M" can increase further with very little change to "v_M". Further increases in "α_M" require huge increases in momentum and energy, per "$\sinh\alpha_M$" and "$\cosh\alpha_M$" functions, respectively. An electron cannot accelerate to the speed-of-light because only a finite quantity of energy is available.

2.5 Adding Hyperbolic-Angles

"$\alpha_{S/M}$" for a bus relative to the roadside is found from speed "$v_{S/M}$".

$$(v_{S/M}/c) = \tanh\alpha_{S/M} \quad ; \quad \alpha_{S/M} = \text{atanh}(v_{S/M}/c)$$

Special Relativity requires time-space hyperbolic-angles be added rather than actual speeds be added.

$$\alpha_S = \alpha_M + \alpha_{S/M}$$

$$\exp(q_x{*}\alpha_M){*}\exp(q_x{*}\alpha_{S/M}) = \exp(q_x{*}(\alpha_M + \alpha_{S/M})) = \exp(q_x{*}\alpha_S)$$

<u>Invariants</u>. Time-space location "$_2r$" is not given a subscript "M" or "S" because "$_2r$" is an "invariant". Also, "$_2p$" is an invariant. The name invariant means the mathematical expression does not change value due to a change in coordinate system, or, specific to relativity, due to a change in the observer's inertial reference frame (which is a change in observer's speed).

To express "$_2r$" and "$_2p$" in terms of "S" rather than in terms of "M", follow a hyperbolic rotation procedure for components and a counter hyperbolic rotation procedure for compound-label-numbers. The net effect is no net change to mathematical invariants "$_2r$" or "$_2p$", such that "$_2r$" or "$_2p$" was multiplied by integer one, "1". The procedure for components and the counter-procedure for compound-label-numbers is called the "Lorentz Transformation".

<u>General Form</u>. In the general form, below, a two-dimensional time-space invariant "$_2r$" is expressed in reference frame "$_M$" and is multiplied by a ratio of exponential functions, "$1 = \exp(q_x{*}\alpha_{S/M})/\exp(q_x{*}\alpha_{S/M})$". "$\exp(q_x{*}\alpha_{S/M})$" in the numerator is multiplied by components to form Lorentz Transformed component values, each multiplied by a simple-label-number. Components can be measured. The other "$\exp(q_x{*}\alpha_{S/M})$" in the denominator is multiplied by compound-label-number "1_M" to form "1_S".

$$\begin{aligned}
_2r &= 1_M{*}c{*}t_B{*}\exp(q_x{*}\alpha_M) \\
&= (1_M{*}c{*}t_B{*}\exp(q_x{*}\alpha_M)){*}(1) \\
&= (1_M{*}c{*}t_B{*}\exp(q_x{*}\alpha_M)){*}(\exp(q_x{*}\alpha_{S/M})/\exp(q_x{*}\alpha_{S/M})) \\
&= (1_M/\exp(q_x{*}\alpha_{S/M})){*}(c{*}t_B{*}\exp(q_x{*}\alpha_M){*}\exp(q_x{*}\alpha_{S/M})) \\
&= (1_M/\exp(q_x{*}\alpha_{S/M})){*}(c{*}t_B{*}\exp(q_x{*}(\alpha_M + \alpha_{S/M}))) \\
&= 1_S{*}c{*}t_B{*}\exp(q_x{*}\alpha_S)
\end{aligned}$$

CHAPTER 2 – PARTICLES

Resulting in:

$1_S = 1_M / \exp(q_x * \alpha_{S/M})$

$c * t_B * \exp(q_x * \alpha_S) = c * t_B * \exp(q_x * \alpha_M) * \exp(q_x * \alpha_{S/M})$

Lorentz Transformation for Components. Component values for "S" are found from component values for "M" by a three-step procedure.

- Change compound-label-numbers to simple-label-numbers
- Apply the exponential function factor with "$\alpha_{S/M}$"
- Sort by label-number factors.

$$\begin{aligned}
c*t_S + q_x*x_S &= (c*t_M + q_x*x_M)*\exp(q_x*\alpha_{S/M}) \\
&= c*t_B*\exp(q_x*\alpha_M)*\exp(q_x*\alpha_{S/M}) \\
&= c*t_B*\exp(q_x*(\alpha_M + \alpha_{S/M})) \\
&= c*t_B*\exp(q_x*\alpha_S) \\
&= c*t_B*\cosh\alpha_S + q_x*c*t_B*\sinh\alpha_S
\end{aligned}$$

$c*t_S = c*t_B*\cosh(\alpha_M + \alpha_{S/M})$; $x_S = c*t_B*\sinh(\alpha_M + \alpha_{S/M})$

And, alternatively

$$\begin{aligned}
c*t_S + q_x*x_S &= (c*t_M + q_x*x_M)*\exp(q_x*\alpha_{S/M}) \\
&= (c*t_M + q_x*x_M)*(\cosh\alpha_{S/M} + q_x*\sinh\alpha_{S/M}) \\
&= (c*t_M*\cosh\alpha_{S/M} + x_M*\sinh\alpha_{S/M}) + q_x*(c*t_M*\sinh\alpha_{S/M} + x_M*\cosh\alpha_{S/M})
\end{aligned}$$

$c*t_S = c*t_M*\cosh\alpha_{S/M} + x_M*\sinh\alpha_{S/M}$; $x_S = c*t_M*\sinh\alpha_{S/M} + x_M*\cosh\alpha_{S/M}$

Similarly

$E_S/c = m_B*c*\cosh(\alpha_M + \alpha_{S/M})$; $p_{xS} = m_B*c*\sinh(\alpha_M + \alpha_{S/M})$

$E_S/c = E_M/c*\cosh\alpha_{S/M} + p_{xM}*\sinh\alpha_{S/M}$; $p_{xS} = E_M/c*\sinh\alpha_{S/M} + p_{xM}*\cosh\alpha_{S/M}$

SPECIAL ALGEBRA FOR SPECIAL RELATIVITY

Lorentz Transformation for Compound-Label-Numbers. The magnitude of each compound-label-number is held constant to the value of one.

$$1_S = 1_M * \exp(-q_x * \alpha_{S/M}) = \exp(-q_x * \varsigma) * \exp(-q_x * \alpha_{S/M})$$
$$= \exp(-q_x * (\alpha_{S/M} + \varsigma))$$

$$q_{xS} = q_{xM} * \exp(-q_x * \alpha_{S/M}) = q_x * \exp(-q_x * \varsigma) * \exp(-q_x * \alpha_{S/M})$$
$$= q_x * \exp(-q_x * (\alpha_{S/M} + \varsigma))$$

$$1_S{}^{*j} = (\exp(-q_x * (\alpha_{S/M} + \varsigma)))^{*j} = \exp((\alpha_{S/M} + \varsigma) * q_x)$$

$$q_{xS}{}^{*j} = (q_x * \exp(-q_x * (\alpha_{S/M} + \varsigma)))^{*j} = (\exp((\alpha_{S/M} + \varsigma) * q_x)) * (-q_x)$$

$$1_S{}^{*j} * 1_S = +1 \ ; \ \ q_{xS}{}^{*j} * q_{xS} = -1 \ ; \ \ 1_S{}^{*j} * q_{xS} = q_x \ ; \ \ q_{xS}{}^{*j} * 1_S = -q_x$$

"$1_S{}^{*j} * 1_S = +1$" is analogous to "$1*1 = +1$", and "$q_{xS}{}^{*j} * q_{xS} = -1$" is analogous to "$q_x{}^{*j} * q_x = -1$". Both "+1" and "-1" are numbers of unit magnitude, as are "$+q_x$" and "$-q_x$", and, therefore, the check is satisfied.

Check on the Lorentz Transformed Invariant. To prove the math was done correctly, the original "M" invariant expression must be derived from the Lorentz Transformed "S" invariant expression.

$$_2r = 1_S * c * t_S + q_{xS} * x_S$$
$$= 1_M * \exp(-q_x * \alpha_{S/M}) * (c * t_S + q_x * x_S)$$
$$= 1_M * \exp(-q_x * \alpha_{S/M}) * (c * t_B * \cosh\alpha_S + q_x * c * t_B * \sinh\alpha_S)$$
$$= 1_M * c * t_B * \exp(-q_x * \alpha_{S/M}) * \exp(+q_x * \alpha_S)$$
$$= 1_M * c * t_B * \exp(q_x * (\alpha_S - \alpha_{S/M}))$$
$$= 1_M * c * t_B * \exp(q_x * \alpha_M)$$
$$= 1_M * (c * t_B * \cosh\alpha_M + q_{xM} * c * t_B * \sinh\alpha_M)$$
$$= 1_M * c * t_M + q_{xM} * x_M$$
$$= {}_2r$$

Constant Magnitude of an Invariant. Proof of independence of the hyperbolic radius (also called magnitude) from the observer is found mathematically by removal of all "M" or "S" subscripts in the expression for the hyperbolic-radius.

$$_2r^{*j} * {}_2r = (1_S * c * t_S + q_{xS} * x_S)^{*j} * (1_S * c * t_S + q_{xS} * x_S)$$
$$= (c * t_S * 1_S{}^{*j} + x_S * q_{xS}{}^{*j}) * (1_S * c * t_S + q_{xS} * x_S)$$

CHAPTER 2 – PARTICLES

$$= c*t_S*(1_S^{*j}*1_S)*c*t_S + c*t_S*(1_S^{*j}*q_{xS})*x_S$$
$$+ x_S*(q_{xS}^{*j}*1_S)*c*t_S + x_S*(q_{xS}^{*j}*q_{xS})*x_S$$

$$= c*t_S*(1)*c*t_S + c*t_S*(q_x)*x_S + x_S*(-q_x)*c*t_S + x_S*(-1)*x_S$$
$$= (c*t_S)^2 - x_S^2 = (c*t_B*\cosh\alpha_S)^2 - (c*t_B*\sinh\alpha_S)^2$$
$$= (c*t_B)^2*(\cosh^2\alpha_S - \sinh^2\alpha_S)$$
$$= (c*t_B)^2$$

2.6 Energy, Time Dilation, Length Contraction

Einstein's Special Theory of Relativity is famous for "$E = m*c^2$", "time dilation", and "length contraction".

"$\underline{E = m*c^2}$" suggests inertia rest mass is formed from the energy of fields. When an object made of matter moves, induced fields increase energy and create momentum.

Time Dilation is "$t_S = t_M*\cosh\alpha_{S/M}$" in which "$\cosh\alpha_{S/M}$" is always greater than "1" for "$\alpha_{S/M}$" mathematically real. It says clock "S" measures time advancing faster when compared to clock "M". A moving clock advances more slowly.

The bus passes the speed limit sign where clocks "t_S" and "t_M" both read zero. An observer on the roadside at a second roadside clock looks at the bus's clock at a later time, and because "$t_S = t_M*\cosh\alpha_{S/M}$" this second clock comparison shows "$t_S > t_M$".

There is reciprocity between "M" and "S" such that the moving object sees itself as stationary and the environment as moving, and it follows time moves slowly for the environment from the perspective of the moving object. To visualize the reciprocity, have two clocks on the bus and one clock on the roadside: A man seated on moving bus M synchronized his clock with a woman standing on roadside S right as he passed her. Later, the man on bus M passed a second S woman further down the road and saw his M clock was slower than this second S woman's clock. Per that visualization, it appears M's perspective on time is relative and, in contrast, S's perspective on time is absolute. But there's a second man further back in bus M who has also synchronized his clock with the first M man. The second M man looked at the first roadside S woman's clock and saw time on the roadside S had slowed. The key to visualizing the reciprocity of special relativity, to really understanding it, was putting this second man on bus M.

SPECIAL ALGEBRA FOR SPECIAL RELATIVITY

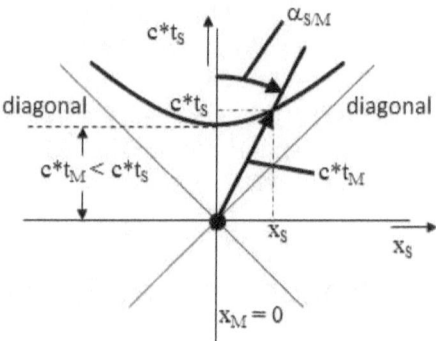

Figure 10. Time goes slower on bus "M" compared to roadside "S".

Length Contraction is most easily shown mathematically using wave-number "$_2k$". Wave-number is a count of evenly spaced nodes over a distance, in measurement units of per-length. A larger wave-number is a shorter length. Wave-number must be an invariant because a spacing of nodes is physically real.

$$_2k = 1_M*0 + q_{xM}*k_{xB} = 1_M*(q_x*k_{xB})*\exp(q_x*0) \qquad (\alpha_M = 0)$$
$$= 1_S*(q_x*k_{xB})*\exp(q_x*\alpha_{S/M})$$
$$= 1_S*k_{xB}*\sinh\alpha_{S/M} + q_{xS}*k_{xB}*\cosh\alpha_{S/M}$$
$$= 1_S*\omega_S/c + q_{xS}*k_{xS}$$

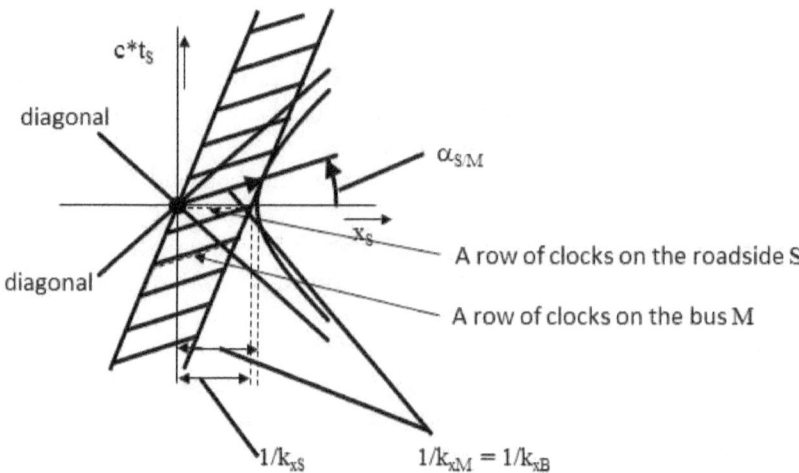

Figure 11. Illustration for why the front of bus "M" is at an earlier time in "S" compared to the back of the bus. On the hypercomplex-plane, lines at an angle appear longer but actually are shorter.

CHAPTER 2 – PARTICLES

"$1_M*0 + q_{xM}*k_{xM}$" states there are "k_{xM}" nodes per distance, for example, one seat per meter on the bus. In "M" there is no frequency "$\omega_M = 0$" because the seats are anchored to the floor of the bus, and, therefore, the seats do not pass a point "x_M" at a rate of frequency.

$$\omega_S/c = k_{xM}*\sinh\alpha_{S/M} \quad ; \quad k_{xS} = k_{xM}*\cosh\alpha_{S/M}$$

For "$\cosh\alpha_{S/M} > 1$" (due to "$\alpha_{S/M} \in R$") lengths are related as "$1/k_{xS} < 1/k_{xM}$". Also, there is a change to the time component, seen as a non-zero frequency term "$1_S*\omega_S/c$", which states seats pass a point "x_S" at a non-zero frequency "ω_S".

Length contraction occurs because time is different from one end to the other end of a moving object. Have everyone front to back on bus "M" synchronize their clocks. Compare, one-to-one, clocks between people on the bus and people on the roadside. At time "t_S", "$t_{M\text{-front}} = 5:00:02$" and "$t_{M\text{-back}} = 5:00:08$". The earlier time at the front of the bus means the front of the bus is not as far forward and, therefore, length is contracted.

For reciprocity, at one time "t_M" everyone looks at corresponding clocks on the roadside "s" and sees an earlier "t_S" in back, and these "s" clocks are spaced in "s" differently than when all the "t_S" clocks had read the same time.

2.7 Space-Like and Time-Like Invariants

Location "$_2r$" and momentum "$_2p$" invariants were each written with a mathematically real hyperbolic-radius, "$c*t_B$" and "m_B*c", respectively.

$$_2r = 1_M*c*t_B*\exp(q_x*\alpha_M) \quad ; \quad _2p = 1_M*m_B*c*\exp(q_x*\alpha_M)$$

For "$\alpha_M = 0$", "$_2r$" and "$_2p$" each have only a time term. It follows that an invariant with a mathematically real hyperbolic-radius is called "time-like". In contrast, the hyperbolic-radius in "$_2k$" had a "q_x" factor, "q_x*k_{xB}", so that "$\alpha_M = 0$" caused "$_2k$" to have only a space term, and that means "$_2k$" was "space-like".

$$_2k = 1_M*(q_x*k_{xB})*\exp(q_x*\alpha_M) = q_{xM}*k_{xB}*\exp(q_x*\alpha_M)$$
$$= 1_M*k_{xB}*\sinh(\alpha_M) + q_{xM}*k_{xB}*\cosh(\alpha_M) = 1_M*\omega_M/c + q_{xM}*k_{xM}$$

Time-like frequency "$_2\omega/c$" partners with space-like "$_2k$".

$$_2\omega/c = 1_M*(\omega_B/c)*\exp(q_x*\alpha_M)$$
$$= 1_M*(\omega_B/c)*\cosh(\alpha_M) + q_{xM}*(\omega_B/c)*\sinh(\alpha_M)$$
$$= 1_M*\omega_M/c + q_{xM}*k_{xM}$$

SPECIAL ALGEBRA FOR SPECIAL RELATIVITY

For "$\alpha_M \in R$" "$_2\omega/c$" has "$\omega_M/c > k_{xM}$" because "$\cosh(\alpha_M) > \sinh(\alpha_M)$", and that contrasts with "$\omega_M/c < k_{xM}$" for "$_2k$".

Similarly, space-like yard-stick "$_2s$" partners with time-like "$_2r$".

$$_2s = 1_M * q_x * s_B * \exp(q_x * \alpha_M) = q_{xM} * s_B * \exp(q_x * \alpha_M)$$

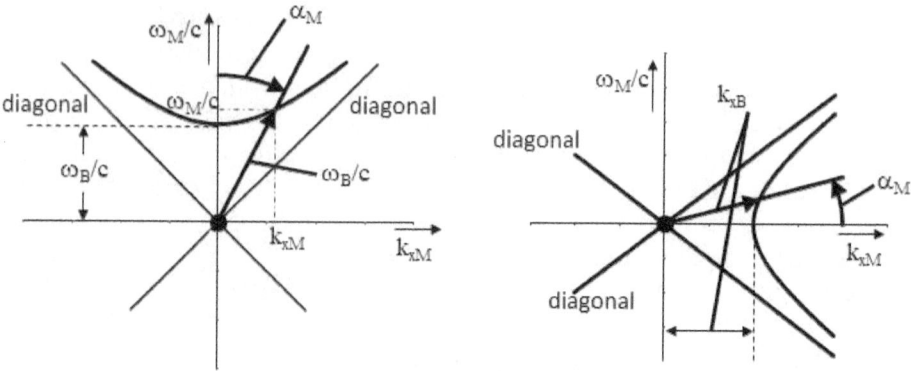

Figure 12. Hypercomplex-plane for frequency and wavenumber, respectively.

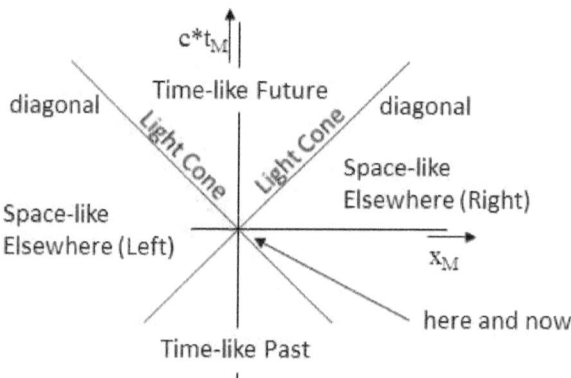

Figure 13. Time-like and space-like.

CHAPTER 2 – PARTICLES

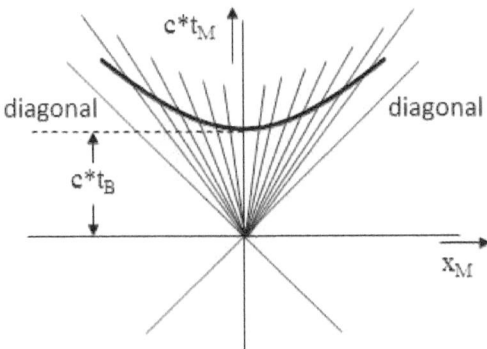

Figure 14. World lines for location in "M" of many baseballs of different speeds "v_M" at the same time "t_B" showing time-like locations on the hyperbola.

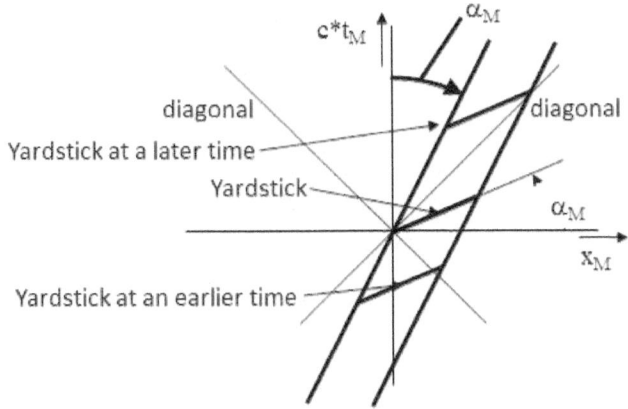

Figure 15. Location in "M" of a yardstick.

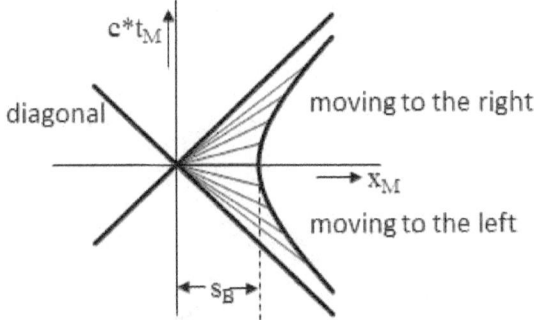

Figure 16. Locations of right ends of many yardsticks, each at a different speed, to illustrate space-like locations.

SPECIAL ALGEBRA FOR SPECIAL RELATIVITY

Lorentz Transformation from Time-Like to Space-Like. $\alpha_{S/M} = (i - j_x)*\pi/2$

$$
\begin{aligned}
{}_2r &= 1_M*(c*t_B)*\exp(q_x*\alpha_M)*1/1 \\
&= 1_M*(c*t_B)*\exp(q_x*\alpha_M)*(\exp(q_x*((i - j_x)*\pi/2))/\exp(q_x*((i - j_x)*\pi/2))) \\
&= 1_M*(c*t_B)*\exp(q_x*\alpha_M)*(\exp(q_x*(i*\pi/2 - j_x*\pi/2))/\exp(q_x*(i*\pi/2 - j_x*\pi/2)))
\end{aligned}
$$

$$
= 1_M*(c*t_B)*\exp(q_x*\alpha_M) \\
(\exp(j_x\pi/2)*\exp(-i*\pi/2))/(\exp(j_x*\pi/2)*\exp(-i*\pi/2))
$$

$$
\begin{aligned}
&= 1_M*(c*t_B)*\exp(q_x*\alpha_M)*(j_x*(-i))/(j_x*(-i)) \\
&= 1_M*(c*t_B)*\exp(q_x*\alpha_M)*(q_x/q_x) \\
&= (1_M/q_x)*((q_x*c*t_B)*\exp(q_x*\alpha_M)) \\
&= 1_S*(q_x*c*t_B)*\exp(q_x*\alpha_M)
\end{aligned}
$$

$$1_S = 1_M/q_x = 1_M*q_x = q_{xM} \quad ; \quad q_{xS} = q_{xM}/q_x = 1_M$$

$$c*t_S + q_x*x_S = (q_x*c*t_B)*\exp(q_x*\alpha_M)$$

$$c*t_S = q_x*c*t_B*q_x*\sinh(\alpha_M) = x_M \quad ; \quad q_x*x_S = q_x*c*t_B*\cosh(\alpha_M) = q_x*c*t_M$$

$$
\begin{aligned}
{}_2r &= 1_S*c*t_S + q_{xS}*x_S \\
&= (q_{xM})*(x_M) + (1_M)*(c*t_M) \\
&= 1_M*c*t_M + q_{xM}*x_M \\
&= {}_2r
\end{aligned}
$$

$$
\begin{aligned}
v_S/c &= \tanh\alpha_S \\
&= \tanh(\alpha_M + (i - j_x)*\pi/2) \\
&= \tanh(\alpha_M + i*\pi/2 - j_x*\pi/2) \\
&= \sinh(\alpha_M + i*\pi/2 - j_x*\pi/2)/\cosh(\alpha_M + i*\pi/2 - j_x*\pi/2)
\end{aligned}
$$

$$
= (\sinh(\alpha_M)*\cosh(i*\pi/2 - j_x*\pi/2) + \cosh(\alpha_M)*\sinh(i*\pi/2 - j_x*\pi/2)) \\
/(\cosh(\alpha_M)*\cosh(i*\pi/2 - j_x*\pi/2) + \sinh(\alpha_M)*\sinh(i*\pi/2 - j_x*\pi/2))
$$

$$
= (\sinh(\alpha_M)*\cosh(i*\pi/2)*\cosh(-j_x*\pi/2) \\
+ \sinh(\alpha_M)*\sinh(i*\pi/2)*\sinh(-j_x*\pi/2) \\
+ \cosh(\alpha_M)*\sinh(i*\pi/2)*\cosh(-j_x*\pi/2) \\
+ \cosh(\alpha_M)*\cosh(i*\pi/2)*\sinh(-j_x*\pi/2)) \\
/(\cosh(\alpha_M)*\cosh(i*\pi/2)*\cosh(-j_x*\pi/2)
$$

CHAPTER 2 – PARTICLES

$$+ \cosh(\alpha_M)*\sinh(i*\pi/2)*\sinh(-j_x*\pi/2)$$
$$+ \sinh(\alpha_M)*\sinh(i*\pi/2)*\cosh(-j_x*\pi/2)$$
$$+ \sinh(\alpha_M)*\cosh(i*\pi/2)*\sinh(-j_x*\pi/2))$$

$$= (\sinh(\alpha_M)*\cos(\pi/2)*\cos(\pi/2) + q_x*\sinh(\alpha_M)*\sin(\pi/2)*\sin(\pi/2)$$
$$+ i*\cosh(\alpha_M)*\sin(\pi/2)*\cos(\pi/2) - j_x*\cosh(\alpha_M)*\cos(\pi/2)*\sin(\pi/2))$$
$$/(\cosh(\alpha_M)*\cos(\pi/2)*\cos(\pi/2) + q_x*\cosh(\alpha_M)*\sin(\pi/2)*\sin(\pi/2)$$
$$+ i*\sinh(\alpha_M)*\sin(\pi/2)*\cos(\pi/2) - j_x*\sinh(\alpha_M)*\cos(\pi/2)*\sin(\pi/2))$$

$$= (q_x*\sinh\alpha_M)/(q_x*\cosh\alpha_M) = \tanh\alpha_M = v_M/c$$

$$v_{S/M}/c = \tanh\alpha_{S/M}$$
$$= \tanh((i - j_x)*\pi/2)$$
$$= \tanh(i*\pi/2 - j_x*\pi/2)$$
$$= \sinh(i*\pi/2 - j_x*\pi/2)/\cosh(i*\pi/2 - j_x*\pi/2)$$

$$= (\sinh(i*\pi/2)*\cosh(-j_x*\pi/2) + \cosh(i*\pi/2)*\sinh(-j_x*\pi/2))$$
$$/(\cosh(i*\pi/2)*\cosh(-j_x*\pi/2) + \sinh(i*\pi/2)*\sinh(-j_x*\pi/2))$$

$$= (i*\sin(\pi/2)*\cos(\pi/2) - j_x*\cos(\pi/2)*\sin(\pi/2))$$
$$/(\cos(\pi/2)*\cos(\pi/2) + q_x*\sin(\pi/2)*\sin(\pi/2))$$

$$= 0/q_x = 0$$

The tangent with a sum of angles identity was not used because it is derived using a division by cosine, and cosine of half pi equals zero.

Bus "M" moves at speed "$v_{S/M}/c = 0$". Observer "S" observes the baseball inside the bus per "$c*t_S = x_M$", "$x_S = c*t_M$" and "$v_S = v_M$" by looking through the bus window and seeing ticks of a clock are nodes along a line and seeing nodes along a line are ticks of a clock. There are no examples in which time and space are swapped, as if a theory of physics prevents it from happening.

"$\alpha_{S/M} = (i - j_x)*\pi/2$" includes a singular-label-number factor "$i - j_x$" so that "$\alpha_{S/M} = (i - j_x)*\pi/2$" is a variation of the number zero. As such, "$\alpha_{S/M} = (i - j_x)*\pi/2$" cannot be in a denominator, which it is not. Also, "$\exp(q_x*((i - j_x)*\pi/2)) = q_x$" must be unit magnitude, which it is, because "q_x" is unit magnitude.

A hypercomplex hyperbolic-angle "$\alpha_{S/M}$" is called (in this book) an "exotic Lorentz Transformation".

SPECIAL ALGEBRA FOR SPECIAL RELATIVITY

2.8 Electric Current Density

Time-like electric current density is a static electric charge.
Space-like electric current density is electric current in a wire.

Time-Like Electric Current Density.

$$_2J = 1_M*(dQ_B/dx_B)*\exp(q_x*\alpha_M) = 1_M*\rho_{xB}*\exp(q_x*\alpha_M)$$

"$\rho_{xB} = dQ_B/dx_B$" (Coulombs per centimeter) refers to rest frame "$_B$" twice: Electric charge "Q_B" and rest space location "x_B".

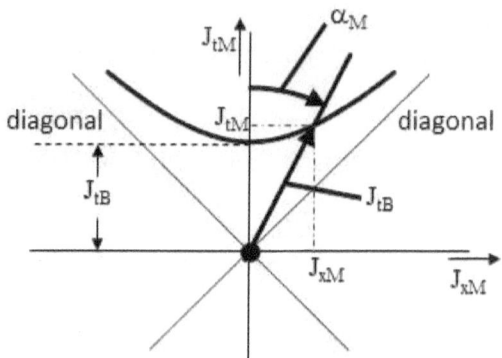

Figure 17. Hypercomplex-plane for charge density "$J_{tB} = \rho_{xB} = dQ_B/dx_B$" as the time-like hyperbolic-radius.

Space-Like Electric Current Density. Inside a wire:

- Stationary positive electric charges "$_2J_+$" with "$\rho_{xB+} > 0$" (in the atomic nuclei), for which "$v_{M+} = 0$", "$\alpha_{M+} = 0$" and "$\exp(q_x*\alpha_{M+}) = 1$"

- Moving negative electric charges "$_2J_-$" with "$\rho_{xB-} < 0$" (valence electrons), for which "$v_{M-}/c = \tanh(\alpha_{M-}) > 0$" and, for a physically real wire, "$v_{M-}/c \approx \alpha_{M-}$".

The sum "$_2J = {_2J_+} + {_2J_-}$" time component in "$_M$" must be zero.

CHAPTER 2 – PARTICLES

$$\begin{aligned}
_2J &= {_2J_+} + {_2J_-} \\
&= 1_M * \rho_{xB+} * \exp(q_x * \alpha_{M+}) + 1_M * \rho_{xB-} * \exp(q_x * \alpha_{M-}) \\
&= 1_M * \rho_{xB+} + 1_M * \rho_{xB-} * \exp(q_x * \alpha_{M-}) \\
&= 1_M * \rho_{xB+} + 1_M * \rho_{xB-} * (\cosh\alpha_{M-} + q_x * \sinh\alpha_{M-}) \\
&= 1_M * (\rho_{xB+} + \rho_{xB-} * \cosh\alpha_{M-}) + q_{xM} * \rho_{xB-} * \sinh\alpha_{M-} \\
&= q_{xM} * \rho_{xB-} * \sinh\alpha_{M-} \\
&= q_{xM} * (-\rho_{xB+}/\cosh\alpha_{M-}) * \sinh\alpha_{M-} = q_{xM} * -\rho_{xB+} * \tanh\alpha_{M-} \\
&= q_{xM} * \rho_{xB+} * (-v_{M-}/c)
\end{aligned}$$

The requirement of a zero time component was satisfied by setting "$\rho_{xB+} + \rho_{xB-} * \cosh\alpha_{M-} = 0$", from which "$\rho_{xB-} = -\rho_{xB+}/\cosh\alpha_{M-}$".

Regardless of "v_{M-}/c" being small, current "$_2J = q_{xM} * \rho_{xB+} * (-v_{M-}/c)$" in a wire produces a very noticeable magnetic field around the wire.

A wire with electric current can, as a piece of metal, move in the positive "x" direction with speed "v_S". With respect to "S", the hyperbolic-radius of the current "$q_x * (\rho_{xB+} * (-v_{M-}/c))$" is space-like. Space-like current density in "S" is not commonly useful.

$$_2J = 1_S * q_x * (\rho_{xB+} * (-v_{M-}/c)) * \exp(q_x * \alpha_{S/M}) = q_{xS} * (\rho_{xB+} * (-v_{M-}/c)) * \exp(q_x * \alpha_{S/M})$$

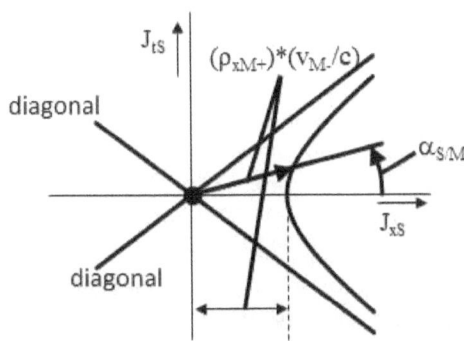

Figure 18. Hypercomplex-plane for a space-like current density.

Space-Like Invariant Complement for Energy-Momentum.
"n_B" relates to rest mass "m_B" as length "s_B" relates to time "$c * t_B$".

$$\begin{aligned}
_2p_{\text{space-like}} &= 1_M * q_x * n_B * c * \exp(q_x * \alpha_M) \\
&= q_{xM} * n_B * c * \exp(q_x * \alpha_M) \\
&= 1_M * n_B * c * \sinh(\alpha_M) + q_{xM} * n_B * c * \cosh(\alpha_M) \\
&= 1_M * (E_M/c) + q_{xM} * p_{xM}
\end{aligned}$$

SPECIAL ALGEBRA FOR SPECIAL RELATIVITY

"$\alpha_{S/M} = (i - j_x)*\pi/2$" has been used to create "$_2p_{space\text{-}like}$" from "$_2p$", but although we can create "$_2p_{space\text{-}like}$", we cannot find a physical example. Can there be momentum without energy?

Angular Momentum Density Invariant. In two-dimensional time-space, the axis of rotation for angular momentum is limited to being in the "x" direction. The visualization is a spinning ring. Use the right-hand-rule: Fingers curl with the direction of rotation and the right-hand thumb points in the direction of angular momentum. If we look to a higher value of "x", then a clockwise rotation (top to the right) has positive angular momentum.

"H_{xB}" is rest-frame angular momentum of a spinning ring. It is similar to rest-frame electric charge "Q_B" because its value does not change with respect to speed of an observer. Like "Q_B", "H_{xB}" is an invariant in itself: Factor "$1 = \cosh(\alpha_M)/\cosh(\alpha_M)$" has the numerator from relativistic mass and the denominator from time dilation.

If "$\rho_{HxB} = dH_{xB}/dx_B$" is density of angular momentum along a particle's length, then the (space-like) invariant is

$$_2\rho_H = 1_M * q_x * \rho_{HxB} * \exp(q_x * \alpha_M)$$
$$= 1_M * \rho_{HxB} * \sinh(\alpha_M) + q_{xM} * \rho_{HxB} * \cosh(\alpha_M)$$
$$= 1_M * \rho_{HtM} + q_{xM} * \rho_{HxM}$$

Space component "$\rho_{HxM} = \rho_{HxB} * \cosh(\alpha_M)$" is positive or negative depending on spin direction of "H_{xB}". Time component "$\rho_{HtM} = \rho_{HxB} * \sinh(\alpha_M)$" is positive or negative depending on "$\sinh(\alpha_M)$" and quantifies angular momentum that passes a point "x_M".

Angular momentum density is similar to electric current density, with the difference being space-like vs time-like, and that difference shows in their respective conservation laws: "$((_2\nabla)^{*/sn}) \mathbf{x} (_2\rho_H) = 0$" as opposed to "$((_2\nabla)^{*/sn}) \bullet (_2J) = 0$". ("sn" will be explained later.)

2.9 Motion Faster than Light

Inside bus "M" a baseball moves toward the front at speed "$v_M < c$", as observed by a person seated on the bus. A person standing on the roadside "S" observes the baseball moving at speed "$v_S > c$".

CHAPTER 2 – PARTICLES

Bus speed hyperbolic-angle "$\alpha_{S/M} = i*\pi/2$" is not real, "$\alpha_{S/M} \notin R$".

$v_{S/M}/c = \tanh\alpha_{S/M} = \tanh(i*\pi/2) = i*\tan(\pi/2) = i*(1/0) = 1/0$

Bus speed "$v_{S/M}/c = 1/0$" is instantaneous movement along "x" so that the bus is at all locations "x" simultaneously, but only for an instant.

General form for "$\alpha_{S/M} = i*\pi/2$" and "$\alpha_M \in R$":

$$\begin{aligned}
_2r &= 1_M*(c*t_B)*\exp(q_x*\alpha_M) \\
&= 1_M*(c*t_B)*\exp(q_x*\alpha_M)*(\exp(q_x*\alpha_{S/M})/\exp(q_x*\alpha_{S/M})) \\
&= 1_M*(c*t_B)*\exp(q_x*\alpha_M)*(\exp(q_x*i*\pi/2)/\exp(q_x*i*\pi/2)) \\
&= 1_M*(c*t_B)*\exp(q_x*\alpha_M)*(\exp(j_x*\pi/2)/(\exp(j_x*\pi/2)) \\
&= 1_M*(c*t_B)*\exp(q_x*\alpha_M)*(j_x/j_x) \\
&= (1_M/j_x)*((j_x*c*t_B)*\exp(q_x*\alpha_M)) \\
&= (-i*1_M/q_x)*((i*q_x*c*t_B)*\exp(q_x*\alpha_M))
\end{aligned}$$

$1_S = -i*1_M/q_x$; $q_{xS} = q_x*i*q_{xM}$
$ = -i*q_{xM}$ $ = -i*1_M$

Hyperbolic radius "$i*q_x*c*t_B$" is space-like and imaginary for "t_B" real.

$c*t_S + q_x*x_S = (i*q_x*c*t_B)*\exp(q_x*\alpha_M)$

$c*t_S = i*q_x*c*t_B*q_x*\sinh(\alpha_M)$; $c*t_S = c*t_M*\cosh\alpha_{S/M} + x_M*\sinh\alpha_{S/M}$
$ = i*x_M$ $ = c*t_M*\cosh(i*\pi/2) + x_M*\sinh(i*\pi/2)$
$ = c*t_M*\cos(\pi/2) + i*x_M*\sin(\pi/2)$
$ = i*x_M$

$q_x*x_S = i*q_x*c*t_B*\cosh(\alpha_M)$; $x_S = c*t_M*\sinh\alpha_{S/M} + x_M*\cosh\alpha_{S/M}$
$ = i*q_x*c*t_M$ $ = c*t_M*\sinh(i*\pi/2) + x_M*\cosh(i*\pi/2)$
$ = i*c*t_M*\sin(\pi/2) + x_M*\cos(\pi/2)$
$ = i*c*t_M$

"$_2r$" is confirmed to be invariant.

$$\begin{aligned}
_2r &= 1_S*c*t_S + q_{xS}*x_S = (-i*q_{xM})*(i*x_M) + (-i*1_M)*(i*c*t_M) \\
&= q_{xM}*x_M + 1_M*c*t_M = 1_M*c*t_M + q_{xM}*x_M = {_2r}
\end{aligned}$$

"$c*t_B$" is real:

SPECIAL ALGEBRA FOR SPECIAL RELATIVITY

$c*t_B = \sqrt{((c*t_S)^2 - x_S^2)} = \sqrt{((i*x_M)^2 - (i*c*t_M)^2)} = \sqrt{((c*t_M)^2 - x_M^2)} = c*t_B$

Time and space (in "M") is swapped for space and time (in "S"), respectively, and that means "α_M" is drawn on the "S" hypercomplex-plane up from the "x_S" axis toward the "$c*t_S$" axis, to illustrate "$\alpha_S = \alpha_M + i*\pi/2$" drawn down from the "$c*t_S$" axis.

$\alpha_S = \alpha_M + \alpha_{S/M} = \alpha_M + i*\pi/2$

$v_S/c = \tanh\alpha_S$
  $= \tanh(\alpha_M + i*\pi/2)$
  $= \sinh(\alpha_M + i*\pi/2)/\cosh(\alpha_M + i*\pi/2)$

  $= (\sinh(\alpha_M)*\cosh(i*\pi/2) + \cosh(\alpha_M)*\sinh(i*\pi/2))/$
    $(\cosh(\alpha_M)*\cosh(i*\pi/2) + \sinh(\alpha_M)*\sinh(i*\pi/2))$

  $= (\sinh(\alpha_M)*\cos(\pi/2) + \cosh(\alpha_M)*i*\sin(\pi/2))/$
    $(\cosh(\alpha_M)*\cos(\pi/2) + \sinh(\alpha_M)*i*\sin(\pi/2))$

  $= (i*\cosh\alpha_M)/(i*\sinh\alpha_M)$
  $= \coth\alpha_M$

For "$\alpha_M \in R$" "$v_S/c = \coth\alpha_M > 1$" so that "$v_S > c$".

"$c*t_S = i*x_M$" and "$x_S = i*c*t_M$" mean "$c*t_S$" and "x_S" are imaginary for "$c*t_M$" and "x_M" real. The "S" hypercomplex-plane requires "$c*t_S$" and "x_S" be real. To force "$c*t_S$" and "x_S" to be real make the hyperbolic-radius (in "B") imaginary.

$t_B = -i*\tau_B$ (tau) ; $\tau_B \in R$; $i*q_x*c*t_B = i*q_x*c*(-i*\tau_B) = q_x*c*\tau_B$

$m_B = -i*\nu_B$ (nu) ; $\nu_B \in R$; $i*q_x*m_B*c = i*q_x*(-i*\nu_B)*c = q_x*\nu_B*c$

The imaginary hyperbolic-radius causes "$c*t_M$" and "x_M" to be imaginary and not plotted on the "M" hypercomplex plane.

Colliding Rods Visualization. Two parallel rods move perpendicular to their length to collide and bounce. All points along the contacting surface coincide everywhere in "x_B" at one instant in time "t_B". Because we are observing a rod collision along "x_B" at one instant "t_B" and not one location "x_B" for all time "t_B", we have "$\alpha_B = i*\pi/2$" and not "$\alpha_B = 0$".

CHAPTER 2 – PARTICLES

The two rods are on the floor of a bus, "$\alpha_M = 0$". Because "$v_{S/M}, \alpha_{S/M} > 0$", the front of the bus is at an earlier time "t_M" compared to the back, for each time "t_S", and that means the rod collision observed from "$_S$" starts in the rear of bus and moves forward at speed "$v_S = c^2/v_{S/M} > c$".

Spooky Action at a Distance. Einstein gave the name "spooky action at a distance" to the polarization of a pair of entangled photons determined simultaneously in two observations a macroscopic distance away. The information of the direction of polarization had to travel from one particle to its entangled partner particle instantaneously over the macroscopic distance. Perhaps this is another example of "$\alpha = i*\pi/2$".

Violation of Cause-and-Effect. A hyper-light-speed signal can arrive before it was emitted. The signal can be two rods colliding.

Set "$\mu_M > 0$" (mu) and "$\alpha_M = \mu_M + i*\pi/2$" so that "$v_M = c*\coth(\mu_M)$". (If "$\mu_M < 0$" then the adder is "$- i*\pi/2$".) Bus "$_M$" moves backward relative to roadside "$_S$" with negative hyperbolic-angle speed-parameter "$\alpha_{S/M} < 0$".

$\mu_M \in R$; $\mu_M > 0$; $\alpha_M = \mu_M + i*\pi/2$; $\alpha_{S/M} \in R$; $\alpha_{S/M} < 0$
$\alpha_S = \alpha_M + \alpha_{S/M} = i*\pi/2 + (\mu_M + \alpha_{S/M})$

$$\begin{aligned}
_2r &= 1_S*(c*t_B)*\exp(q_x*\alpha_S) \\
&= 1_S*(-i*c*\tau_B)*\exp(q_x*\alpha_S) \\
&= 1_S*(-i*c*\tau_B)*\exp(q_x*(\alpha_M + \alpha_{S/M})) \\
&= 1_S*(-i*c*\tau_B)*\exp(q_x*(i*\pi/2 + \mu_{M1} + \alpha_{S/M})) \\
&= 1_S*(-i*c*\tau_B)*\exp(q_x*i*\pi/2)*\exp(q_x*(\mu_M + \alpha_{S/M})) \\
&= 1_S*(-i*c*\tau_B)*j_x*\exp(q_x*(\mu_M + \alpha_{S/M})) \\
&= q_{xS}*(i*-i*c*\tau_B)*\exp(q_x*(\mu_M + \alpha_{S/M})) \\
&= q_{xS}*(c*\tau_B)*\exp(q_x*(\mu_M + \alpha_{S/M}))
\end{aligned}$$

$c*t_S = c*\tau_B*\sinh(\mu_M + \alpha_{S/M})$ $x_S = c*\tau_B*\cosh(\mu_M + \alpha_{S/M})$

Space-like equations for "$c*t_S$" and "x_S" depend on angle "$\mu_M + \alpha_{S/M}$", which is positive for "$\alpha_{S/M} = 0$" and negative for "$\alpha_{S/M} < -\mu_M$".

SPECIAL ALGEBRA FOR SPECIAL RELATIVITY

- For "$\alpha_{S/M} = 0$", the colliding rods move forward inside the bus at speed "$v_{M\text{-rods}} = c*\tanh(\mu_M) < c$" so that their contact point moves at speed "$v_S = c*\coth(\mu_M) > c$" and "$c*t_S > 0$" and "$x_S > 0$".

- In contrast, for "$\alpha_{S/M} < -\mu_M$" the rods move backward at speed "$-c < v_{S\text{-rods}} = c*\tanh(\mu_M + \alpha_{S/M}) < 0$" so that their contact point moves backward at speed "$v_S = c*\coth(\mu_M + \alpha_{S/M}) < -c$". "$c*t_S < 0$" with "$x_S > 0$" means increasing values of "x_S" occur for decreasing values of "$c*t_S$" and that properly corresponds to negative direction movement because "$_S$" time is only seen moving positively.

The contact point was seen moving forward on the bus but moving backward from the roadside so that past locations of the contact point on the bus are future locations of the contact point from the roadside. Time goes backward.

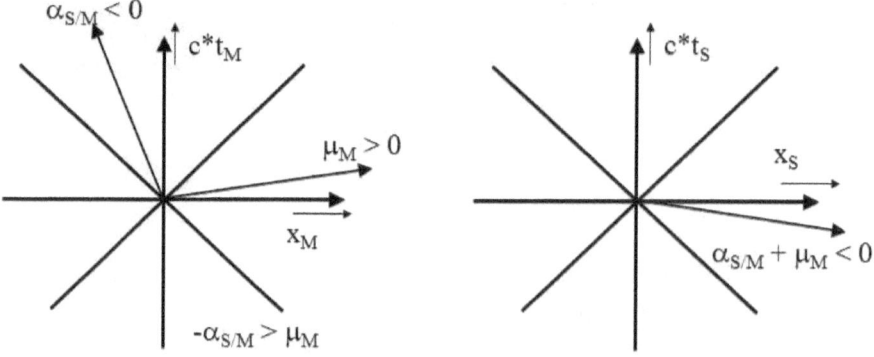

Figure 19. Illustration of a violation of cause-and-effect for motion faster than the speed-of-light. The observer sees time as moving up vertically. The arrow on the far-right figure points right, but the "S" observer sees the activity as moving left, toward the source of the emission.

<u>Visualization of the Violation of Cause-and-Effect</u>. A solar system has two small planets in counter-orbits. Reference frame "$_M$" is Planet M. "$_S$" is Planet S. At "$c*t_M = 0$" and "$c*t_S = 0$" the two planets are a few days passed each other with Planet M on the left and moving left. Therefore, "$v_{S/M}/c = \tanh\alpha_{S/M}$" with "$\alpha_{S/M} < 0$" and "$v_{S/M} < 0$" (and "$v_{M/S} > 0$").

CHAPTER 2 – PARTICLES

Planet M sends a hyper-light-speed signal to Planet S at speed "$v_M/c = \coth \mu_M$" with "$\mu_M > 0$" and "$v_M > 0$" / "$v_M > c$". Planet S received it at "$c*t_M = 7$" and "$x_M = t_M*v_M = c*t_M*\coth(\mu_M)$" ("$x_M > 0$").

$$_2r = 1_M*(-i*c*\tau_B)*\exp(q_x*\alpha_M)$$
$$= 1_M*(-i*c*\tau_B)*\exp(q_x*(\mu_M + i*\pi/2))$$
$$= 1_M*(-i*c*\tau_B)*(j_x)*\exp(q_x*\mu_M)$$
$$= 1_M*(c*\tau_B)*(q_x)*\exp(q_x*\mu_M)$$

$c*t_M = c*\tau_B*\sinh(\mu_M)$; $x_M = c*\tau_B*\cosh(\mu_M)$
$\qquad\qquad\qquad\qquad\qquad\qquad = (c*t_M/\sinh(\mu_M))*\cosh(\mu_M)$
$\qquad\qquad\qquad\qquad\qquad\qquad = c*t_M*\coth(\mu_M)$
$\qquad\qquad\qquad\qquad\qquad\qquad = t_M*v_M$

Note that "$x_M > 0$" and "$x_M > c*t_M$" because "$\mu_M \in R$".

Relative to Planet S, the hyper-light-speed signal time-space location components "$c*t_S$" and "x_S" are found using the below component equations, as given previously.

$c*t_S = c*\tau_B*\sinh(\mu_M + \alpha_{S/M})$; $x_S = c*\tau_B*\cosh(\mu_M + \alpha_{S/M})$

If "$\mu_M = 0$", such that the signal arrived instantly (as would happen for colliding rods), then "$c*t_S < 0$", and the signal was received on Planet S prior to being emitted, per "$c*t_S = c*\tau_B*\sinh(\mu_M + \alpha_{S/M})$".

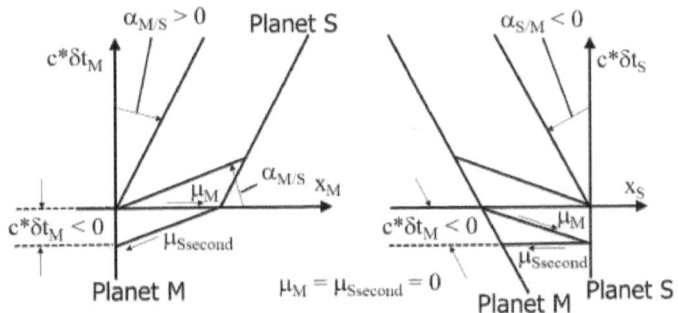

Figure 20. Instant signal sent from Planet M to Planet S and then another instant signal is sent back to Planet M. The second signal arrives prior to the first signal being emitted because "$\mu_M = 0$" and "$\mu_{S(second)} = 0$" with the two planets moving apart.

SPECIAL ALGEBRA FOR SPECIAL RELATIVITY

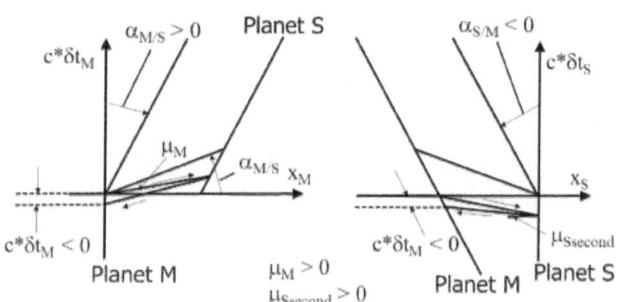

Figure 21. Hyper light speed signal from Planet M to Planet S and then another signal back to Planet M. The second signal arrives prior to the first signal being emitted because "$\mu_M - \mu_{S(second)} < -\alpha_{S/M}$" with the two planets moving apart.

But people on Planet S were unaware of the emission time of the signal because it occurred on Planet M, far away from them, and so are unaware the signal moved backward in time. To make everyone aware of the reverse passage of time, Planet S sent a second hyper-light-speed signal to Planet M the instant they received the signal from Planet M. For this second signal, "$\mu_{S(second)} < 0$" because the signal traveled from Planet S to the left towards Planet M. "-" in "-$i*\pi/2$" is because the second signal is on the left side of the "$c*t_S$" axis on the hypercomplex-plane.

$$\alpha_{S(second)} = \mu_{S(second)} - i*\pi/2 \quad ; \quad \mu_{S(second)} < 0$$

$$\begin{aligned}
2r_{(second)} &= 1_S*(-i*c*\tau_{B(second)})*\exp(q_x*\alpha_{S(second)}) \\
&= 1_S*(-i*c*\tau_{B(second)})*\exp(q_x*(\mu_{S(second)} - i*\pi/2)) \\
&= 1_S*(-i*c*\tau_{B(second)})*(-j_x)*\exp(q_x*\mu_{S(second)}) \\
&= 1_S*(c*\tau_{B(second)})*(-q_x)*\exp(q_x*\mu_{S(second)})
\end{aligned}$$

$$c*t_{S(second)} = -c*\tau_{B(second)}*\sinh(\mu_{S(second)})$$
$$x_{S(second)} = -c*\tau_{B(second)}*\cosh(\mu_{S(second)})$$

"$\mu_{S(second)} < 0$" in "$c*t_{S(second)}$" and "$x_{S(second)}$" component equations means "$c*t_{S(second)} > 0$" for "$c*\tau_{B(second)} > 0$". And "$x_{S(second)} < 0$" for left motion.

Relative to Planet S, the total (first signal plus second signal) time elapsed, and the total distance traveled, are given below.

CHAPTER 2 – PARTICLES

$$c*t_{S(total)} = c*t_S + c*t_{S(second)}$$
$$= c*\tau_B*\sinh(\mu_M + \alpha_{S/M}) - c*\tau_{B(second)}*\sinh(\mu_{S(second)})$$

$$x_{S(total)} = x_S + x_{S(second)}$$
$$= c*\tau_B*\cosh(\mu_M + \alpha_{S/M}) - c*\tau_{B(second)}*\cosh(\mu_{S(second)})$$

The general form of the Lorentz Transformation is now applied to find the arrival time of this second signal on Planet M.

$$2^{r_{(second)}}$$
$$= 1_S*(c*\tau_{B(second)})*(-q_x)*\exp(q_x*\mu_{S(second)}))*(\exp(q_x*\alpha_{M/S})/\exp(q_x*\alpha_{M/S}))$$

$$c*t_{M(second)} = -c*\tau_{B(second)}*\sinh(\mu_{S(second)} + \alpha_{M/S})$$

$$x_{M(second)} = -c*\tau_{B(second)}*\cosh(\mu_{S(second)} + \alpha_{M/S})$$

$$c*t_{M(total)} = c*t_M + c*t_{M(second)}$$
$$= c*\tau_B*\sinh(\mu_M) - c*\tau_{B(second)}*\sinh(\mu_{S(second)} + \alpha_{M/S})$$
$$= c*\tau_B*\sinh(\mu_M) - c*\tau_{B(second)}*\sinh(\mu_{S(second)} - \alpha_{S/M})$$

$$x_{M(total)} = x_M + x_{M(second)}$$
$$= c*\tau_B*\cosh(\mu_M) - c*\tau_{B(second)}*\cosh(\mu_{S(second)} - \alpha_{S/M})$$

"$x_{M(total)} = 0$" when "$c*\tau_B*\cosh(\mu_M) = c*\tau_{B(second)}*\cosh(\mu_{S(second)} - \alpha_{S/M})$".

$$c*\tau_{B(second)} = c*\tau_B*\cosh(\mu_M)/\cosh(\mu_{S(second)} + \alpha_{M/S})$$

$$c*t_{M(total)} = c*\tau_B*\sinh(\mu_M) - c*\tau_{B(second)}*\sinh(\mu_{S(second)} - \alpha_{S/M})$$
$$= c*\tau_B*(\sinh(\mu_M) - \cosh(\mu_M)*\tanh(\mu_{S(second)} - \alpha_{S/M}))$$

For the second signal to be received by Planet M prior to the first signal being emitted, "$c*t_{M(total)} < 0$", and that requires "$\sinh(\mu_M) < \cosh(\mu_M)*\tanh(\mu_{S(second)} - \alpha_{S/M})$", or "$\tanh(\mu_M) < \tanh(\mu_{S(second)} - \alpha_{S/M})$" and that means "$\mu_M < \mu_{S(second)} - \alpha_{S/M}$". It was specified that "$\mu_M > 0$", "$\alpha_{S/M} < 0$" and "$\mu_{S(second)} < 0$", therefore "$\mu_M - \mu_{S(second)} > 0$". If both "$\mu_M$" and "$\mu_{S(second)}$" are small enough in magnitude (so that the signal speeds are very fast), then "$\mu_M - \mu_{S(second)} < -\alpha_{S/M}$" to the result "$c*t_{M(total)} < 0$".

SPECIAL ALGEBRA FOR SPECIAL RELATIVITY

If the first signal is a weapon, then the weapon is countered by a second weapon counterattack that goes backward in time to destroy the enemy prior to their initial attack. But then the attack is not initiated, and, therefore, the counterattack is not initiated. To avoid such a strange condition, a rule of nature prevents controlled or prescribed information from being transmitted faster than the speed-of-light, to not violate cause-and-effect. The hypothetical violation of cause-and-effect for hyper-light-speed signals is a classic feature of Special Relativity.

Matter-waves. An electron's matter-wave moves at phase speed "v_p".

$$v_{pM}/c = \omega_M/k_{xM} = (\hbar*\omega_M)/(\hbar*k_{xM}) = (E_M)/(p_{xM}) = \cosh\alpha_M/\sinh\alpha_M = c/v_M$$

Per the above equation (which assumes no potential energy for the electron) phase speed "v_{pM}/c" is the reciprocal of group speed of the electron, "c/v_M". If group speed "$v_M = 0$" (for a stopped electron), then phase speed "v_{pM}" equals reciprocal of zero. If all of space cycled as a wave per "$T = \exp(i*\omega_M*t_M)$", then information of the value of "T" can be thought of as being transmitted in both directions of "x" instantly because it goes to the extremes of "x" and back instantly.

The wave carries no controlled information and so is not a signal. Controlled information is formed from the interference pattern of the waves. The interference pattern moves at the group speed. The group forms the probability function of particle location, and the group speed is slower than the speed-of-light.

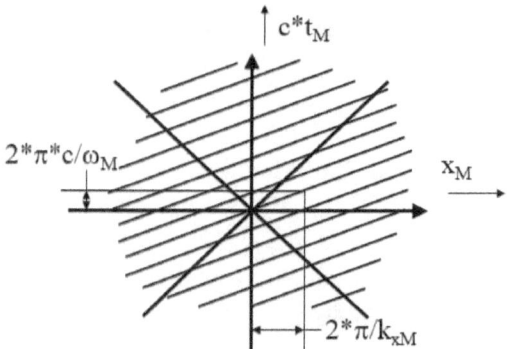

Figure 22. Wavelength node points for a matter-wave, with "$v_{pM} > c$".

CHAPTER 2 – PARTICLES

2.10 Anti-Matter

Anti-matter was first proposed because of the alternative/opposite electron identified in the solution of the Dirac Equation. A few years later, the anti-matter electron, called the positron, was discovered experimentally. Many years later, Feynman proposed that anti-matter is matter that moves backward in time.

(The below proposed use of Special Relativity as the basis of a theory for anti-matter might be new with this book. A search didn't find it anywhere.)

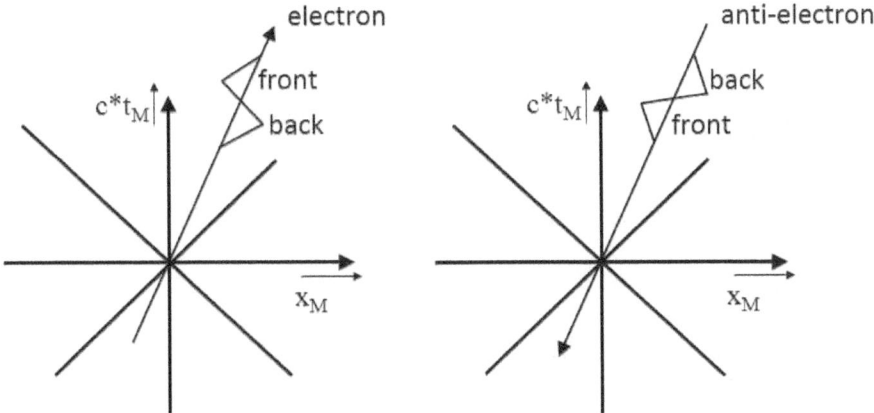

Figure 23. Back and front are swapped for anti-matter. But more than that, anti-matter is turned inside-out, like a glove, such that the left glove appears to be a right glove. Momentum for matter is on the left figure and is to the right. Momentum for anti-matter is on the right figure and is to the left. An observer made of matter feels the push of the momentum of anti-matter as if it moves to the right, because the observed momentum is a reaction from the push that drove the anti-matter backward in time and space.

An observer sits inside bus "M" and observes a baseball with speed "v_M" pass their seat toward the front of the bus, "$\alpha_M, v_M > 0$".

Bus speed is represented by "$\alpha_{S/M}$".

$\alpha_{S/M} = (N + 1/2)*i*2*\pi$; $N \in Z$

"$\alpha_{S/M} = (N + 1/2)*i*2*\pi$" is written simplified to "$\alpha_{S/M} = \pm i*\pi$".

$v_{S/M}/c = \tanh\alpha_{S/M} = \tanh(\pm i*\pi) = \pm i*\tan(\pi) = \pm i*0 = 0$

SPECIAL ALGEBRA FOR SPECIAL RELATIVITY

"$\alpha_{S/M} = \pm i*\pi$" has no effect on speed. Total speed "v_S" of the baseball inside the bus is found by adding hyperbolic-angles.

$\alpha_S = \alpha_M + \alpha_{S/M} = \alpha_M \pm i*\pi$

$v_S/c = \tanh\alpha_S = \tanh(\alpha_M \pm i*\pi) = \sinh(\alpha_M \pm i*\pi)/\cosh(\alpha_M \pm i*\pi)$

$\quad = (\sinh(\alpha_M)*\cosh(\pm i*\pi) + \cosh(\alpha_M)*\sinh(\pm i*\pi))$
$\quad\quad /(\cosh(\alpha_M)*\cosh(\pm i*\pi) + \sinh(\alpha_M)*\sinh(\pm i*\pi))$

$\quad = (\sinh(\alpha_M)*\cos(\pi) \pm \cosh(\alpha_M)*i*\sin(\pi))$
$\quad\quad /(\cosh(\alpha_M)*\cos(\pi) \pm \sinh(\alpha_M)*i*\sin(\pi))$

$\quad = (-\sinh\alpha_M)/(-\cosh\alpha_M) = \tanh\alpha_M$

"$v_S = v_M = c*\tanh\alpha_M$" even though "$\alpha_S \neq \alpha_M$". General form:

$_2r = 1_M*(c*t_B)*\exp(q_x*\alpha_M)$
$\quad = 1_M*(c*t_B)*\exp(q_x*\alpha_M)*(\exp(q_x*\pm i*\pi)/\exp(q_x*\pm i*\pi))$
$\quad = 1_M*(c*t_B)*\exp(q_x*\alpha_M)*(\exp(\pm j_x*\pi)/\exp(\pm j_x*\pi))$
$\quad = 1_M*(c*t_B)*\exp(q_x*\alpha_M)*(-1/-1)$
$\quad = (-1_M)*((-c*t_B)*\exp(q_x*\alpha_M))$

$c*t_S = c*t_M*\cosh\alpha_{S/M} + x_M*\sinh\alpha_{S/M}$
$\quad = c*t_M*\cosh(\pm i*\pi) + x_M*\sinh(\pm i*\pi)$
$\quad = c*t_M*\cos(\pi) \pm i*x_M*\sin(\pi)$
$\quad = c*t_M*(-1) \pm i*x_M*0$
$\quad = -c*t_M$

$c*t_S = c*t_B*\cosh\alpha_S$
$\quad = c*t_B*\cosh(\alpha_M + \alpha_{S/M})$
$\quad = c*t_B*\cosh(\alpha_M + \pm i*\pi)$
$\quad = c*t_B*(\cosh(\alpha_M)*\cosh(\pm i*\pi) + \sinh(\alpha_M)*\sinh(\pm i*\pi))$
$\quad = c*t_B*(\cosh(\alpha_M)*\cos(\pm\pi) + i*\sinh(\alpha_M)*\sin(\pm\pi))$
$\quad = -c*t_B*\cosh(\alpha_M)$
$\quad = -c*t_M$

CHAPTER 2 – PARTICLES

$$x_S = c*t_M*\sinh\alpha_{S/M} + x_M*\cosh\alpha_{S/M}$$
$$= c*t_M*\sinh(\pm i*\pi) + x_M*\cosh(\pm i*\pi)$$
$$= \pm i*c*t_M*\sin(\pi) + x_M*\cos(\pi)$$
$$= i*c*t_M*0 + x_M*(-1)$$
$$= -x_M$$

$$x_S = c*t_B*\sinh\alpha_S$$
$$= c*t_B*\sinh(\alpha_M + \alpha_{S/M})$$
$$= c*t_B*\sinh(\alpha_M + \pm i*\pi)$$
$$= c*t_B*(\sinh(\alpha_M)*\cosh(\pm i*\pi) + \cosh(\alpha_M)*\sinh(\pm i*\pi))$$
$$= c*t_B*(\sinh(\alpha_M)*\cos(\pm\pi) + \cosh(\alpha_M)*\sin(\pm\pi))$$
$$= -c*t_B*\sinh(\alpha_M)$$
$$= -x_M$$

$$1_S = 1_M*\exp(-q_x*\alpha_{S/M}) = 1_M*\exp(\mp q_x*i*\pi) = 1_M*\exp(\mp j_x*\pi) = 1_M*(-1) = -1_M$$

$$q_{xS} = q_{xM}*\exp(-q_x*\alpha_{S/M}) = -q_{xM}$$

$$_2r = 1_S*c*t_S + q_{xS}*x_S$$
$$= (-1_M)*(-c*t_M) + (-q_{xM})*(-x_M)$$
$$= 1_M*c*t_M + q_{xM}*x_M$$
$$= {_2r}$$

Observer "S" sees time "t_S" go forward and, due to "$c*t_S = -c*t_M$", sees time going backward for what observer "M" sees going forward. To visualize this: Observer "S" watches time pass "$t_{B\text{-matter}}$" on his clock and through the bus window sees a clock for "$t_{B\text{-antimatter}}$" made of anti-matter and recognizes "$t_{B\text{-antimatter}} = -t_{B\text{-matter}}$". "$_{B\text{-matter}}$" is the rest frame ("$\alpha_{B\text{-matter}} = 0$") for matter. In contrast, "$\alpha_{B\text{-antimatter}} = \pm i*\pi$" creates the negative. (It is one-in-the-same with "$\alpha_{S/M} = \pm i*\pi$" for the bus, with the difference being how reference frames are defined.) For anti-matter insert a negative.

$$c*t_{B\text{-matter}} = +\sqrt{((c*t_S)^2 - x_S^2)}$$
$$c*t_{B\text{-antimatter}} = -\sqrt{((c*t_S)^2 - x_S^2)}$$

$$t_{B\text{-antimatter}}*\cosh(\alpha_{B\text{-antimatter}}) = -t_{B\text{-matter}}*-\cosh(\alpha_{B\text{-matter}}) = t_{B\text{-matter}}*\cosh(\alpha_{B\text{-matter}})$$

"$c*t_S = -c*t_M$" is complemented by "$x_S = -x_M$". Measuring tape on the floor of the bus has increasing numbers that are negative of the tape on the roadside, and therefore the bus points backward. The clock held by the anti-matter person

68
Special Algebra for Special Relativity

"M" is observed by person "S" to be moving to smaller numbers and those numbers are reversed left-to-right so that the clock hand moves clockwise, just like the matter clock.

Imagine the baseball is rolling on the floor toward the front of the bus (to the right) inside a little toy car frame. Person "M" seated in the bus sees head-lights in front and tail-lights in back. Person "S" standing on the roadside sees the little car frame moving to the right, too, per "$v_S = v_M$", but with tail-lights leading and the headlights following. Person "S" looks at the whole bus and sees the back faces positive "x_S" (to the right) and the front faces negative "x_S" and concludes "$v_S = v_M$" because of the double negative, time and space.

Frequency. Person "M" seated in the bus hears clock ticks at frequency "$\omega_M = \omega_B$". Person "S" on the roadside hears tick frequency "$\omega_S = -\omega_M$" because "$\cosh\alpha_S = -\cosh\alpha_M$". The hyperbolic-radius of the frequency invariant for anti-matter is negative.

$\omega_{B\text{-matter}} = -\omega_{B\text{-antimatter}}$ same as $c*t_{B\text{-matter}} = -c*t_{B\text{-antimatter}}$

$\omega_S = \omega_{B\text{-antimatter}}*\cosh\alpha_S = \omega_{B\text{-antimatter}}*\cosh(\alpha_M + \pm i*\pi) = \omega_{B\text{-antimatter}}*\cosh(\pm i*\pi)$
$= \omega_{B\text{-antimatter}}*-1 = -\omega_{B\text{-antimatter}} = \omega_{B\text{-matter}} = \omega_{B\text{-matter}}*1 = \omega_{B\text{-matter}}*\cosh(0)$
$= \omega_{B\text{-matter}}*\cosh\alpha_M$

Our observation of frequency "ω_S" is positive, both for observing the anti-matter baseball's clock ticks and for observing the matter baseball's clock ticks. Therefore "$\omega_M = \omega_{B\text{-antimatter}}$" is negative.

Energy and Momentum. "$E_S/c = -E_M/c$" and "$p_{xS} = -p_{xM}$". Person "S" is impacted by an anti-matter baseball and it feels the same as if it were matter. The difference is that the matter baseball was received by person "S", and the anti-matter baseball was launched by person "S". Person "S" felt the reaction force from launching it backward in time and space.

To help visualize matter and anti-matter feeling the same, imagine the baseball re-composes itself into an electric field. The moving electric field induces a magnetic field and therefore has inertia same as mass. Anti-matter has the electric field reversed because anti-matter is what is formed when matter is subtracted from a vacuum. Regardless of the direction of the electric field and its induced magnetic field, the momentum will feel the same when it impacts person "S".

CHAPTER 2 – PARTICLES

$E_{S(\text{of antimatter})}/c = +E_{S(\text{of matter})}/c$

$p_{xS(\text{of antimatter})} = +p_{xS(\text{of matter})}$

Two particles move down the road and pass person "S". The matter particle has "$\alpha_{S(\text{of matter})} \in R$", "$\alpha_{S(\text{of matter})} = \alpha_M$", and "$m_{B\text{-matter}} > 0$".

$E_{S(\text{of matter})}/c = m_{B\text{-matter}}*c*\cosh(\alpha_{S(\text{of matter})}) = m_{B\text{-matter}}*c*\cosh(\alpha_M)$

$p_{xS(\text{of matter})} = m_{B\text{-matter}}*c*\sinh(\alpha_{S(\text{of matter})}) = m_{B\text{-matter}}*c*\sinh(\alpha_M)$

Anti-matter particle has "$\alpha_{S(\text{of antimatter})} = \alpha_M \pm i*\pi$", and "$m_{B\text{-antimatter}} < 0$".

$E_{S(\text{of antimatter})}/c = m_{B\text{-antimatter}}*c*\cosh(\alpha_{S(\text{of antimatter})})$
$\qquad = m_{B\text{-antimatter}}*c*\cosh(\alpha_M \pm i*\pi) = -m_{B\text{-antimatter}}*c*\cosh(\alpha_M)$
$\qquad = m_{B\text{-matter}}*c*\cosh(\alpha_M) = E_{S(\text{of matter})}/c$

$p_{xS(\text{of antimatter})} = m_{B\text{-antimatter}}*c*\sinh(\alpha_{S(\text{of antimatter})})$
$\qquad = m_{B\text{-antimatter}}*c*\sinh(\alpha_M \pm i*\pi) = -m_{B\text{-antimatter}}*c*\sinh(\alpha_M)$
$\qquad = m_{B\text{-matter}}*c*\sinh(\alpha_M) = p_{xS(\text{of matter})}$

Newtonian Mechanics with Anti-Matter. "$m_{B\text{-antimatter}} = -m_{B\text{-matter}}$" should not be put into the context of mass times acceleration equals force, but rather into the context of force equals the time derivative of momentum, as Newton originally presented his second law. Per "$p_{xS(\text{of antimatter})} = +p_{xS(\text{of matter})}$" the negative mass of anti-matter is inconsequential.

Negative rest mass also affects Newton's Law of Gravity, but General Relativity, which uses energy and not mass, supersedes it, and per "$E_{S(\text{of antimatter})}/c = +E_{S(\text{of matter})}/c$" the negative mass of anti-matter is inconsequential. Note that anti-matter has not been produced in a quantity large enough to measure force due to gravity.

Reverse Parity of Anti-Matter. A bus in reverse parity is perhaps best visualized using a right-hand glove. Fingers point to positive "x". Pull fingers through the open end to turn it inside out. Fingers now point to negative "x" and the glove looks like a left-hand glove. That same operation is not possible with a bus or particle. Rather, imagine the bus is an illustration on a sheet of paper as

SPECIAL ALGEBRA FOR SPECIAL RELATIVITY

one page in a book sitting flat on a table. The page is turned by lifting it up and placing it upside down on the other side of the book to create a mirror image. If ever anti-matter is made from its matter counterpart, then it will have been rotated through another dimension, just as the page had to be lifted out of the plane of the table. This page-turning visualization is a classic interpretation of the reverse parity of anti-matter.

Anti-matter Electric Current Density. Electric charge "Q_B" is in a space derivative (with respect to "x_B" as a second reference to reference frame "$_B$") to form electric charge density "$\rho_{B\text{-matter}} = dQ_B/dx_B$" as the hyperbolic-radius of the current density invariant "$_2J$".

$$\rho_{B\text{-matter}} = dQ_{B\text{-matter}}/dx_{B\text{-matter}} \quad ; \quad _2J = 1_M*\rho_{B\text{-matter}}*\exp(q_x*\alpha_M)$$

The ratio of anti-matter electric charge to location, formed as a derivative, has a double negative, and, therefore, no negative.

$$Q_{B\text{-antimatter}} = -Q_{B\text{-matter}} \quad ; \quad dQ_{B\text{-antimatter}} = -dQ_{B\text{-matter}}$$
$$x_{B\text{-antimatter}} = -x_{B\text{-matter}} \quad ; \quad dx_{B\text{-antimatter}} = -dx_{B\text{-matter}}$$

$$dQ_{B\text{-antimatter}}/dx_{B\text{-antimatter}} = -dQ_{B\text{-matter}}/-dx_{B\text{-matter}} = dQ_{B\text{-matter}}/dx_{B\text{-matter}}$$
$$\rho_{B\text{-antimatter}} = +\rho_{B\text{-matter}}$$

The Lorentz Transformation from matter observed in bus "M" to anti-matter observed from the roadside "S" uses "$\alpha_{S/M} = \pm i*\pi$".

$$\begin{aligned}
_2J &= 1_M*J_{tM} + q_{xM}*J_{xM} \\
&= 1_M*\rho_{B\text{-matter}}*\exp(q_x*\alpha_M) \\
&= 1_M*\rho_{B\text{-matter}}*\exp(q_x*\alpha_M)*\exp(q_x*\pm i*\pi)/\exp(q_x*\pm i*\pi)) \\
&= 1_M*\rho_{B\text{-matter}}*\exp(q_x*\alpha_M)*(-1)/(-1) \\
&= (1_M*/(-1))*(-\rho_{B\text{-matter}})*\exp(q_x*\alpha_M) \\
&= 1_S*(-\rho_{B\text{-matter}})*\cosh\alpha_M + q_{xS}*(-\rho_{B\text{-matter}})*\sinh\alpha_M \\
&= 1_S*(-\rho_{B\text{-matter}})*\cosh\alpha_M + q_{xS}*(-\rho_{B\text{-matter}})*\sinh\alpha_M \\
&= 1_S*(-J_{tM}) + q_{xS}*(-J_{xM}) \\
&= 1_S*J_{tS} + q_{xS}*J_{xS}
\end{aligned}$$

"$J_{tS} = -J_{tM}$" and "$J_{xS} = -J_{xM}$" state the electric charge density and the electric current density become negative. Because "$\rho_{B\text{-antimatter}} = \rho_{B\text{-matter}}$":

$$J_{tS\text{(of antimatter)}} = -J_{tS\text{(of matter)}} \quad ; \quad J_{xS\text{(of antimatter)}} = -J_{xS\text{(of matter)}}$$

CHAPTER 2 – PARTICLES

(The analogous equations for energy-momentum did not have the negative.)
Two particles pass person "S". Matter particle: "$\alpha_{S(of\ matter)} \in R$", "$\alpha_{S(of\ matter)} = \alpha_M$", and "$\rho_{B\text{-matter}} > 0$".

$J_{tS(of\ matter)} = \rho_{B\text{-matter}} * c * \cosh(\alpha_{S(of\ matter)}) = \rho_{B\text{-matter}} * c * \cosh(\alpha_M)$
$J_{xS(of\ matter)} = \rho_{B\text{-matter}} * c * \sinh(\alpha_{S(of\ matter)}) = \rho_{B\text{-matter}} * c * \sinh(\alpha_M)$

For the anti-matter particle "$\alpha_{S(of\ antimatter)} = \alpha_M \pm i*\pi$", "$\rho_{B\text{-antimatter}} > 0$".

$J_{tS(of\ antimatter)} = \rho_{B\text{-antimatter}} * c * \cosh(\alpha_{S(of\ antimatter)})$
$\quad = \rho_{B\text{-antimatter}} * c * \cosh(\alpha_M \pm i*\pi) = \rho_{B\text{-antimatter}} * c * -\cosh(\alpha_M)$
$\quad = \rho_{B\text{-matter}} * c * -\cosh(\alpha_M) = -J_{tS(of\ matter)}$

$J_{xS(of\ antimatter)} = \rho_{B\text{-antimatter}} * c * \sinh(\alpha_{S(of\ antimatter)})$
$\quad = \rho_{B\text{-antimatter}} * c * \sinh(\alpha_M \pm i*\pi) = \rho_{B\text{-antimatter}} * c * -\sinh(\alpha_M)$
$\quad = \rho_{B\text{-matter}} * c * -\sinh(\alpha_M) = -J_{xS(of\ matter)}$

"$J_{tS(of\ antimatter)} = -J_{tS(of\ matter)}$" says anti-matter electric charge density is the negative of matter. Specifically, for the electron, the anti-electron (called the positron) is observed in "$_S$" having a positive electric charge (compared to negative charge of the electron).

"$J_{xS(of\ antimatter)} = -J_{xS(of\ matter)}$" states anti-matter electric current density is the negative of matter. A flow of positively charged particles is the negative of a flow of negatively charged particles.

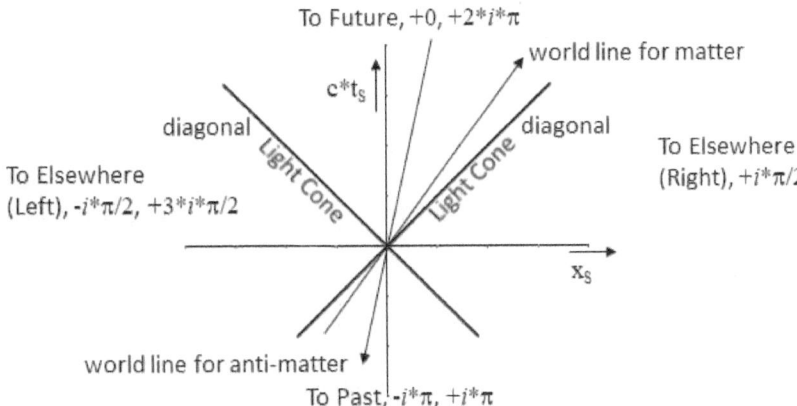

Figure 24. Anti-matter moving from future to past contrasted with matter moving from past to future, for "$\alpha_{S(matter)} = \alpha_M$" and "$\alpha_{S(anti\text{-}matter)} = \alpha_M \pm i*\pi$".

Matter-waves of Anti-matter. A factory makes anti-matter one particle at a time and assembles an anti-matter bus, an android, and a rubber bounce ball.

The (anti-matter) rubber ball bounces off walls of the (anti-matter) bus with smaller and smaller bounces losing energy to friction and eventually stops. Time appeared to go forward, just like for us in the factory made of matter. To explain this, we can propose that the direction of time is determined by the particle that makes the observation (which is us made of matter), and not by the particle being observed (which is the bouncing ball). This would make sense in terms of quantum mechanics because the "collapse of the wave function" in which particle properties get specified, is an observation.

What would the android made of anti-matter observe of us? How would it view the two slit experiment? Can time move in both directions in our cosmological model of the universe? What experiments are needed?

2.11 Distributed Material Theory

A distributed material is a field of point particles, each point with an infinitesimal quantity. Examples are a fluid, a solid, an electromagnetic field, and a distributed electric charge. Each is spread through space as a continuum and varies in time. Differential calculus applies.

Space-Negative. Time-space differential gradient operator invariant "$_4\nabla^{sn}$" (del) is the mathematical tool for distributed material.

$$_4\nabla = \nabla_{tM}*\mathbf{i}_{tM} + \nabla_{xM}*\mathbf{i}_{xM} + \nabla_{yM}*\mathbf{i}_{yM} + \nabla_{zM}*\mathbf{i}_{zM}$$

translates to all number expression

$$_4\nabla^{sn} = 1_M^{sn}*\nabla_{tM} + q_{xM}^{sn}*\nabla_{xM} + q_{yM}^{sn}*\nabla_{yM} + q_{zM}^{sn}*\nabla_{zM}$$

Space-negative operator "^{sn}" has two aspects: Space compound-label-numbers "q_{xM}^{sn}", "q_{yM}^{sn}" and "q_{zM}^{sn}" are negative, and the Lorentz Transformation is inverted. The inverted Lorentz Transformation compensates for the lack of a negative on "1_M". A space-negative is necessary because time and space are in the denominator for "$_4\nabla$".

A long rod has temperature "$_1T$" along its length. Invariant "$_1T$" can loosely be called a "scalar" field. The compound-label-number associated with a scalar field is integer "1". It is devoid of a reference to the unspecified-speed-parameter "ς", but is "compound", regardless.

CHAPTER 2 – PARTICLES

$_1T(t_M, x_M) = C + a*t_M + b*x_M$

The rod is mounted inside bus "M" and moves with the bus. "$_1T$" has slope "b" (measured in degrees centigrade per meter) and increases at rate "a" (measured in degrees centigrade per second) when measured by a person seated on bus "M".

$\partial T/\partial ct_M = a/c$; $\partial T/\partial x_M = b$

The two gradients are placed into a time-space gradient invariant.

$_2\nabla^{sn}*_1T = 1_M^{sn}*\partial T/\partial ct_M + q_{xM}^{sn}*\partial T/\partial x_M = 1_M^{sn}*(a/c) + q_{xM}^{sn}*b$

The time-space gradient operator is

$_2\nabla^{sn} = 1_M^{sn}*\partial/\partial ct_M + q_{xM}^{sn}*\partial/\partial x_M$

Multiplication symbol "*" after operator "$_2\nabla^{sn}$" tells us to think of "$_2\nabla^{sn}$" as an invariant.

Bus "M" moves at speed "$v_{S/M}$" relative to roadside "S". Gradients "$\partial T/\partial ct_S$" and "$\partial T/\partial x_S$" are measured.

$_2\nabla^{sn}*_1T = 1_S^{sn}*\partial T/\partial ct_S + q_{xS}^{sn}*\partial T/\partial x_S$; $_2\nabla^{sn} = 1_S^{sn}*\partial/\partial ct_S + q_{xS}^{sn}*\partial/\partial x_S$

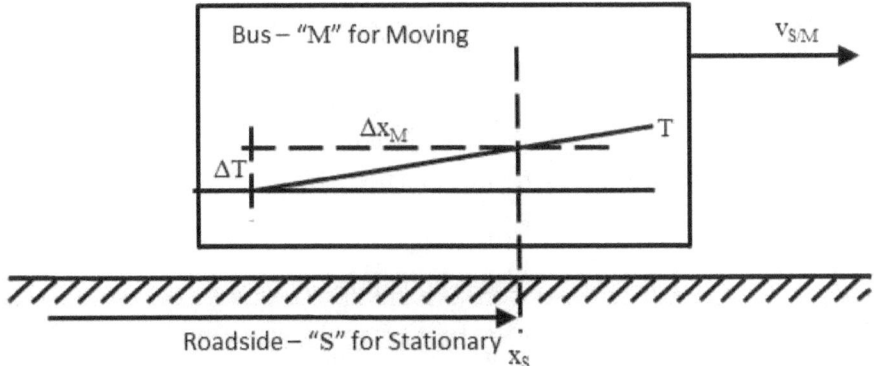

Figure 25. Temperature "T" at point "x_S" decreases when the bus moves forward at speed "$v_{S/M} > 0$". Therefore, "$\partial T_S/\partial ct_S < 0$" for "$\partial T_M/\partial x_M > 0$".

SPECIAL ALGEBRA FOR SPECIAL RELATIVITY

Bus "M" moves forward ("$v_{S/M} > 0$") to present colder and colder temperature to location "x_S" if "$a = 0$" and "$b > 0$", so that "$\partial T/\partial ct_S < 0$". Colder temperature requires a negative.

$$\partial T/\partial ct_S = (\partial T/\partial ct_M)*(\cosh\alpha_{S/M}) + (\partial T/\partial x_M)*(-\sinh\alpha_{S/M})$$
$$= (a/c)*(\cosh\alpha_{S/M}) + (b)*(-\sinh\alpha_{S/M})$$

$$\partial T/\partial x_S = (\partial T/\partial ct_M)*(-\sinh\alpha_{S/M}) + (\partial T/\partial x_M)*(\cosh\alpha_{S/M})$$
$$= (a/c)*(-\sinh\alpha_{S/M}) + (b)*(\cosh\alpha_{S/M})$$

To better see the negative, consider a non-relativistic speed by replacing "$\sinh\alpha_{S/M}$" with "$v_{S/M}/c$" and "$\cosh\alpha_{S/M}$" with "1".

$$\partial T/\partial ct_S = -b*v_{S/M}/c \qquad\qquad \partial T/\partial x_S = b$$

Using the above "$\partial T/\partial ct_S$" and "$\partial T/\partial x_S$" information, one-component invariant "$_1T$" can be expressed in terms of "$c*t_S$" and "x_S".

$$_1T(t_S, x_S) = C + (\partial T/\partial ct_S)*(c*t_S) + (\partial T/\partial x_S)*(x_S)$$

$$= C + ((\partial T/\partial ct_M)*(\cosh\alpha_{S/M}) + (\partial T/\partial x_M)*(-\sinh\alpha_{S/M}))*(c*t_S)$$
$$+ ((\partial T/\partial ct_M)*(-\sinh\alpha_{S/M}) + (\partial T/\partial x_M)*(\cosh\alpha_{S/M}))*(x_S)$$

To prove this is correct, derive "$_1T(t_M, x_M)$" from "$_1T(t_S, x_S)$".

$$_1T(t_S, x_S) = C + ((\partial T/\partial ct_M)*(\cosh\alpha_{S/M}) + (\partial T/\partial x_M)*(-\sinh\alpha_{S/M}))$$
$$*((c*t_M)*(\cosh\alpha_{S/M}) + (x_M)*(\sinh\alpha_{S/M}))$$
$$+ ((\partial T/\partial ct_M)*(-\sinh\alpha_{S/M}) + (\partial T/\partial x_M)*(\cosh\alpha_{S/M}))$$
$$*((c*t_M)*(\sinh\alpha_{S/M}) + (x_M)*(\cosh\alpha_{S/M}))$$

$$= C + (\partial T/\partial ct_M)*c*t_M + (\partial T/\partial x_M)*x_M$$
$$= {_1T}(t_M, x_M)$$

Gradient operators based on those equations are below.

$$\partial/\partial ct_S = (\partial/\partial ct_M)*(\cosh\alpha_{S/M}) + (\partial/\partial x_M)*(-\sinh\alpha_{S/M})$$
$$\partial/\partial x_S = (\partial/\partial ct_M)*(-\sinh\alpha_{S/M}) + (\partial/\partial x_M)*(\cosh\alpha_{S/M})$$

CHAPTER 2 – PARTICLES

$\nabla_{tS} = \nabla_{tM} * \cosh\alpha_{S/M} - \nabla_{xM} * \sinh\alpha_{S/M}$
$\nabla_{xS} = -\nabla_{tM} * \sinh\alpha_{S/M} + \nabla_{xM} * \cosh\alpha_{S/M}$

The critical concept here is the "-" sign in front of "$\sinh\alpha_{S/M}$". This "-" sign makes the Lorentz Transformation for "$_2\nabla^{sn}$" the opposite (or inverse) of the Lorentz Transformation for "$_2r$", given below.

$c*t_S = (c*t_M)*(\cosh\alpha_{S/M}) + (x_M)*(\sinh\alpha_{S/M})$
$x_S = (c*t_M)*(\sinh\alpha_{S/M}) + (x_M)*(\cosh\alpha_{S/M})$

There is no negative sign in the above "$_2r$" invariant equations. Negatives make the two Lorentz Transformations inverses of each other. A matrix multiplied by its inverse equals one.

$\cosh\alpha_{S/M}$	$\sinh\alpha_{S/M}$		$\cosh\alpha_{S/M}$	$-\sinh\alpha_{S/M}$		1	0
$\sinh\alpha_{S/M}$	$\cosh\alpha_{S/M}$	*	$-\sinh\alpha_{S/M}$	$\cosh\alpha_{S/M}$	=	0	1

Including "-" with "$\sinh\alpha_{S/M}$" makes the Lorentz Transformation for "$_2\nabla^{sn}$" special. To show it is special, it is given symbol "^{sn}".

Space Negative on Other Invariants. "^{sn}" applies to other invariants.
"$_2k = (\omega_M/c)*\mathbf{i}_{tM} + k_{xM}*\mathbf{i}_{xM}$" translates to "$_2k = 1_M*(\omega_M/c) + q_{xM}*k_{xM}$".

$\omega_S/c = (\omega_M/c)*(\cosh\alpha_{S/M}) + (k_{xM})*(\sinh\alpha_{S/M})$
$k_{xS} = (\omega_M/c)*(\sinh\alpha_{S/M}) + (k_{xM})*(\cosh\alpha_{S/M})$

"$_2k = (\omega_M/c)*\mathbf{i}_{tM} - k_{xM}*\mathbf{i}_{xM}$" translates to "$_2k^{sn} = 1_M^{sn}*(\omega_M/c) + q_{xM}^{sn}*k_{xM}$".

$\omega_S/c = (\omega_M/c)*(\cosh\alpha_{S/M}) - (k_{xM})*(\sinh\alpha_{S/M})$
$k_{xS} = -(\omega_M/c)*(\sinh\alpha_{S/M}) + (k_{xM})*(\cosh\alpha_{S/M})$

"$_2k^{sn}$" is abnormal (because of the negative before "$\mathbf{i}_{xM}$"). When we use "$_2k^{sn}$", we are modelling a wave-number in a special way, for example, for anti-matter, if we choose that anti-matter moves to the left for positive "α_M" when matter moves to the right for positive "α_M", as will be done in the chapter on waves. "$_2k^{sn}$" being abnormal is unlike the time-space gradient operator "$_2\nabla^{sn}$", because the space-negative on the gradient operator "$_2\nabla^{sn}$" is normal (because of the positive before "$\mathbf{i}_{xM}$").

SPECIAL ALGEBRA FOR SPECIAL RELATIVITY

Multiplication Operation with the Space-Negative. Because of the inverted matrix, space label-numbers are negative. Also, multiplication needs a hypercomplex-conjugate, for another negative.

$_2\nabla^{sn*}{}_1 T = 1_M{}^{sn}*\partial T/\partial ct_M + q_{xM}{}^{sn}*\partial T/\partial x_M$
$(_2\nabla^{sn*}{}_1 T)^{*j} = \partial T/\partial ct_M * 1_M{}^{*jsn} + \partial T/\partial x_M * q_{xM}{}^{*jsn}$

$_2 r = 1_M * c * t_M + q_{xM} * x_M$

$_1 T(t_M, x_M) - C = ((_2\nabla^*{}_1 T)^{*jsn}) \bullet (_2 r)$
$= (\partial T/\partial ct_M * ((1_M)^{*jsn}) + (\partial T/\partial x_M * (q_{xM})^{*jsn})) \bullet (1_M * c * t_M + q_{xM} * x_M)$
$= (\partial T/\partial ct_M * ((1_M)^{*jsn}) * (1_M * c * t_M) + (\partial T/\partial x_M * (q_{xM})^{*jsn}) * (q_{xM} * x_M)$
$= (\partial T/\partial ct_M) * (c * t_M) * ((1_M)^{*jsn}) * (1_M) + (\partial T/\partial x_M) * (x_M) * ((q_{xM})^{*jsn}) * (q_{xM})$
$= (\partial T/\partial ct_M) * (c * t_M) + (\partial T/\partial x_M) * (x_M)$

"$(\partial T/\partial ct_M) * (c * t_M) + (\partial T/\partial x_M) * (x_M)$" has a "+" sign between the two terms. It is different from "$((_2 k)^{*j}) \bullet (_2 r)$", as given below, because "$((_2 k)^{*j}) \bullet (_2 r)$" has a negative sign "-" between the two terms.

$((_2 k)^{*j}) \bullet (_2 r) = (\omega_M/c) * ((1_M)^{*j}) + (k_{xM} * (q_{xM})^{*j})) \bullet (1_M * c * t_M + q_{xM} * x_M)$
$= (\omega_M/c) * ((1_M)^{*j}) * (1_M * c * t_M) + (k_{xM} * (q_{xM})^{*j}) * (q_{xM} * x_M)$
$= (\omega_M/c) * (c * t_M) * ((1_M)^{*j}) * (1_M) + (k_{xM}) * (x_M) * ((q_{xM})^{*j}) * (q_{xM})$
$= \omega_M * t_M - k_{xM} * x_M$

To create the "+" sign between the two terms in "$((_2\nabla^{sn*}{}_1 T)^{*j}) \bullet (_2 r)$", a negative was on the space compound-label-numbers introduced by the hypercomplex-conjugate operation, and another negative was introduced by the space-negative operator, for a net result of a positive. ("p_{xM}" and "k_{xM}" will be discussed later.)

$(1_M)^{sn} = 1_M$; $(q_{xM})^{sn} = -q_{xM}$; $(q_{yM})^{sn} = -q_{yM}$; $(q_{zM})^{sn} = -q_{zM}$
$(1_M)^{*jsn} = 1_M$; $(q_{xM})^{*jsn} = -q_{xM}{}^{*j}$; $(q_{yM})^{*jsn} = -q_{yM}{}^{*j}$; $(q_{zM})^{*jsn} = -q_{zM}{}^{*j}$
$((1)^{*jsn}) * 1_M = 1$; $((q_{xM})^{*jsn}) * q_{xM} = +1$
$((q_{yM})^{*jsn}) * q_{yM} = +1$; $((q_{zM})^{*jsn}) * q_{zM} = +1$
$((1_M)^{*jsn}) * q_{xM} = ((q_{xM})^{*jsn}) * 1_M = p_{xM}$
$((q_{yM})^{*jsn}) * q_{zM} = -((q_{zM})^{*jsn}) * q_{yM} = -i * p_{xM} = -k_{xM}$

CHAPTER 2 – PARTICLES

Examples of Gradient Operations with Space-Negative. A conservation law typically uses the divergence operator, "$((_4\nabla)^{*/sn})\bullet$".

$((_4\nabla)^{*/sn})\bullet(_4\rho) = (\partial\rho_{tM}/\partial ct_M) + (\partial\rho_{xM}/\partial x_M) + (\partial\rho_{yM}/\partial y_M) + (\partial\rho_{zM}/\partial z_M) = 0$
$((_4\nabla)^{*/sn})\bullet(_4J) = (\partial J_{tM}/\partial ct_M) + (\partial J_{xM}/\partial x_M) + (\partial J_{yM}/\partial y_M) + (\partial J_{zM}/\partial z_M) = 0$
$((_4\nabla)^{*/sn})\bullet(_4V) = (\partial V_{tM}/\partial ct_M) + (\partial V_{xM}/\partial x_M) + (\partial V_{yM}/\partial y_M) + (\partial V_{zM}/\partial z_M) = 0$

"$_4\rho$" is a density of particles (for example, a count of gravel particles in the back of a truck). "$_4J$" is electric current density and is a special case of "$_4\rho$". "$_4V$" is voltage from the next chapter.

Curl operator, "$((_4\nabla)^{*/sn})\mathbf{x}$":

$((_4\nabla)^{*/sn})\mathbf{x}(_4J) = -_6G$ $((_4\nabla)^{*/sn})\mathbf{x}(_4V) = -_6E$

"$_6E$" is the electromagnetic field invariant from the next chapter.

What is special about the four-dimensional time-space curl operator "$((_4\nabla)^{*/sn})\mathbf{x}$" is the lack of a negative between time term and space term. This special feature is shown in the two-dimensional time-space simplification given below.

$((_2\nabla)^{*/sn})\mathbf{x}(_2V) =$
$= (\partial/\partial ct_M*((1_M)^{*/sn}) + (\partial/\partial x_M*(q_{xM})^{*/sn}))\mathbf{x}(1_M*V_{tM} + q_{xM}*V_{xM})$
$= (\partial/\partial ct_M)*((1_M)^{*/sn})*(q_{xM}*V_{xM}) + (\partial/\partial x_M)*((q_{xM})^{*/sn})*(1_M*V_{tM})$
$= (\partial V_{xM}/\partial ct_M)*((1_M)^{*/sn})*(q_{xM}) + (\partial V_{tM}/\partial x_M)*((q_{xM})^{*/sn})*(1_M)$
$= (\partial V_{xM}/\partial ct_M + \partial V_{tM}/\partial x_M)*(p_{xM})$
$= -E_{xM}*p_{xM}$

$-E_{xM} = \partial V_{xM}/\partial ct_M + \partial V_{tM}/\partial x_M$; $p_{xM} = ((1_M)^{*/sn})*(q_{xM}) = ((q_{xM})^{*/sn})*(1_M)$

d'Alembert operator, what is called the harmonic operator:

$((_4\nabla)^{sn})*((_4\nabla)^{*/sn}) = ((_4\nabla)^{sn})\bullet((_4\nabla)^{*/sn})$
$= \nabla_{tM}^2 - (\nabla_{xM}^2 + \nabla_{yM}^2 + \nabla_{zM}^2)$
$= (_4\nabla)^2$

$_4\nabla^2*_4V = {_4J}$; $_4\nabla^2*_6E = {_6G}$

SPECIAL ALGEBRA FOR SPECIAL RELATIVITY

<u>Theory-Development-Algebra Mathematics</u>. "$((_4\nabla)^{*/sn})\bullet(_4\rho) = 0$" is the same as fluid mass conservation from engineering class, where it is written "$\partial\rho/\partial t + {}_3\nabla\bullet(\rho*_3v) = 0$" for which "$_3v = v_x*i_x + v_y*i_y + v_z*i_z$" is velocity. In engineering class, it was derived very pragmatically by equating what goes into and out of a cube to what is inside.

"$\partial\rho/\partial t + {}_3\nabla\bullet(\rho*_3v) = 0$" appeared strange because "+" is between time and space terms, and not "−" as in "$((_2k)^{*/})\bullet(_2r) = \omega*t - k_x*x$". The "+" is now explained by use of the much less efficient nomenclature of the space-negative in the all-number theory-development-algebra. Theory-development-algebra is much more symbol intensive and is being used because our purpose for math has changed from engineering calculations to theory development, with the benefit being we see how mass conservation fits into the larger scheme of patterns in our world.

Tensor notation calculus of General Relativity is a form of differential geometry and also is an engineering-calculation-algebra. Covariant variables time and space are in the denominator of a derivative with the result being a contravariant vector. A contravariant vector is the tensor notation calculus analogy to the space-negative and special rules state where to insert negatives. The challenge is to create a theory-development-algebra for General Relativity that can replace tensor notation calculus. The challenge looks difficult.

<u>Example of Particle Count Conservation</u>. A pile of gravel moves in the positive "x" direction. Particle count starts on the left of "x_B" in "$dCount_B/dx_B = A*\exp(-(x_B*k_{xB})^2)$".

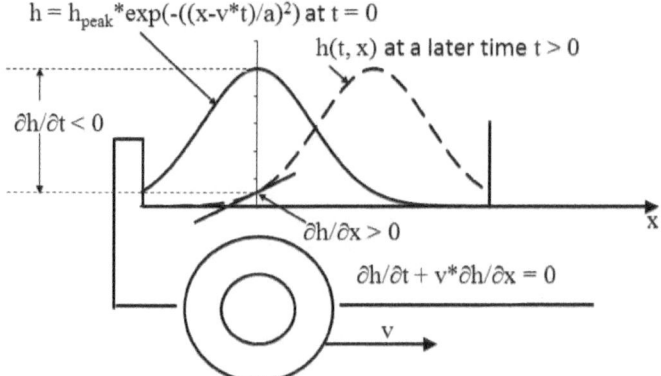

Figure 26. A moving pile of gravel of height "h".

CHAPTER 2 – PARTICLES

Subscript "$_B$" identifies the rest frame of the pile of gravel and is the same as "$_M$". "$dCount_B/dx_B$" is placed into a time-space invariant "$_2\rho_{lump}$" by multiplying it by the compound-label-number "1_B".

$$_2\rho_{lump} = 1_B*(\ dCount_B/dx_B\ ) = 1_B*(\ A*exp(-(x_B*k_{xB})^2)\ ) = 1_B*\rho_{lumptB}$$

Use a Lorentz Transformation to derive "ρ_{lumptS}" and "ρ_{lumpxS}".

$$\begin{aligned}
_2\rho_{lump} &= 1_B*(\ dCount_B/dx_B\ ) \\
&= 1_S*(\ dCount_B/dx_B\ )*exp(q_x*\alpha_{S/B}) \\
&= 1_S*(\ A*exp(-(x_B*k_{xB})^2)\ )*exp(q_x*\alpha_{S/B}) \\
&= 1_S*(\ A*exp(\ -((x_S - v_S*t_S)*cosh\alpha_M*k_{xB})^2\ )\ )*exp(q_x*\alpha_{S/B}) \\
&= 1_S*(\ A*exp(\ -((x_S - v_S*t_S)*cosh\alpha_M*k_{xB})^2\ )\ )*exp(q_x*\alpha_S) \\[4pt]
&= 1_S*(\ A*exp(\ -((x_S - v_S*t_S)*cosh\alpha_S*k_{xB})^2\ )\ )*cosh\alpha_S \\
&\quad + q_{xS}*(\ A*exp(\ -((x_S - v_S*t_S)*cosh\alpha_S*k_{xB})^2\ )\ )*sinh\alpha_S \\[4pt]
&= 1_S*\rho_{lumptS} + q_{xS}*\rho_{lumpxS}
\end{aligned}$$

$$\rho_{lumptS} = (\ A*exp(\ -((x_S - v_S*t_S)*cosh\alpha_S*k_{xB})^2\ )\ )*cosh\alpha_S$$

$$\rho_{lumpxS} = (\ A*exp(\ -((x_S - v_S*t_S)*cosh\alpha_S*k_{xB})^2\ )\ )*sinh\alpha_S$$

$$1_S = 1_B/exp(q_x*\alpha_{S/B})$$

Substitution "$x_B = (x_S - v_S*t_S)*cosh\alpha_S$" was derived from the Lorentz Transformation from "B" to "S".

$$c*t_S = c*t_B*cosh\alpha_S + x_B*sinh\alpha_S\ ;\quad c*t_B = c*t_S*cosh\alpha_S - x_S*sinh\alpha_S$$
$$x_S = c*t_B*sinh\alpha_S + x_B*cosh\alpha_S\ ;\quad x_B = -c*t_S*sinh\alpha_S + x_S*cosh\alpha_S$$

$$\begin{aligned}
x_B &= -c*t_S*sinh\alpha_S + x_S*cosh\alpha_S & ;\quad c*t_B &= c*t_S*cosh\alpha_S - x_S*sinh\alpha_S \\
&= x_S*cosh\alpha_S - c*t_S*sinh\alpha_S & &= c*t_S*cosh\alpha_S - x_S*sinh\alpha_S \\
&= x_S*cosh\alpha_S - c*t_S*tanh\alpha_S*cosh\alpha_S & &= c*t_S*cosh\alpha_S - x_S*tanh\alpha_S*cosh\alpha_S \\
&= (x_S - tanh\alpha_S*c*t_S)*cosh\alpha_S & &= (c*t_S - tanh\alpha_S*x_S)*cosh\alpha_S \\
&= (x_S - v_S*t_S)*cosh\alpha_S & &= c*(t_S - v_S*x_S/c^2)*cosh\alpha_S
\end{aligned}$$

The conservation law is applied.

SPECIAL ALGEBRA FOR SPECIAL RELATIVITY

$_2\nabla^{sn} = 1_S{}^{sn}*\nabla_{tS} + q_{xS}{}^{sn}*\nabla_{xS}$; $_2\nabla^{*jsn} = \nabla_{tS}*1_S{}^{*jsn} + \nabla_{xS}*q_{xS}{}^{*jsn}$
$= 1_S{}^{sn}*\partial/\partial ct_S + q_{xS}{}^{sn}*\partial/\partial x_S$

$_2\rho_{lump} = 1_S*(dCount_B/dx_B)*\cosh\alpha_S + q_{xS}*(dCount_B/dx_B)*\sinh\alpha_S$

$1_S{}^{*jsn}*1_S = 1$; $q_{xS}{}^{*jsn}*q_{xS} = 1$

$_2\nabla^{*jsn} \bullet {}_2\rho_{lump} = (\nabla_{tS}*1_S{}^{*jsn})*(1_S*(dCount_B/dx_B)*\cosh\alpha_S)$
$\qquad\qquad\qquad + (\nabla_{xS}*q_{xS}{}^{*jsn})*(q_{xS}*(dCount_B/dx_B)*\sinh\alpha_S)$

$\qquad = \nabla_{tS}*((dCount_B/dx_B)*\cosh\alpha_S) + \nabla_{xS}*((dCount_B/dx_B)*\sinh\alpha_S)$

For the example above

$\nabla_{tS}*((dCount_B/dx_B)*\cosh\alpha_S)$
$= (\partial/\partial ct_S)*((A*\exp(- ((x_S - v_S*t_S)*(\cosh\alpha_S)*k_{xB})^2))*\cosh\alpha_S)$

$= (-\partial((x_S - v_S*t_S)^2)/\partial ct_S)*(k_{xB}*\cosh\alpha_S)^2$
$\quad *((A*\exp(- ((x_S - v_S*t_S)*(k_{xB}*\cosh\alpha_S))^2))*\cosh\alpha_S)$

$= -2*(\partial(x_S - v_S*t_S)/\partial ct_S)*(x_S - v_S*t_S)*(k_{xB}*\cosh\alpha_S)^2$
$\quad *((A*\exp(- ((x_S - v_S*t_S)*(k_{xB}*\cosh\alpha_S))^2))*\cosh\alpha_S)$

$= -2*(-v_S/c)*(x_S - v_S*t_S)*(k_{xB}{}^2*\cosh^2\alpha_S)$
$\quad *A*\exp(-((x_S - v_S*t_S)*(k_{xB}*\cosh\alpha_S))^2)*\cosh\alpha_S$

$= 2*(\sinh\alpha_S/\cosh\alpha_S)*(x_S - v_S*t_S)*(k_{xB}{}^2*\cosh^2\alpha_S)$
$\quad *A*\exp(-((x_S - v_S*t_S)*(k_{xB}*\cosh\alpha_S))^2)*\cosh\alpha_S$

$= 2*(x_S - v_S*t_S)*(k_{xB}{}^2*\cosh^2\alpha_S)$
$\quad *A*\exp(-((x_S - v_S*t_S)*(k_{xB}*\cosh\alpha_S))^2)*\sinh\alpha_S$

$\nabla_{xS}*((dCount_B/dx_B)*\sinh\alpha_S)$

$= (\partial/\partial x_S)*((A*\exp(- ((x_S - v_S*t_S)*(k_{xB}*\cosh\alpha_S))^2))*\sinh\alpha_S)$

$= (-\partial((x_S - v_S*t_S)^2)/\partial x_S)*(k_{xB}*\cosh\alpha_S)^2$
$\quad *((A*\exp(- ((x_S - v_S*t_S)*(k_{xB}*\cosh\alpha_S))^2))*\sinh\alpha_S)$

CHAPTER 2 – PARTICLES

$$= -2*(\partial(x_S - v_S*t_S)/\partial x_S)*(x_S - v_S*t_S)*(k_{xB}*\cosh\alpha_S)^2$$
$$*((\ A*\exp(-\ ((x_S - v_S*t_S)*(k_{xB}*\cosh\alpha_S))^2\ )\ )*\sinh\alpha_S)$$

$$= -2*(x_S - v_S*t_S)*(k_{xB}^2*\cosh^2\alpha_S)$$
$$*A*\exp(-((x_S - v_S*t_S)*(k_{xB}*\cosh\alpha_S))^2)*\sinh\alpha_S$$

The above analysis showed

$$_2\nabla^{*/sn} \bullet\ _2\rho_{lump} = 0$$

Conservation Law with a Space-Negative Invariant. Abnormal space-negative invariant "$_2\rho_{anti-lump}{}^{sn}$" has subscript "$_{anti}$" to indicate it applies to anti-matter because if matter moves to the right, then the equivalent anti-matter is chosen to move to the left (if observed by the same observer), and that is because the space-negative operator means the negative of space, or, with respect to movement, the opposite direction.

"$_2\rho_{anti-lump}{}^{sn}$" and "$_2k^{sn}$" invariants are abnormal because the geometric translation has "$+i_t$" and "$-i_x$", same as the abnormal "$_2\nabla$". This is in contrast to "$+i_x$" of the geometric translation for normal invariants "$_2\rho_{lump}$", "$_2k$" and "$_2\nabla^{sn}$".

For matter invariant "$_2\rho_{lump}$", movement in the positive "x_M" direction is identified by "$x_B = x_M - v_M*t_M$". In contrast, space-negative anti-matter invariant "$_2\rho_{anti-lump}{}^{sn}$" includes "$x_B = x_M + v_M*t_M$". The "+" rather than "-" means movement is in the opposite direction.

"$x_B = x_M - v_M*t_M$" was derived from the Lorentz Transformation. In contrast, "$x_B = x_M + v_M*t_M$" for the space-negative is derived from the inverse matrix Lorentz Transformation for "$_2r^{sn}$", rather than for "$_2r$". (The inverse matrix has the negatives on the "$\sinh\alpha_{S/M}$".) Alternatively, "$x_B = x_M + v_M*t_M$" can be derived from "$-_2k^{sn*/} \bullet\ _2r$" rather than from "$-_2k^{*/} \bullet\ _2r$" if speed "v_M" is frequency divided by wavenumber.

The overall movement of electric charge particles is the same for matter to the right or for anti-matter to the left: For matter, if "$q = 1$" is the quantity of electric charge per particle, then the count that passes point "x_S" moving right is proportional to the total electric charge and is positive. For anti-matter the count is negative because "$q = -1$". Using the space-negative, those negative particles move left past point "x_S" and so get subtracted, for a double negative.

"$_2\nabla^{*/sn} \bullet (_2\rho_{anti-lump})^{sn} = 0$" is shown below to be valid:

$$dCount_B/dx_B = -A*\exp(\ -(x_B*k_{xB})^2\ )$$

SPECIAL ALGEBRA FOR SPECIAL RELATIVITY

$$_2\rho_{\text{anti-lump}}{}^{sn} = 1_S{}^{sn}*(d\text{Count}_B/dx_B)*\cosh\alpha_S + q_{xS}{}^{sn}*(d\text{Count}_B/dx_B)*\sinh\alpha_S$$
$$= 1_S{}^{sn}*(d\text{Count}_B/dx_B)*\exp(-q_x*\alpha_S)$$
$$= 1_S{}^{sn}*(-A*\exp(-(x_B*k_{xB})^2))*\exp(-q_x*\alpha_S)$$
$$= 1_S{}^{sn}*(-A*\exp(-((x_S + v_S*t_S)*k_{xB}*\cosh\alpha_S)^2))*\exp(-q_x*\alpha_S)$$

$$= 1_S{}^{sn}*(-A*\exp(-((x_S + v_S*t_S)*(k_{xB}*\cosh\alpha_S))^2))*\cosh\alpha_S$$
$$+ q_{xS}{}^{sn}*(-A*\exp(-((x_S + v_S*t_S)*(k_{xB}*\cosh\alpha_S))^2))*\sinh\alpha_S$$

$c*t_S = c*t_B*\cosh\alpha_S - x_B*\sinh\alpha_S$; $c*t_B = c*t_S*\cosh\alpha_S + x_S*\sinh\alpha_S$
$x_S = -c*t_B*\sinh\alpha_S + x_B*\cosh\alpha_S$; $x_B = c*t_S*\sinh\alpha_S + x_S*\cosh\alpha_S$

$$x_B = c*t_S*\sinh\alpha_S + x_S*\cosh\alpha_S = x_S*\cosh\alpha_S + c*t_S*\sinh\alpha_S$$
$$= x_S*\cosh\alpha_S + c*t_S*\tanh\alpha_S*\cosh\alpha_S = (x_S + \tanh\alpha_S*c*t_S)*\cosh\alpha_S$$
$$= (x_S + v_S*t_S)*\cosh\alpha_S$$

$$_2\nabla^{sn} = 1_S{}^{sn}*\nabla_{tS} + q_{xS}{}^{sn}*\nabla_{xS} \qquad ; \qquad _2\nabla^{*/sn} = \nabla_{tS}*1_S{}^{*/sn} + \nabla_{xS}*q_{xS}{}^{*/sn}$$
$$= 1_S{}^{sn}*\partial/\partial ct_S + q_{xS}{}^{sn}*\partial/\partial x_S$$

$(1_S{}^{*/sn})*(1_S{}^{sn}) = (1_S{}^{*j})*(1_S) = (1)*(1) = 1$
$(q_{xS}{}^{*/sn})*(q_{xS}{}^{sn}) = (-q_{xS}{}^{*j})*(-q_{xS}) = (-q_x)*(q_x) = -1$

$$_2\nabla^{*/sn} \bullet {}_2\rho_{\text{anti-lump}}{}^{sn} = (\nabla_{tS}*1_S{}^{*/sn})*(1_S{}^{sn}*(d\text{Count}/dx_B)*\cosh\alpha_S)$$
$$+ (\nabla_{xS}*q_{xS}{}^{*/sn})*(q_{xS}{}^{sn}*(d\text{Count}_B/dx_B)*\sinh\alpha_S)$$

$$= \nabla_{tS}*((d\text{Count}_B/dx_B)*\cosh\alpha_S) - \nabla_{xS}*((d\text{Count}_B/dx_B)*\sinh\alpha_S)$$

$$\nabla_{tS}*((d\text{Count}_B/dx_B)*\cosh\alpha_S)$$
$$= (\partial/\partial ct_S)*((-A*\exp(-((x_S + v_S*t_S)*(k_{xB}*\cosh\alpha_S))^2))*\cosh\alpha_S)$$

$$= (-\partial((x_S + v_S*t_S)^2)/\partial ct_S)*(k_{xB}*\cosh\alpha_S)^2$$
$$*((-A*\exp(-((x_S + v_S*t_S)*(k_{xB}*\cosh\alpha_S))^2))*\cosh\alpha_S)$$

$$= -2*(\partial(x_S + v_S*t_S)/\partial ct_S)*(x_S + v_S*t_S)*(k_{xB}*\cosh\alpha_S)^2$$
$$*((-A*\exp(-((x_S + v_S*t_S)*(k_{xB}*\cosh\alpha_S))^2))*\cosh\alpha_S)$$

$$= -2*(v_S/c)*(x_S + v_S*t_S)*(k_{xB}{}^2*\cosh^2\alpha_S)$$
$$*-A*\exp(-((x_S + v_S*t_S)*(k_{xB}*\cosh\alpha_S))^2)*\cosh\alpha_S$$

CHAPTER 2 – PARTICLES

$$= -2*(\sinh\alpha_S/\cosh\alpha_S)*(x_S + v_S*t_S)*(k_{xB}{}^2*\cosh^2\alpha_S)$$
$$*\text{-}A*\exp(-((x_S + v_S*t_S)*(k_{xB}*\cosh\alpha_S))^2)*\cosh\alpha_S$$

$$= -2*(x_S + v_S*t_S)*(k_{xB}{}^2*\cosh^2\alpha_S)$$
$$*\text{-}A*\exp(-((x_S + v_S*t_S)*(k_{xB}*\cosh\alpha_S))^2)*\sinh\alpha_S$$

$$\nabla_{xS}*((dCount_B/dx_B)*\sinh\alpha_S)$$

$$= (\partial/\partial x_S)*((\text{-}A*\exp(-((x_S + v_S*t_S)*(k_{xB}*\cosh\alpha_S))^2))*\sinh\alpha_S)$$

$$= (-\partial((x_S + v_S*t_S)^2)/\partial x_S)*(k_{xB}*\cosh\alpha_S)^2$$
$$*((\text{-}A*\exp(-((x_S + v_S*t_S)*(k_{xB}*\cosh\alpha_S))^2))*\sinh\alpha_S)$$

$$= -2*(\partial(x_S + v_S*t_S)/\partial x_S)*(x_S + v_S*t_S)*(k_{xB}*\cosh\alpha_S)^2$$
$$*((\text{-}A*\exp(-((x_S + v_S*t_S)*(k_{xB}*\cosh\alpha_S))^2))*\sinh\alpha_S)$$

$$= -2*(x_S + v_S*t_S)*(k_{xB}{}^2*\cosh^2\alpha_S)$$
$$*\text{-}A*\exp(-((x_S + v_S*t_S)*(k_{xB}*\cosh\alpha_S))^2)*\sinh\alpha_S$$

2.12 Exercises

<u>Check on Text Comprehension</u>.

1) For "$c*t = 13$" and "$x = 12$" calculate "$r_{hyperbolic} = \sqrt{(c*t)^2 - x^2}$", "$\cosh\alpha = c*t/r_{hyperbolic}$", "$\sinh\alpha = x/r_{hyperbolic}$", and "$\tanh\alpha = x/(c*t)$". Find "$\alpha = \text{atanh}(x/(c*t))$", confirm "$1 = \sqrt{\cosh^2\alpha - \sinh^2\alpha}$". Write "$_2r = 1*c*t + q_x*x$" and "$_2r = r_{hyperbolic}*\exp(q_x*\alpha)$" using numbers. Plot "$(c*t, x) = (13, 12)$" on the hypercomplex-plane. Draw a straight line at hyperbolic-angle "$\alpha = \text{atanh}(12/13)$" and hyperbola with hyperbolic-radius "$r_{hyperbolic} = \sqrt{13^2 - 12^2}$".

2) For "$c*t_M = 11$", "$\alpha_M = 5$" and "$\alpha_{S/M} = 3$" find "$c*t_S$" and "x_S". For "$E_M/c = 7$" find "E_S/c" and "p_{xS}". What is "$c*t_B$" and what is "m_B*c"?

3) How fast "$v_{S/M}/c$" must bus "M" move so that its seats have half the spacing of a stationary bus? At that speed, what is the ratio of time between ticks of the bus's clock to the roadside's clock? What is the energy ratio of the moving bus to a stationary bus?

SPECIAL ALGEBRA FOR SPECIAL RELATIVITY

4) A hyper-light-speed signal of speed "$v_{M(first)} = c*\coth(\mu_{M(first)})$" is sent from Person M in the positive x-direction to Person S who walks away from Person M at a speed "$v_{M/S}$" equal to positive seven feet per second. The instant the signal is received, Person S sends a signal back to the source at the same speed but in the opposite direction, "$v_{S(second)} = -v_{M(first)}$". At what speed "$v_{M(first)}$" must the signal travel so that the second signal arrives at the same instant the first signal is emitted? Draw the two signals on the hypercomplex-plane for Person M stationary and then, again, on the hypercomplex-plane for Person S stationary.

5) An anti-matter electron moves on an anti-matter bus "M" at a speed "$v_M = c*\tanh\alpha_M = 11$ meters per second" with "$\alpha_M \in R$". What is its electric charge "$Q_{B\text{-antimatter}}$"? What is its rest mass "$m_{B\text{-antimatter}}$"? The bus moves relative to the road "S" with hyperbolic angle "$\alpha_{S/M} = \pm i*\pi$". What is the electron's energy "E_S" and momentum "p_{xS}"?

6) A wavy pattern for static electric charge is across the roof of a bus: "$dQ_B/dx_B = \rho_{Qwavy} = A*\sin(x_B/a)$". "$dQ_B/dx_B$" is hyperbolic-radius "$\rho_{Qwavy}$" of the current density invariant "$_2J$". Show electric charge is conserved for any speed of the bus by showing "$_2\nabla^{*/sn} \bullet _2J = 0$". Write "$_2J^{sn}$" (to represent an anti-matter car), by analogy using the example in the text above, and show "$_2\nabla^{*/sn} \bullet _2J^{sn} = 0$".

7) A row of spinning disks has a density "$\rho_{HxB+} = dH_{Bx+}/dx_B$" with "$H_{Bx+} > 0$". In parallel, a moving row of spinning disks has a density "$\rho_{HxB-} = dH_{Bx-}/dx_B$" with "$H_{Bx-} < 0$". This second row moves at speed "$v_{M-} > 0$" (to the right), so that "$\alpha_{M-} > 0$". The total angular momentum of both rows together is constrained to be zero. For both rows together, what is the hyperbolic radius of the time-like angular momentum invariant for the rate of angular momentum that passes point "x_M".

8) Angular momentum density "$_2\rho_H = 1_M*q_x*dH_{Bx}/dx_B*\exp(q_x*\alpha_M)$" is used in conservation law "$((_2\nabla)^{*/sn})\mathbf{x}(_2\rho_H) = 0$" for which no torque is applied to change angular momentum. For "$_2\rho_H$" below, verify the conservation law.

$$_2\rho_H = 1_M*q_x*(dH_{Bx}/dx_B)*\exp(q_x*\alpha_M)$$
$$= 1_M*q_x*(A*\exp(-((x_B)/a)^2))*\exp(q_x*\alpha_M)$$

CHAPTER 2 – PARTICLES

Answers to Select Exercises.

1) $r_{hyperbolic} = \sqrt{(13^2 - 12^2)} = 5$; $\alpha = \operatorname{atanh}(12/13) = 1.60943...$
   $\cosh\alpha = 13/5 = 2.6$; $(2.6)^2 - (2.4)^2 = 6.76 - 5.76 = 1$
   $\sinh\alpha = 12/5 = 2.4$; $_2r = 1*13 + q_x*12$
   $\tanh\alpha = 12/13 = 0.092307...$; $_2r = 5*\exp(q_x*1.60943...)$

2) $c*t_B = c*t_M/\cosh(\alpha_M) = 11/\cosh(5) = 0.14822...$
   $x_M = c*t_B*\sinh(\alpha_M) = 0.14822...*\sinh(5) = 10.99900...$

   $c*t_S = c*t_B*\cosh(\alpha_M + \alpha_{S/M}) = 0.14822...*\cosh(5 + 3) = 220.9309...$
   $x_S = c*t_B*\sinh(\alpha_M + \alpha_{S/M}) = 0.14822...*\sinh(5 + 3) = 220.9308...$

   $c*t_S = c*t_M*\cosh(\alpha_{S/M}) + x_M*\sinh(\alpha_{S/M})$
   $= 11*\cosh(3) + 10.99900...*\sinh(3) = 220.9309...$

   $x_S = c*t_M*\sinh(\alpha_{S/M}) + x_M*\cosh(\alpha_{S/M})$
   $= 11*\sinh(3) + 10.99900...*\cosh(3) = 220.9308...$

   $m_B*c = (E_M/c)/\cosh(\alpha_M) = 7/\cosh(5) = 0.09432...$
   $p_{xM} = m_B*c*\sinh(\alpha_M) = 0.09432...*\sinh(5) = 6.99930...$

   $E_S/c = m_B*c*\cosh(\alpha_M + \alpha_{S/M}) = 0.09432...*\cosh(8) = 140.59239...$
   $p_{xS} = m_B*c*\sinh(\alpha_M + \alpha_{S/M}) = 0.09432...*\sinh(8) = 140.59235...$

   $E_S/c = (E_M/c)*\cosh(\alpha_{S/M}) + p_{xM}*\sinh(\alpha_{S/M})$
   $= 7*\cosh(3) + 6.99930...*\sinh(3) = 140.59239...$

   $p_{xS} = (E_M/c)*\sinh(\alpha_{S/M}) + p_{xM}*\cosh(\alpha_{S/M})$
   $= 7*\sinh(3) + 6.99930...*\cosh(3) = 140.59235...$

3) $k_{xS}/k_{xM} = 2$; $\alpha_{S/M} = \operatorname{acosh}(k_{xS}/k_{xM}) = \operatorname{acosh}(2) = 1.31695...$
   $v_{S/M}/c = \tanh(\alpha_{S/M}) = \sqrt{3}/2$; $t_S/t_M = E_S/E_M = \cosh\alpha_{S/M} = 2$

4) $\mu_{M(first)} = \alpha_{M/S}/2 = \operatorname{atanh}(v_{M/S}/c)/2$

5) and 6) solution not given

SPECIAL ALGEBRA FOR SPECIAL RELATIVITY

7) $_2\rho_H = {_2\rho_{H+}} + {_2\rho_{H-}}$
$= 1_M * q_x * \rho_{HxB+} * \exp(q_x * \alpha_{M+}) + 1_M * q_x * \rho_{HxB-} * \exp(q_x * \alpha_{M-})$
$= 1_M * q_x * \rho_{HxB+} + 1_M * q_x * \rho_{HxB-} * \exp(q_x * \alpha_{M-})$
$= 1_M * q_x * \rho_{HxB+} + 1_M * q_x * \rho_{HxB-} * (\cosh\alpha_{M-} + q_x * \sinh\alpha_{M-})$
$= 1_M * q_x * (\rho_{HxB+} + \rho_{HxB-} * \cosh\alpha_{M-}) + 1_M * \rho_{HxB-} * \sinh\alpha_{M-}$
$= 1_M * \rho_{HxB-} * \sinh\alpha_{M-} = 1_M * (-\rho_{HxB+}/\cosh\alpha_{M-}) * \sinh\alpha_{M-}$
$= 1_M * -\rho_{HxB+} * \tanh\alpha_{M-} = 1_M * \rho_{HxB+} * (-v_{M-}/c)$

For both rows together, the hyperbolic radius of the time-like angular momentum invariant for the rate of angular momentum that passes point "x_M" is "$\rho_{HxB+} * (-v_{M-}/c)$".

8) $_2\rho_H = 1_M * q_x * (dH_{Bx}/dx_B) * \cosh\alpha_M + q_{xM} * q_x * (dH_{Bx}/dx_B) * \sinh\alpha_M$
$= 1_M * q_x * (dH_{Bx}/dx_B) * \exp(q_x * \alpha_M)$
$= 1_M * q_x * (A * \exp(-((x_B)/a)^2)) * \exp(q_x * \alpha_M)$
$= 1_M * q_x * (A * \exp(-(((x_M - v_M * t_M) * \cosh\alpha_M)/a)^2)) * \exp(q_x * \alpha_M)$

$= 1_M * q_x * (A * \exp(-((x_M - v_M * t_M) * (\cosh\alpha_M/a))^2)) * \cosh\alpha_M$
$+ q_{xM} * q_x * (A * \exp(-((x_M - v_M * t_M) * (\cosh\alpha_M/a))^2)) * \sinh\alpha_M$

$x_B = -c * t_M * \sinh\alpha_M + x_M * \cosh\alpha_M = x_M * \cosh\alpha_M - c * t_M * \sinh\alpha_M$
$= x_M * \cosh\alpha_M - c * t_M * \tanh\alpha_M * \cosh\alpha_M = (x_M - \tanh\alpha_M * c * t_M) * \cosh\alpha_M$
$= (x_M - v_M * t_M) * \cosh\alpha_M$

$_2\nabla^{sn} = 1_M^{sn} * \nabla_{tM} + q_{xM}^{sn} * \nabla_{xM}$; $_2\nabla^{*/sn} = \nabla_{tM} * 1_M^{*/sn} + \nabla_{xM} * q_{xM}^{*/sn}$
$= 1_M^{sn} * \partial/\partial ct_M + q_{xM}^{sn} * \partial/\partial x_M$

$_2\nabla^{*/sn} \mathbf{x} _2\rho_H = (\nabla_{tM} * 1_M^{*/sn}) * (1_M * q_x * (dH_{xB}/dx_B) * \cosh\alpha_M)$
$+ (\nabla_{xM} * q_{xM}^{*/sn}) * (q_{xM} * q_x * (dH_{xB}/dx_B) * \sinh\alpha_M)$

$= (\nabla_{tM} * ((dH_{xB}/dx_B) * \cosh\alpha_M) + \nabla_{xM} * ((dH_{xB}/dx_B) * \sinh\alpha_M)) * q_x$

$\nabla_{tM} * ((dH_{xB}/dx_B) * \cosh\alpha_M)$
$= (\partial/\partial ct_M) * ((A * \exp(- ((x_M - v_M * t_M) * (\cosh\alpha_M/a))^2)) * \cosh\alpha_M)$

$= (-\partial((x_M - v_M * t_M)^2)/\partial ct_M) * (\cosh\alpha_M/a)^2$
$* ((A * \exp(- ((x_M - v_M * t_M) * (\cosh\alpha_M/a))^2)) * \cosh\alpha_M)$

CHAPTER 2 – PARTICLES

$= -2*(\partial(x_M - v_M*t_M)/\partial ct_M)*(x_M - v_M*t_M)*(\cosh\alpha_M/a)^2$
$\quad *((A*\exp(- ((x_M - v_M*t_M)*(\cosh\alpha_M/a))^2))*\cosh\alpha_M)$

$= -2*(-v_M/c)*(x_M - v_M*t_M)*(\cosh^2\alpha_M/a^2)$
$\quad *A*\exp(-((x_M - v_M*t_M)*(\cosh\alpha_M/a))^2)*\cosh\alpha_M$

$= 2*(\sinh\alpha_M/\cosh\alpha_M)*(x_M - v_M*t_M)*(\cosh^2\alpha_M/a^2)$
$\quad *A*\exp(-((x_M - v_M*t_M)*(\cosh\alpha_M/a))^2)*\cosh\alpha_M$

$= 2*(x_M - v_M*t_M)*(\cosh^2\alpha_M/a^2)$
$\quad *A*\exp(-((x_M - v_M*t_M)*(\cosh\alpha_M/a))^2)*\sinh\alpha_M$

$\nabla_{xM}*((dH_{xB}/dx_B)*\sinh\alpha_M)$

$= (\partial/\partial x_M)*((A*\exp(- ((x_M - v_M*t_M)*(\cosh\alpha_M/a))^2))*\sinh\alpha_M)$

$= (-\partial((x_M - v_M*t_M)^2)/\partial x_M)*(\cosh\alpha_M/a)^2$
$\quad *((A*\exp(- ((x_M - v_M*t_M)*(\cosh\alpha_M/a))^2))*\sinh\alpha_M)$

$= -2*(\partial(x_M - v_M*t_M)/\partial x_M)*(x_M - v_M*t_M)*(\cosh\alpha_M/a)^2$
$\quad *((A*\exp(- ((x_M - v_M*t_M)*(\cosh\alpha_M/a))^2))*\sinh\alpha_M)$

$= -2*(x_M - v_M*t_M)*(\cosh^2\alpha_M/a^2)$
$\quad *A*\exp(-((x_M - v_M*t_M)*(\cosh\alpha_M/a))^2)*\sinh\alpha_M$

$_2\nabla^{*jsn}\mathbf{x}_2\rho_H = 0$

SPECIAL ALGEBRA FOR SPECIAL RELATIVITY

Questions for Further Thought.

1) If unspecified-speed-parameter "ς" is a rational number, then it cannot be infinite, has equal probability of being positive or negative, and has equal probability of magnitude less than one or greater than one. What alternative assumption might we make for the structure of "ς"? Numerator only, for equal probability of being any number on the number-line?

2) Assume "$\alpha_{S/M} = i*\theta$" and find "$c*t_S$", "x_S", "$v_{S/M}/c$" and "v_S/c". If "$\theta \neq z*\pi/2$" for "$z \in Z$", what problem is there? Should we make a rule (as a theory of physics), so that we mimic nature?

3) Is an invariant with hyperbolic-radius "dm_B/dx_B" or "Q_B" useful?

4) What does "N" in "$\alpha_{S/M} = i*\pi*(2*N + 1)$" represent physically?

5) What might an anti-matter person observe of us with respect to entropy increase, cause and effect, and successive collapses of matter-wave functions? Does the anti-matter person see us moving backward relative to their sense of time? Propose a cosmological model of the universe.

6) Rewrite the section on anti-matter electric current but substitute angular momentum for electric current.

7) Explain why "$\partial\rho/\partial t_M + {}_3\nabla\bullet(\rho*_3v) = 0$" has the two terms added, in contrast to the subtraction in "$({}_2r^{*j})\bullet({}_2r) = (c*t_M)^2 - (x_M^2 + y_M^2 + z_M^2)$".

8) Assume information transfer between detectors in the EPR experiment (see an Appendix) occurs instantly. Try to design an experiment in which detectors move away from each other so that the detectors prove time can step backward.

9) Is the angular momentum density invariant the missing space-like energy-momentum invariant as measurement units imply it is? If time-like energy-momentum is formed from induced electromagnetic fields (as explained in the next chapter), what fields could produce the space-like energy-momentum invariant?

CHAPTER 3 – FIELDS

Chapter 3 – Fields

Using his equations, Maxwell derived that the speed for light was independent of the observer's speed. But the path of light between two people sitting side-by-side on a bus is longer when observed by a person on the roadside because the path has a component in the direction of bus travel. To keep the speed for light constant, Einstein created Special Relativity to explain that a clock-tower shows faster passage of time compared to a clock mounted inside the bus.

3.1 Geometric-Vector Notation

The four materials in Maxwell's Equations are physical entities in our geometric world. **Bold** indicates a geometric-vector. Non-bold a scalar.

Electric field $_3$**E** ; Current density $_3$**j**
Magnetic field $_3$**B** ; Charge density ρ

Each exists as a distribution in time and space, as does voltage (see Panofsky, Wolfgang and Philips, Melba: *Classical Electricity and Magnetism*, Addison-Wesley Publishing Company, Inc.; 1955).

Vector voltage $_3$**A** ; Scalar voltage ϕ

A ground for voltage is analogous to an origin for location and to an inertial reference frame for speed or momentum.

Maxwell evolved his equations and settled on component equations. Hamilton worked with Maxwell and attempted to apply quaternions, but that effort failed, probably because, per a read of Hamilton's 1843 paper, he saw quaternions as a replacement for "*i*", and so could not combine quaternions with "*i*" to formulate the correct properties of these factors. A little before year 1900, Heaviside and others applied geometric-unit-vectors "**i**$_x$", "**i**$_y$", and "**i**$_z$", dot-product "•", cross-product "**x**", and gradient operator "$_3\nabla$" to Maxwell's Equations. Geometric-unit-vectors explicitly placed components of electro-magnetism into physical space.

Three constants for measurement unit conversion:

$\mu = 4\pi * 10^{-7}$ tesla*meter/amp $\approx 1.256637 * 10^{-6}$ tesla*meter/amp

$\partial \approx 8.854188 * 10^{-12}$ coulomb^2/(newton*meter^2)

$c \approx 2.99792458 * 10^8$ meters/second

SPECIAL ALGEBRA FOR SPECIAL RELATIVITY

The three constants were selected so that "$c^2 * ə * \mu = 1$".

A coulomb is electric charge from a specific finite quantity of protons, or that same quantity of electrons times negative one.

1 coulomb = $-6.24150965(16) * 10^{18}$ electron charges
$Q_{B\text{-electron}} = -1.6021176462 * 10^{-19}$ coulomb

Alternate units of measure are found from:

1 tesla*meter/amp = 1 newton/amp^2 = 1 volt*second/(amp*meter)

1 coulomb^2/(newton*meter^2) = 1 farad/meter
$\qquad\qquad\qquad\qquad\qquad$ = 1 coulomb/(volt*meter)

Maxwell's Equations in geometric-vector notation:

$_3\mathbf{E} = E_x * \mathbf{i}_x + E_y * \mathbf{i}_y + E_z * \mathbf{i}_z$ $\qquad$; $\qquad$ $_3\mathbf{j} = j_x * \mathbf{i}_x + j_y * \mathbf{i}_y + j_z * \mathbf{i}_z$
$_3\mathbf{B} = B_x * \mathbf{i}_x + B_y * \mathbf{i}_y + B_z * \mathbf{i}_z$
$_3\mathbf{\nabla} = \nabla_x * \mathbf{i}_x + \nabla_y * \mathbf{i}_y + \nabla_z * \mathbf{i}_z = \partial/\partial x * \mathbf{i}_x + \partial/\partial y * \mathbf{i}_y + \partial/\partial z * \mathbf{i}_z$

$_3\mathbf{\nabla} \bullet {_3\mathbf{E}} = \rho/ə$ $\qquad\qquad\qquad\qquad$; $\qquad$ $_3\mathbf{\nabla} \bullet {_3\mathbf{B}} = 0$
$_3\mathbf{\nabla} \mathbf{x} {_3\mathbf{B}} = \mu * {_3\mathbf{j}} + (\partial({_3\mathbf{E}})/\partial t)/c^2$ $\quad$; $\qquad$ $_3\mathbf{\nabla} \mathbf{x} {_3\mathbf{E}} = -\partial({_3\mathbf{B}})/\partial t$

$\partial E_x/\partial x + \partial E_y/\partial y + \partial E_z/\partial z = \rho/ə$ $\qquad$; $\qquad$ $\partial B_x/\partial x + \partial B_y/\partial y + \partial B_z/\partial z = 0$
$\partial B_z/\partial y - \partial B_y/\partial z = (\partial E_x/\partial t)/c^2 + \mu * j_x$ $\;$; $\qquad$ $\partial E_z/\partial y - \partial E_y/\partial z = -\partial B_x/\partial t$
$\partial B_x/\partial z - \partial B_z/\partial x = (\partial E_y/\partial t)/c^2 + \mu * j_y$ $\;$; $\qquad$ $\partial E_x/\partial z - \partial E_z/\partial x = -\partial B_y/\partial t$
$\partial B_y/\partial x - \partial B_x/\partial y = (\partial E_z/\partial t)/c^2 + \mu * j_z$ $\;$; $\qquad$ $\partial E_y/\partial x - \partial E_x/\partial y = -\partial B_z/\partial t$

Electric Charge Conservation in geometric-vector notation:

$\qquad \partial\rho/\partial t + {_3\mathbf{\nabla}} \bullet {_3\mathbf{j}} = 0$ $\qquad$; $\qquad$ $\partial\rho/\partial t + \partial j_x/\partial x + \partial j_y/\partial y + \partial j_z/\partial z = 0$

Voltage equations in geometric-vector notation:

$\qquad _3\mathbf{A} = A_x * \mathbf{i}_x + A_y * \mathbf{i}_y + A_z * \mathbf{i}_z$; $\;$ $_3\mathbf{B} = {_3\mathbf{\nabla}} \mathbf{x} {_3\mathbf{A}}$ $\;$; $\;$ $_3\mathbf{E} = -{_3\mathbf{\nabla}}\phi - \partial({_3\mathbf{A}})/\partial t$

$\qquad \partial A_z/\partial y - \partial A_y/\partial z = B_x$ $\qquad$; $\qquad$ $-\partial\phi/\partial x - \partial A_x/\partial t = E_x$
$\qquad \partial A_x/\partial z - \partial A_z/\partial x = B_y$ $\qquad$; $\qquad$ $-\partial\phi/\partial y - \partial A_y/\partial t = E_y$
$\qquad \partial A_y/\partial x - \partial A_x/\partial y = B_z$ $\qquad$; $\qquad$ $-\partial\phi/\partial z - \partial A_z/\partial t = E_z$

CHAPTER 3 – FIELDS

$_3\nabla \bullet {_3}\mathbf{A} + (\partial\phi/\partial t)/c^2 = 0$; $\partial A_x/\partial x + \partial A_y/\partial y + \partial A_z/\partial z + (\partial\phi/\partial t)/c^2 = 0$

$(_3\nabla \bullet {_3}\nabla)\phi - \partial^2(\phi)/\partial t^2/c^2 = -\rho/\vartheta$; $(_3\nabla \bullet {_3}\nabla)_3\mathbf{A} - \partial^2(_3\mathbf{A})/\partial t^2/c^2 = -\mu^*{_3}\mathbf{j}$

$\partial^2\phi/\partial x^2 + \partial^2\phi/\partial y^2 + \partial^2\phi/\partial z^2 - (\partial^2\phi/\partial t^2)/c^2 = -\rho/\vartheta$
$\partial^2 A_x/\partial x^2 + \partial^2 A_x/\partial y^2 + \partial^2 A_x/\partial z^2 - (\partial^2 A_x/\partial t^2)/c^2 = -\mu^* j_x$
$\partial^2 A_y/\partial x^2 + \partial^2 A_y/\partial y^2 + \partial^2 A_y/\partial z^2 - (\partial^2 A_y/\partial t^2)/c^2 = -\mu^* j_y$
$\partial^2 A_z/\partial x^2 + \partial^2 A_z/\partial y^2 + \partial^2 A_z/\partial z^2 - (\partial^2 A_z/\partial t^2)/c^2 = -\mu^* j_z$

Gauss's Law for Electricity. "$_3\nabla \bullet {_3}\mathbf{E} = \rho/\vartheta$" states divergence of an electric field is proportional to electric charge density. In other words, electric charge is the source of an electric field.

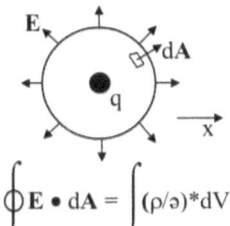

$\oint \mathbf{E} \bullet d\mathbf{A} = \int (\rho/\vartheta)^* dV$

Figure 27. Gauss's Law for Electricity. Electric field times area of a closed surface (a sphere) equals the electric charge inside. The electric field at the radius of the sphere is calculated from "$E_{radial}*(r^2*\pi*4/3) = q/\vartheta$".

Gauss's Law for Magnetism. "$_3\nabla \bullet {_3}\mathbf{B} = 0$" states there are no magnetic field charges (in contrast to the existence of electric charges).

Faraday's Law of Electric Field Induction. "$_3\nabla \times {_3}\mathbf{E} = -\partial(_3\mathbf{B})/\partial t$" states a time varying magnetic field is a source of vorticity for an induced electric field.

SPECIAL ALGEBRA FOR SPECIAL RELATIVITY

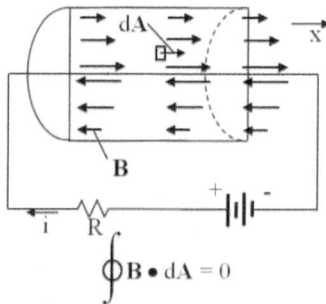

Figure 28. Gauss's Law of Magnetism. Magnetic field times area of a closed surface equals zero. Ends and the curved surface of the cylinder cut in half have "**B**•d**A** = 0". The "**B**•d**A**" of the upper rectangle equals the negative of the "**B**•d**A**" of the lower rectangle, so that all the magnetic flux out of the identified volume equals all the flux into the volume, for a net total of zero.

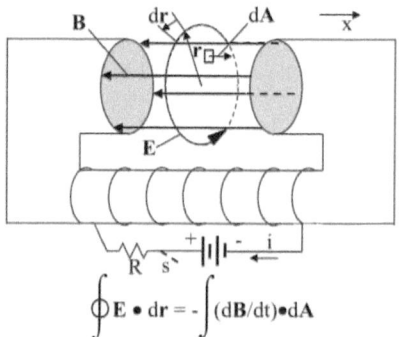

Figure 29. Faraday's Law of Induction. Electric field times length along a closed curve (a circle) equals the rate of change of the magnetic field times area enclosed by the closed curve. Switch "s" closes to initiate a flow of current "i", to create an iron magnet inside wire coils. The gap of cross-section area "A" in the iron has increasing magnitude magnetic field "dB_x/dt", which is negative. Changing magnetic field induces an electric field per "$E_{circumferential}*(2*\pi*r) = -(dB_x/dt)*A$", for "$r > \sqrt{A/\pi}$" (for "r" outside the gap).

Ampere's Law of Magnetic Field Induction. "$_3\nabla \mathbf{x}_3\mathbf{B} = \mu *_3\mathbf{j} + (\partial(_3\mathbf{E})/\partial t)/c^2$" states vorticity of an induced magnetic field is created by a time varying electric field and/or by electric current density.

CHAPTER 3 – FIELDS

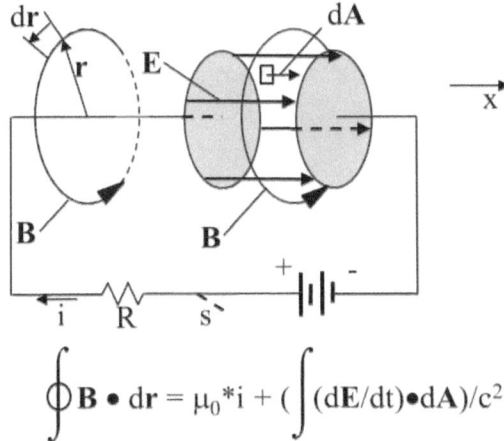

$$\oint \mathbf{B} \bullet d\mathbf{r} = \mu_0 * i + \left(\int (d\mathbf{E}/dt) \bullet d\mathbf{A} \right)/c^2$$

Figure 30. Ampere's Law of Induction. Magnetic field times length along a closed curve (a circle) equals the rate of change of the electric field times area enclosed by the closed curve, plus the electric current through the area. Switch "s" closes to initiate a current that induces a magnetic field around the wire of strength "$B_{circumferential} = \mu_0 * i/(2*\pi*r)$". The magnetic field continues along the wire so that it also exists around the capacitor of area "A", per "$B_{circumferential} * (2*\pi*r) = (dE_x/dt) * A/c^2$" for the radius outside the area of the capacitor.

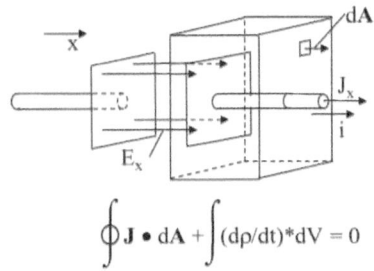

$$\oint \mathbf{J} \bullet d\mathbf{A} + \int (d\rho/dt) * dV = 0$$

Figure 31. Electric Charge Conservation. The flow of electric charge out a closed surface (a cube), calculated as electric current density times area of the surface, equals the negative change in electric charge inside. In other words, electric charge cannot be created or destroyed. The box on the right contains one of the two capacitor plates. The change in electric charge on the capacitor plate inside the box is "$dq/dt < 0$" as noticed by the increase in the electric field "E_x". The box is penetrated by the wire out the right for which the current "i" equals the current density "J_x" times the cross-section area of the wire. The conservation equation reduces to "$i + dq/dt = 0$".

SPECIAL ALGEBRA FOR SPECIAL RELATIVITY

Electric Charge Conservation. "$\partial \rho / \partial t + {}_3\mathbf{\nabla} \bullet \mathbf{j} = 0$", rewritten "$-\partial \rho / \partial t = {}_3\mathbf{\nabla} \bullet \mathbf{j}$", states the divergence of an electric current density equals the negative of the change in electric charge density.

3.2 All-Number Notation

New component symbols apply to the second step's translation of geometry into all-number algebra.

$V_t = \phi$; $J_t = \rho/\partial = \rho*(\mu*c^2)$
$V_x = A_x*c$; $K_x = B_x*c$; $J_x = j_x*(\mu*c) = j_x/(\partial*c)$
$V_y = A_y*c$; $K_y = B_y*c$; $J_y = j_y*(\mu*c) = j_y/(\partial*c)$
$V_z = A_z*c$; $K_z = B_z*c$; $J_z = j_z*(\mu*c) = j_z/(\partial*c)$

"$\mu*c$" and "$\partial*c$" are near a numerical value of one.

Simple-Label-Numbers. Using components specified above, Maxwell's Equations, the electric charge conservation equation and the voltage equation may be written using label-numbers.

$j_x*j_y = -j_y*j_x = j_z$; $i*j_x = j_x*i$
$j_y*j_z = -j_z*j_y = j_x$; $i*j_y = j_y*i$
$j_z*j_x = -j_x*j_z = j_y$; $i*j_z = j_z*i$

$j_x^2 = j_y^2 = j_z^2 = i^2 = -1$; $q_x^2 = q_y^2 = q_z^2 = 1^2 = +1$

$q_x = j_x/i$; $q_y = j_y/i$; $q_z = j_z/i$

$q_x*j_y = -j_y*q_x = -q_y*j_x = j_x*q_y = q_z$; $q_x*q_y = -q_y*q_x = -i*q_z = -j_z$
$q_y*j_z = -j_z*q_y = -q_z*j_y = j_y*q_z = q_x$; $q_y*q_z = -q_z*q_y = -i*q_x = -j_x$
$q_z*j_x = -j_x*q_z = -q_x*j_z = j_z*q_x = q_y$; $q_z*q_x = -q_x*q_z = -i*q_y = -j_y$

Unspecified-Label-Number. There is an unspecified-speed-parameter "ς", and there is an unspecified-label-number "κ" (kappa) restricted to one of "q_x", "q_y" or "q_z". "κ" is unknown and unknowable.

CHAPTER 3 – FIELDS

"ς" and "κ" are placed into two exponential functions: One for each side, left and right, of a simple-label-number "1", "q_x", "q_y" or "q_z". Each is a square root by dividing the argument by two.

$1_M = \exp(-\kappa*\varsigma/2)*1*\exp(-\kappa*\varsigma/2) = \exp(-\kappa*\varsigma)$
$q_{xM} = \exp(-\kappa*\varsigma/2)*q_x*\exp(-\kappa*\varsigma/2)$
$q_{yM} = \exp(-\kappa*\varsigma/2)*q_y*\exp(-\kappa*\varsigma/2)$
$q_{zM} = \exp(-\kappa*\varsigma/2)*q_z*\exp(-\kappa*\varsigma/2)$

Hypothetically: "$\kappa = q_x$" for "$1_M = \exp(-q_x*\varsigma)$", "$q_{xM} = q_x*\exp(-q_x*\varsigma)$", "$q_{yM} = q_y$" and "$q_{zM} = q_z$" from the previous chapter. The concept of using a square root factor on both sides, which takes advantage of the anti-commute law "$q_x*q_y = -q_y*q_x$", is a classic concept in Special Relativity.

$1_M = \exp(-\kappa*\varsigma/2)*1*\exp(-\kappa*\varsigma/2)$
$\quad = \exp(-q_x*\varsigma/2)*1*\exp(-q_x*\varsigma/2) \quad$ "$\kappa = q_x$"
$\quad = \exp(-q_x*\varsigma/2)*\exp(-q_x*\varsigma/2) = \exp(-q_x*\varsigma)$

$q_{xM} = \exp(-\kappa*\varsigma/2)*q_x*\exp(-\kappa*\varsigma/2)$
$\quad = \exp(-q_x*\varsigma/2)*(q_x*\exp(-q_x*\varsigma/2)) \quad$ "$\kappa = q_x$"
$\quad = \exp(-q_x*\varsigma/2)*(q_x*(\cosh(\varsigma/2) - q_x*\sinh(\varsigma/2)))$
$\quad = \exp(-q_x*\varsigma/2)*((\cosh(\varsigma/2) - q_x*\sinh(\varsigma/2))*q_x)$
$\quad = \exp(-q_x*\varsigma/2)*(\exp(-q_x*\varsigma/2)*q_x)$
$\quad = \exp(-q_x*\varsigma)*q_x$

$q_{yM} = \exp(-\kappa*\varsigma/2)*q_y*\exp(-\kappa*\varsigma/2)$
$\quad = \exp(-q_x*\varsigma/2)*(q_y*\exp(-q_x*\varsigma/2)) \quad$ "$\kappa = q_x$"
$\quad = \exp(-q_x*\varsigma/2)*(q_y*(\cosh(\varsigma/2) - q_x*\sinh(\varsigma/2)))$
$\quad = \exp(-q_x*\varsigma/2)*((\cosh(\varsigma/2) + q_x*\sinh(\varsigma/2))*q_y)$
$\quad = \exp(-q_x*\varsigma/2)*(\exp(q_x*\varsigma/2)*q_y)$
$\quad = q_y$

$q_{zM} = \exp(-\kappa*\varsigma/2)*q_z*\exp(-\kappa*\varsigma/2)$
$\quad = \exp(-q_x*\varsigma/2)*(q_z*\exp(-q_x*\varsigma/2)) \quad$ "$\kappa = q_x$"
$\quad = \exp(-q_x*\varsigma/2)*(q_z*(\cosh(\varsigma/2) - q_x*\sinh(\varsigma/2)))$
$\quad = \exp(-q_x*\varsigma/2)*((\cosh(\varsigma/2) + q_x*\sinh(\varsigma/2))*q_z)$
$\quad = \exp(-q_x*\varsigma/2)*(\exp(q_x*\varsigma/2)*q_z)$
$\quad = q_z$

SPECIAL ALGEBRA FOR SPECIAL RELATIVITY

<u>Invariants</u>. Compound-label-numbers combine with components to form invariants. "$_4\nabla^{sn}$" is a space-negative.

$$_4\mathbf{V} = V_{tM}*\mathbf{i}_t + c*A_{xM}*\mathbf{i}_x + c*A_{yM}*\mathbf{i}_y + c*A_{zM}*\mathbf{i}_z$$
$$_4\mathbf{V} = 1_M*V_{tM} + q_{xM}*V_{xM} + q_{yM}*V_{yM} + q_{zM}*V_{zM}$$

$$_4\mathbf{J} = \rho*(\mu*c^2)*\mathbf{i}_t + j_x*(\mu*c)*\mathbf{i}_x + j_y*(\mu*c)*\mathbf{i}_y + j_z*(\mu*c)*\mathbf{i}_z$$
$$_4\mathbf{J} = 1_M*J_{tM} + q_{xM}*J_{xM} + q_{yM}*J_{yM} + q_{zM}*J_{zM}$$

$$_3\mathbf{E} = E_{xM}*\mathbf{i}_x + E_{yM}*\mathbf{i}_y + E_{zM}*\mathbf{i}_z$$
$$c*_3\mathbf{B} = c*B_{xM}*\mathbf{i}_x + c*B_{yM}*\mathbf{i}_y + c*B_{zM}*\mathbf{i}_z$$

$$_6\mathrm{E} = p_{xM}*E_{xM} + p_{yM}*E_{yM} + p_{zM}*E_{zM}$$
$$+ k_{xM}*K_{xM} + k_{yM}*K_{yM} + k_{zM}*K_{zM}$$

$$_4\nabla = \nabla_{tM}*\mathbf{i}_t + \nabla_{xM}*\mathbf{i}_x + \nabla_{yM}*\mathbf{i}_y + \nabla_{zM}*\mathbf{i}_z$$
$$_4\nabla^{sn} = 1_M^{sn}*\nabla_{tM} + q_{xM}^{sn}*\nabla_{xM} + q_{yM}^{sn}*\nabla_{yM} + q_{zM}^{sn}*\nabla_{zM}$$
$$= 1_M^{sn}*(\partial/\partial ct_M) + q_{xM}^{sn}*\partial/\partial x_M + q_{yM}^{sn}*\partial/\partial y_M + q_{zM}^{sn}*\partial/\partial z_M$$

<u>E-M-Compound-Label-Numbers</u>. Two new sets of compound-label-numbers ("p_{xM}", "p_{yM}" and "p_{zM}" and "k_{xM}", "k_{yM}" and "k_{zM}") have been introduced for the electromagnetic field "$_6\mathrm{E}$".

$$p_{xM} = \exp(\kappa*\varsigma/2)*q_x*\exp(-\kappa*\varsigma/2)$$
$$= 1_M^{*jsn}*q_{xM} = -q_{xM}^{*jsn}*1_M = -q_{yM}^{*jsn}*q_{zM}/i = q_{zM}^{*jsn}*q_{yM}/i = k_{xM}/i$$

$$p_{yM} = \exp(\kappa*\varsigma/2)*q_y*\exp(-\kappa*\varsigma/2)$$
$$= 1_M^{*jsn}*q_{yM} = -q_{yM}^{*jsn}*1_M = -q_{zM}^{*jsn}*q_{xM}/i = q_{xM}^{*jsn}*q_{zM}/i = k_{yM}/i$$

$$p_{zM} = \exp(\kappa*\varsigma/2)*q_z*\exp(-\kappa*\varsigma/2)$$
$$= 1_M^{*jsn}*q_{zM} = -q_{zM}^{*jsn}*1_M = -q_{xM}^{*jsn}*q_{yM}/i = q_{yM}^{*jsn}*q_{xM}/i = k_{zM}/i$$

$$k_{xM} = \exp(\kappa*\varsigma/2)*j_x*\exp(-\kappa*\varsigma/2) = i*p_{xM}$$
$$= 1_M^{*jsn}*j_{xM} = j_{xM}^{*jsn}*1_M = -q_{yM}^{*jsn}*q_{zM} = q_{zM}^{*jsn}*q_{yM} = j_{yM}^{*jsn}*j_{zM} = -j_{zM}^{*jsn}*j_{yM}$$

$$k_{yM} = \exp(\kappa*\varsigma/2)*j_y*\exp(-\kappa*\varsigma/2) = i*p_{yM}$$
$$= 1_M^{*jsn}*j_{yM} = j_{yM}^{*jsn}*1_M = -q_{zM}^{*jsn}*q_{xM} = q_{xM}^{*jsn}*q_{zM} = j_{zM}^{*jsn}*j_{xM} = -j_{xM}^{*jsn}*j_{zM}$$

CHAPTER 3 – FIELDS

$$k_{zM} = \exp(\kappa^*\varsigma/2)^*j_z{}^*\exp(-\kappa^*\varsigma/2) = i^*p_{zM} = -j_{yM}{}^{*jsn}{}^*j_{xM}$$
$$= 1_M{}^{*jsn}{}^*j_{zM} = j_{zM}{}^{*jsn}{}^*1_M = -q_{xM}{}^{*jsn}{}^*q_{yM} = q_{yM}{}^{*jsn}{}^*q_{xM} = j_{xM}{}^{*jsn}{}^*j_{yM}$$

$$1_M = q_{xM}{}^*p_{xM} = q_{yM}{}^*k_{zM}{}^*p_{xM} = q_{yM}{}^*p_{yM}$$
$$= q_{yM}{}^*p_{zM}{}^*k_{xM} = 1_M{}^*1 = 1_M{}^*p_{xM}{}^*p_{xM}$$
$$= 1_M{}^*1 = -1_M{}^*k_{xM}{}^*k_{xM} = q_{xM}{}^*p_{xM}$$

$$q_{zM} = 1_M{}^*p_{zM} = 1_M{}^*k_{xM}{}^*p_{yM} = 1_M{}^*p_{zM} = 1_M{}^*p_{xM}{}^*k_{yM} = j_{xM}{}^*p_{yM}$$
$$= -q_{yM}{}^*p_{zM}{}^*p_{yM} = q_{xM}{}^*k_{yM} = q_{yM}{}^*k_{zM}{}^*k_{yM} = 1_M{}^*p_{zM}$$
$$= q_{zM}{}^*p_{zM}{}^*p_{zM} = 1_M{}^*i^*k_{zM} = q_{zM}{}^*k_{zM}{}^*k_{zM}$$

$$1_M{}^*i = q_{yM}{}^*k_{yM} = q_{yM}{}^*k_{zM}{}^*k_{xM} = j_{yM}{}^*p_{xM} = -q_{yM}{}^*p_{zM}{}^*p_{xM}$$
$$= 1_M{}^*p_{xM}{}^*k_{xM} = 1_M{}^*k_{xM}{}^*p_{xM} = q_{xM}{}^*k_{xM}$$

$$j_{zM} = j_{xM}{}^*k_{yM} = 1_M{}^*k_{xM}{}^*k_{yM} = 1_M{}^*k_{zM} = -1_M{}^*p_{xM}{}^*p_{yM} = -q_{xM}{}^*p_{yM}$$
$$= -q_{yM}{}^*k_{zM}{}^*p_{yM} = j_{xM}{}^*k_{yM} = -q_{yM}{}^*p_{zM}{}^*k_{yM} = 1_M{}^*k_{zM} = q_{zM}{}^*k_{zM}{}^*p_{zM}$$
$$= 1_M{}^*k_{yM} = q_{zM}{}^*p_{zM}{}^*k_{yM}$$

$$1_M = \exp(-\kappa^*\varsigma/2)^*1^*\exp(-\kappa^*\varsigma/2) = \exp(-\kappa^*\varsigma)$$
$$q_{xM} = \exp(-\kappa^*\varsigma/2)^*q_x{}^*\exp(-\kappa^*\varsigma/2)$$
$$q_{yM} = \exp(-\kappa^*\varsigma/2)^*q_y{}^*\exp(-\kappa^*\varsigma/2)$$
$$q_{zM} = \exp(-\kappa^*\varsigma/2)^*q_z{}^*\exp(-\kappa^*\varsigma/2)$$

$$p_{xM} = \exp(\kappa^*\varsigma/2)^*q_x{}^*\exp(-\kappa^*\varsigma/2)$$
$$p_{yM} = \exp(\kappa^*\varsigma/2)^*q_y{}^*\exp(-\kappa^*\varsigma/2)$$
$$p_{zM} = \exp(\kappa^*\varsigma/2)^*q_z{}^*\exp(-\kappa^*\varsigma/2)$$
$$k_{xM} = \exp(\kappa^*\varsigma/2)^*j_x{}^*\exp(-\kappa^*\varsigma/2) = i^*p_{xM}$$
$$k_{yM} = \exp(\kappa^*\varsigma/2)^*j_y{}^*\exp(-\kappa^*\varsigma/2) = i^*p_{yM}$$
$$k_{zM} = \exp(\kappa^*\varsigma/2)^*j_z{}^*\exp(-\kappa^*\varsigma/2) = i^*p_{zM} = j_{xM}{}^{*jsn}{}^*j_{yM}$$

Quaternion hypercomplex-conjugate operation "$*j$" reverses the sign of each "j", and, therefore, also reverses the sign of each "q", and reverses order of factors.

$$i^{*j} = i \quad ; \quad 1^{*j} = 1 \qquad\qquad j_x{}^{*j} = -j_x \;\; ; \;\; j_y{}^{*j} = -j_y \;\; ; \;\; j_z{}^{*j} = -j_z$$
$$q_x{}^{*j} = -q_x \;\; ; \;\; q_y{}^{*j} = -q_y \;\; ; \;\; q_z{}^{*j} = -q_z$$

SPECIAL ALGEBRA FOR SPECIAL RELATIVITY

$$j_{xM}{}^{*j} = \exp(\kappa^*\varsigma/2)*(-j_x)*\exp(\kappa^*\varsigma/2) \quad ; \quad k_{xM}{}^{*j} = -k_{xM}$$
$$j_{yM}{}^{*j} = \exp(\kappa^*\varsigma/2)*(-j_y)*\exp(\kappa^*\varsigma/2) \quad ; \quad k_{yM}{}^{*j} = -k_{yM}$$
$$j_{zM}{}^{*j} = \exp(\kappa^*\varsigma/2)*(-j_z)*\exp(\kappa^*\varsigma/2) \quad ; \quad k_{zM}{}^{*j} = -k_{zM}$$

$$1_M{}^{*j} = \exp(\kappa^*\varsigma/2)*(1)*\exp(\kappa^*\varsigma/2)$$
$$q_{xM}{}^{*j} = \exp(\kappa^*\varsigma/2)*(-q_x)*\exp(\kappa^*\varsigma/2) \quad ; \quad p_{xM}{}^{*j} = -p_{xM}$$
$$q_{yM}{}^{*j} = \exp(\kappa^*\varsigma/2)*(-q_y)*\exp(\kappa^*\varsigma/2) \quad ; \quad p_{yM}{}^{*j} = -p_{yM}$$
$$q_{zM}{}^{*j} = \exp(\kappa^*\varsigma/2)*(-q_z)*\exp(\kappa^*\varsigma/2) \quad ; \quad p_{zM}{}^{*j} = -p_{zM}$$

Conjugate Form of Invariants.

$$_4V^{*j} = V_{tM}*1_M{}^{*j} + V_{xM}*q_{xM}{}^{*j} + V_{yM}*q_{yM}{}^{*j} + V_{zM}*q_{zM}{}^{*j}$$

$$_4J^{*j} = J_{tM}*1_M{}^{*j} + J_{xM}*q_{xM}{}^{*j} + J_{yM}*q_{yM}{}^{*j} + J_{zM}*q_{zM}{}^{*j}$$

$$_6E^{*j} = E_{xM}*p_{xM}{}^{*j} + E_{yM}*p_{yM}{}^{*j} + E_{zM}*p_{zM}{}^{*j}$$
$$\quad + K_{xM}*k_{xM}{}^{*j} + K_{yM}*k_{yM}{}^{*j} + K_{zM}*k_{zM}{}^{*j}$$

$$_4\nabla^{*jsn} = \nabla_{tM}*1_M{}^{*jsn} + \nabla_{xM}*q_{xM}{}^{*jsn} + \nabla_{yM}*q_{yM}{}^{*jsn} + \nabla_{zM}*q_{zM}{}^{*jsn}$$

Governing Equations

Voltage Equation $\quad\quad _4\nabla^{*jsn}\mathbf{x}(_4V) = -_6E$

Lorenz Condition $\quad\quad _4\nabla^{*jsn}\bullet(_4V) = 0$

Maxwell's Equations $\quad _4\nabla^{sn}*_4\nabla^{*jsn}*(_4V) = _4\nabla^{sn}*(-_6E)$
$$_4\nabla^{sn}*(-_6E) = _4J$$

Electric Charge Conservation Equation

$$_4\nabla^{*jsn}\bullet(_4\nabla^{sn}*_4\nabla^{*jsn}*(_4V)) = _4\nabla^{*jsn}\bullet(_4\nabla^{sn}*(-_6E)) = _4\nabla^{*jsn}\bullet(_4J) = 0$$

<u>Decomposition</u>. Invariants in the above equations separate into pieces per nomenclature given below. Pieces alone are not invariants. Therefore, pieces are specific to an inertial reference frame even though the "M" or "S" subscript may be dropped to simplify what is written.

CHAPTER 3 – FIELDS

$$_4V = {_1V} + {_3V} \quad ; \quad {_1V} = 1_M * V_{tM} \quad ; \quad {_3V} = q_{xM} * V_{xM} + q_{yM} * V_{yM} + q_{zM} * V_{zM}$$

$$_4J = {_1J} + {_3J} \quad ; \quad {_1J} = 1_M * J_{tM} \quad ; \quad {_3J} = q_{xM} * J_{xM} + q_{yM} * J_{yM} + q_{zM} * J_{zM}$$

$$_4\nabla^{sn} = {_1\nabla^{sn}} + {_3\nabla^{sn}} \quad ; \quad {_1\nabla^{sn}} = 1_M{}^{sn} * \nabla_{tM}$$

$$_3\nabla^{sn} = q_{xM}{}^{sn} * \nabla_{xM} + q_{yM}{}^{sn} * \nabla_{yM} + q_{zM}{}^{sn} * \nabla_{zM}$$

$$_6E = {_3E} + {_3K}$$

$$_3E = p_{xM} * E_{xM} + p_{yM} * E_{yM} + p_{zM} * E_{zM} \quad ; \quad {_3K} = k_{xM} * K_{xM} + k_{yM} * K_{yM} + k_{zM} * K_{zM}$$

Multiplication operation "*" decomposes into dot-product "•" and cross-product "**x**".

$$q_x * q_y = q_x \bullet q_y + q_x \mathbf{x} q_y \qquad ; \qquad q_x * q_x = q_x \bullet q_x + q_x \mathbf{x} q_x$$
$$= q_x \mathbf{x} q_y \qquad \qquad \qquad \qquad = q_x \bullet q_x$$

$$q_x \bullet q_y = 0 \qquad ; \qquad q_x \mathbf{x} q_x = 0$$

Lorenz Condition "$_4\nabla^{*jsn} \bullet ({_4V}) = 0$".

$$_4\nabla^{*jsn} * ({_4V}) = {_4\nabla^{*jsn}} \bullet ({_4V}) + {_4\nabla^{*jsn}} \mathbf{x} ({_4V})$$

$$_4\nabla^{*jsn} \bullet ({_4V}) = {_1\nabla^{*jsn}} * ({_1V}) + {_3\nabla^{*jsn}} \bullet ({_3V}) = 0$$

$$_4\nabla^{*jsn} \bullet ({_4V}) = \nabla_{tM} * V_{tM} * 1_M{}^{*jsn} * 1_M + \nabla_{xM} * V_{xM} * q_{xM}{}^{*jsn} * q_{xM}$$
$$+ \nabla_{yM} * V_{yM} * q_{yM}{}^{*jsn} * q_{yM} + \nabla_{zM} * V_{zM} * q_{zM}{}^{*jsn} * q_{zM}$$

$$= \nabla_{tM} * V_{tM} + \nabla_{xM} * V_{xM} + \nabla_{yM} * V_{yM} + \nabla_{zM} * V_{zM} = 0$$

Electromagnetic Field Voltage Equation. Cross-product "$_4\nabla^{*jsn}\mathbf{x}({_4V})$" results in negative of the electromagnetic field.

$$_4\nabla^{*jsn} \mathbf{x} ({_4V}) = {_1\nabla^{*jsn}} * ({_3V}) + {_3\nabla^{*jsn}} * ({_1V}) + {_3\nabla^{*jsn}} \mathbf{x} ({_3V}) = -{_6E}$$

$$_1\nabla^{*jsn} * ({_3V}) + {_3\nabla^{*jsn}} * ({_1V}) = -{_3E} \qquad ; \qquad {_3\nabla^{*jsn}} \mathbf{x} ({_3V}) = -{_3K}$$

SPECIAL ALGEBRA FOR SPECIAL RELATIVITY

Electric Field Voltage Equation.

$$_1\nabla^{*jsn}*(_3V) + {_3}\nabla^{*jsn}*(_1V) = -_3E$$

$$\nabla_{tM}*V_{xM}*1_M{}^{*jsn}*q_{xM} + \nabla_{xM}*V_{tM}*q_{xM}{}^{*jsn}*1_M$$
$$= \nabla_{tM}*V_{xM}*p_{xM} + \nabla_{xM}*V_{tM}*p_{xM}$$
$$= -p_{xM}*E_{xM}$$

$$-\nabla_{tM}*V_{xM} - \nabla_{xM}*V_{tM} = E_{xM}$$
$$-\nabla_{tM}*V_{yM} - \nabla_{yM}*V_{tM} = E_{yM}$$
$$-\nabla_{tM}*V_{zM} - \nabla_{zM}*V_{tM} = E_{zM}$$

Magnetic Field Voltage Equation.

$$_3\nabla^{*jsn}\mathbf{x}(_3V) = -_3K$$

$$\nabla_{yM}*V_{zM}*q_{yM}{}^{*jsn}*q_{zM} + \nabla_{zM}*V_{yM}*q_{zM}{}^{*jsn}*q_{yM}$$
$$= -\nabla_{yM}*V_{zM}*k_{xM} + \nabla_{zM}*V_{yM}*k_{xM}$$
$$= -k_{xM}*K_{xM}$$

$$q_{yM}{}^{*jsn}*q_{zM} = -q_{yM}{}^{*j}*q_{zM} = j_{yM}{}^{*j}*j_{zM} = -k_{xM}$$

$$-\nabla_{yM}*V_{zM} + \nabla_{zM}*V_{yM} = -K_{xM}$$
$$-\nabla_{zM}*V_{xM} + \nabla_{xM}*V_{zM} = -K_{yM}$$
$$-\nabla_{xM}*V_{yM} + \nabla_{yM}*V_{xM} = -K_{zM}$$

Triple-Vector-Product and Remnant-Product.
Dot-product and/or cross-product do not apply to six-component invariants. Instead, use the triple-vector-product "■" and the remnant-product "♦".

$$_4\nabla^{sn}*(_4\nabla^{*jsn}*(_4V)) = {_4}\nabla^{sn}*(_4\nabla^{*jsn}\bullet(_4V)) + {_4}\nabla^{sn}*(_4\nabla^{*jsn}\mathbf{x}(_4V))$$

$$_4\nabla^{sn}*(_4\nabla^{*jsn}\mathbf{x}(_4V)) = {_4}\nabla^{sn}\blacksquare(_4\nabla^{*jsn}\mathbf{x}(_4V)) + {_4}\nabla^{sn}\blacklozenge(_4\nabla^{*jsn}\mathbf{x}(_4V))$$
$$= {_4}\nabla^{sn}\blacksquare(-_6E) + {_4}\nabla^{sn}\blacklozenge(-_6E)$$

CHAPTER 3 – FIELDS

<u>Triple-Vector-Product Gradient Identities</u>.

$$_4\nabla^{sn}\blacksquare(_4\nabla^{*jsn}\mathbf{x}(_4V)) = {}_1\nabla^{sn}*(_3\nabla^{*jsn}\mathbf{x}(_3V)) + {}_3\nabla^{sn}\mathbf{x}(_1\nabla^{*jsn}*(_3V))$$
$$+ {}_3\nabla^{sn}\mathbf{x}(_3\nabla^{*jsn}*(_1V)) + {}_3\nabla^{sn}\bullet(_3\nabla^{*jsn}\mathbf{x}(_3V))$$

$$_4\nabla^{sn}\blacksquare(-_6E) = {}_1\nabla^{sn}*(-_3K) + {}_3\nabla^{sn}\mathbf{x}(-_3E) + {}_3\nabla^{sn}\bullet(-_3K)$$

"$_4\nabla^{sn}\blacksquare(_4\nabla^{*jsn}\mathbf{x}(_4V))$" includes identities.

$$_1\nabla^{sn}*(_3\nabla^{*jsn}\mathbf{x}(_3V)) + {}_3\nabla^{sn}\mathbf{x}(_1\nabla^{*jsn}*(_3V)) \equiv 0$$
$$_3\nabla^{sn}\mathbf{x}(_3\nabla^{*jsn}*(_1V)) \equiv 0$$
$$_3\nabla^{sn}\bullet(_3\nabla^{*jsn}\mathbf{x}(_3V)) \equiv 0 \quad ; \quad {}_4\nabla^{sn}\blacksquare(_4\nabla^{*jsn}\mathbf{x}(_4V)) \equiv 0$$

Electromagnetic components substitute into the sum of the first two identities to result in the first of four of Maxwell's Equations.

$$_1\nabla^{sn}*(_3\nabla^{*jsn}\mathbf{x}(_3V)) + (_3\nabla^{sn}\mathbf{x}(_1\nabla^{*jsn}*(_3V)) + {}_3\nabla^{sn}\mathbf{x}(_3\nabla^{*jsn}*(_1V))) \equiv 0$$

$$_1\nabla^{sn}*(-_3K) + {}_3\nabla^{sn}\mathbf{x}(-_3E) = 0$$

$$-\nabla_{tM}*K_{xM}*1_M^{sn}*k_{xM} - \nabla_{yM}*E_{zM}*q_{yM}^{sn}*p_{zM} - \nabla_{zM}*E_{yM}*q_{zM}^{sn}*p_{yM} = 0$$

$$-\nabla_{tM}*K_{xM}*j_{xM} - \nabla_{yM}*E_{zM}*j_{xM} + \nabla_{zM}*E_{yM}*j_{xM} = 0$$

$$-\nabla_{tM}*K_{xM} - \nabla_{yM}*E_{zM} + \nabla_{zM}*E_{yM} = 0$$

Electromagnetic components substitute into "$_3\nabla^{sn}\bullet(_3\nabla^{*jsn}\mathbf{x}(_3V)) \equiv 0$" to result in the second of four of Maxwell's Equations.

$$_3\nabla^{sn}\bullet(_3\nabla^{*jsn}\mathbf{x}(_3V)) \equiv 0 \quad ; \quad {}_3\nabla^{sn}\bullet(-_3K) = 0$$

$$-\nabla_{xM}*K_{xM}*q_{xM}^{sn}*k_{xM} - \nabla_{yM}*K_{yM}*q_{yM}^{sn}*k_{yM} - \nabla_{zM}*K_{zM}*q_{zM}^{sn}*k_{zM} = 0$$

$$\nabla_{xM}*K_{xM} + \nabla_{yM}*K_{yM} + \nabla_{zM}*K_{zM} = 0$$

First with second of the four Maxwell's Equations are given below.

$$_4\nabla^{sn}\blacksquare(_4\nabla^{*jsn}\mathbf{x}(_4V)) = {}_4\nabla^{sn}\blacksquare(-_6E) = 0$$

Special Algebra for Special Relativity

Remnant-Product Results.

$$_4\nabla^{sn}\blacklozenge(_4\nabla^{*jsn}\mathbf{x}(_4V)) = {}_4\nabla^{sn}*(_4\nabla^{*jsn}\mathbf{x}(_4V)) - {}_4\nabla^{sn}\blacksquare(_4\nabla^{*jsn}\mathbf{x}(_4V))$$

$$= {}_1\nabla^{sn}*(_1\nabla^{*jsn}*(_3V)) + {}_1\nabla^{sn}*(_3\nabla^{*jsn}*(_1V)) + {}_3\nabla^{sn}\mathbf{x}(_3\nabla^{*jsn}\mathbf{x}(_3V))$$
$$+ {}_3\nabla^{sn}\bullet(_1\nabla^{*jsn}*(_3V)) + {}_3\nabla^{sn}\bullet(_3\nabla^{*jsn}*(_1V))$$

Third of four of Maxwell's Equations:

$$_1\nabla^{sn}*(_1\nabla^{*jsn}*(_3V)) + {}_1\nabla^{sn}*(_3\nabla^{*jsn}*(_1V)) + {}_3\nabla^{sn}\mathbf{x}(_3\nabla^{*jsn}\mathbf{x}(_3V))$$
$$= {}_1\nabla^{sn}*(-_3E) + {}_3\nabla^{sn}\mathbf{x}(-_3K)$$
$$= {}_3J$$

$$-\nabla_{tM}*E_{xM}*1_M{}^{sn}*p_{xM} - \nabla_{yM}*K_{zM}*q_{yM}{}^{sn}*k_{zM} - \nabla_{zM}*K_{yM}*q_{zM}{}^{sn}*k_{yM} = q_{xM}*J_{xM}$$

$$-\nabla_{tM}*E_{xM}*q_{xM} + \nabla_{yM}*K_{zM}*q_{xM} - \nabla_{zM}*K_{yM}*q_{xM} = q_{xM}*J_{xM}$$

$$-\nabla_{tM}*E_{xM} + \nabla_{yM}*K_{zM} - \nabla_{zM}*K_{yM} = J_{xM}$$

Fourth of four of Maxwell's Equations:

$$_3\nabla^{sn}\bullet(_1\nabla^{*jsn}*(_3V)) + {}_3\nabla^{sn}\bullet(_3\nabla^{*jsn}*(_1V)) = {}_3\nabla^{sn}\bullet(-_3E) = {}_1J$$

$$-\nabla_{xM}*E_{xM}*q_{xM}{}^{sn}*p_{xM} - \nabla_{yM}*E_{yM}*q_{yM}{}^{sn}*p_{yM} - \nabla_{zM}*E_{zM}*q_{zM}{}^{sn}*p_{zM} = 1_M*J_{tM}$$

$$\nabla_{xM}*E_{xM} + \nabla_{yM}*E_{yM} + \nabla_{zM}*E_{zM} = J_{tM}$$

Third with fourth of Maxwell's Equations:

$$_4\nabla^{sn}\blacklozenge(_4\nabla^{*jsn}\mathbf{x}(_4V)) = {}_4\nabla^{sn}\blacklozenge(-_6E) = {}_4J$$

CHAPTER 3 – FIELDS

Maxwell's Equations. $\quad _4\nabla^{sn}*(_4\nabla^{*jsn}\mathbf{x}(_4V)) = {_4}\nabla^{sn}*(-_6E) = {_4}J + 0$

$-\nabla_{tM}*K_{xM} - \nabla_{yM}*E_{zM} + \nabla_{zM}*E_{yM} = 0 \;;\; -\nabla_{tM}*E_{xM} + \nabla_{yM}*K_{zM} - \nabla_{zM}*K_{yM} = J_{xM}$
$-\nabla_{tM}*K_{yM} - \nabla_{zM}*E_{xM} + \nabla_{xM}*E_{zM} = 0 \;;\; -\nabla_{tM}*E_{yM} + \nabla_{zM}*K_{xM} - \nabla_{xM}*K_{zM} = J_{yM}$
$-\nabla_{tM}*K_{zM} - \nabla_{xM}*E_{yM} + \nabla_{yM}*E_{xM} = 0 \;;\; -\nabla_{tM}*E_{zM} + \nabla_{xM}*K_{yM} - \nabla_{yM}*K_{xM} = J_{zM}$

$\nabla_{xM}*K_{xM} + \nabla_{yM}*K_{yM} + \nabla_{zM}*K_{zM} = 0 \;;\; \nabla_{xM}*E_{xM} + \nabla_{yM}*E_{yM} + \nabla_{zM}*E_{zM} = J_{tM}$

Electric Charge Conservation.

$$_4\nabla^{*jsn}*{_4}\nabla^{sn}*{_4}\nabla^{*jsn}*({_4}V) = {_4}\nabla^{*jsn}*({_4}\nabla^{sn}*({_4}\nabla^{*jsn}*({_4}V)))$$
$$= {_4}\nabla^{*jsn}*({_4}\nabla^{sn}*({_4}\nabla^{*jsn}\bullet({_4}V))) + {_4}\nabla^{*jsn}*({_4}\nabla^{sn}*({_4}\nabla^{*jsn}\mathbf{x}({_4}V)))$$

Lorenz Condition "$_4\nabla^{*jsn}\bullet(_4V) = 0$" and other identities apply.

$$_4\nabla^{*jsn}*{_4}\nabla^{sn}*{_4}\nabla^{*jsn}*({_4}V)$$
$$= {_4}\nabla^{*jsn}\bullet({_4}\nabla^{sn}\blacklozenge({_4}\nabla^{*jsn}\mathbf{x}({_4}V))) + {_4}\nabla^{*jsn}\mathbf{x}({_4}\nabla^{sn}\blacklozenge({_4}\nabla^{*jsn}\mathbf{x}({_4}V)))$$

$$_4\nabla^{*jsn}\bullet({_4}\nabla^{sn}\blacklozenge({_4}\nabla^{*jsn}\mathbf{x}({_4}V))) = {_3}\nabla^{*jsn}\bullet({_1}\nabla^{sn}*({_1}\nabla^{*jsn}*({_3}V) + {_3}\nabla^{*jsn}*({_1}V)))$$
$$+ {_3}\nabla^{*jsn}\bullet({_3}\nabla^{sn}\mathbf{x}({_3}\nabla^{*jsn}\mathbf{x}({_3}V)))$$
$$+ {_1}\nabla^{*jsn}*({_3}\nabla^{sn}\bullet({_1}\nabla^{*jsn}*({_3}V) + {_3}\nabla^{*jsn}*({_1}V)))$$

$_3\nabla^{*jsn}\bullet({_1}\nabla^{sn}*({_1}\nabla^{*jsn}*({_3}V))) + {_1}\nabla^{*jsn}*({_3}\nabla^{sn}\bullet({_1}\nabla^{*jsn}*({_3}V))) \equiv 0$
$_3\nabla^{*jsn}\bullet({_1}\nabla^{sn}*({_3}\nabla^{*jsn}*({_1}V))) + {_1}\nabla^{*jsn}*({_3}\nabla^{sn}\bullet({_3}\nabla^{*jsn}*({_1}V))) \equiv 0$
$_3\nabla^{*jsn}\bullet({_3}\nabla^{sn}\mathbf{x}({_3}\nabla^{*jsn}\mathbf{x}({_3}V))) \equiv 0$

$_4\nabla^{*jsn}\bullet({_4}\nabla^{sn}\blacklozenge({_4}\nabla^{*jsn}\mathbf{x}({_4}V))) \equiv 0$

$_4\nabla^{*jsn}\bullet({_4}J) = 0$

$$\nabla_{tM}*J_{tM}*1_M*{^{*jsn}}*1_M + \nabla_{xM}*J_{xM}*q_{xM}{^{*jsn}}*q_{xM}$$
$$+ \nabla_{yM}*J_{yM}*q_{yM}{^{*jsn}}*q_{yM} + \nabla_{zM}*J_{zM}*q_{zM}{^{*jsn}}*q_{zM} = 0$$

$\nabla_{tM}*J_{tM} + \nabla_{xM}*J_{xM} + \nabla_{yM}*J_{yM} + \nabla_{zM}*J_{zM} = 0$

SPECIAL ALGEBRA FOR SPECIAL RELATIVITY

Summary of Component Equations.

Lorenz Condition: $\quad _4\nabla^{*jsn}\bullet(_4V) = 0$

$$\nabla_{tM}*V_{tM} + \nabla_{xM}*V_{xM} + \nabla_{yM}*V_{yM} + \nabla_{zM}*V_{zM} = 0$$

Electric Field Voltage Equation:

$_1\nabla^{*jsn}*(_3V) + _3\nabla^{*jsn}*(_1V) = -_3E$

$-\nabla_{tM}*V_{xM} - \nabla_{xM}*V_{tM} = E_{xM}$
$-\nabla_{tM}*V_{yM} - \nabla_{yM}*V_{tM} = E_{yM}$
$-\nabla_{tM}*V_{zM} - \nabla_{zM}*V_{tM} = E_{zM}$

Magnetic Field Voltage Equation:

$_3\nabla^{*jsn}\mathbf{x}(_3V) = -_3K \quad ;$

$-\nabla_{yM}*V_{zM} + \nabla_{zM}*V_{yM} = -K_{xM}$
$-\nabla_{zM}*V_{xM} + \nabla_{xM}*V_{zM} = -K_{yM}$
$-\nabla_{xM}*V_{yM} + \nabla_{yM}*V_{xM} = -K_{zM}$

Maxwell's Induced Electric Field Equation:

$_1\nabla^{sn}*(-_3K) + _3\nabla^{sn}\mathbf{x}(-_3E) = 0$

$-\nabla_{tM}*K_{xM} - \nabla_{yM}*E_{zM} + \nabla_{zM}*E_{yM} = 0$
$-\nabla_{tM}*K_{yM} - \nabla_{zM}*E_{xM} + \nabla_{xM}*E_{zM} = 0$
$-\nabla_{tM}*K_{zM} - \nabla_{xM}*E_{yM} + \nabla_{yM}*E_{xM} = 0$

Maxwell's Zero Magnetic Charge Equation: $\quad _3\nabla^{sn}\bullet(-_3K) = 0$

$$\nabla_{xM}*K_{xM} + \nabla_{yM}*K_{yM} + \nabla_{zM}*K_{zM} = 0$$

Maxwell's Induced Magnetic Field Equation:

$_1\nabla^{sn}*(-_3E) + _3\nabla^{sn}\mathbf{x}(-_3K) = _3J$

$-\nabla_{tM}*E_{xM} + \nabla_{yM}*K_{zM} - \nabla_{zM}*K_{yM} = J_{xM}$
$-\nabla_{tM}*E_{yM} + \nabla_{zM}*K_{xM} - \nabla_{xM}*K_{zM} = J_{yM}$
$-\nabla_{tM}*E_{zM} + \nabla_{xM}*K_{yM} - \nabla_{yM}*K_{xM} = J_{zM}$

Maxwell's Electric Charge Equation: $\quad _3\nabla^{sn}\bullet(-_3E) = _1J$

$$\nabla_{xM}*E_{xM} + \nabla_{yM}*E_{yM} + \nabla_{zM}*E_{zM} = J_{tM}$$

Electric Charge Conservation Equation: $\quad _4\nabla^{*jsn}\bullet(_4J) = 0$

$$\nabla_{tM}*J_{tM} + \nabla_{xM}*J_{xM} + \nabla_{yM}*J_{yM} + \nabla_{zM}*J_{zM} = 0$$

CHAPTER 3 – FIELDS

3.3 Gauges and Super-Potentials

Super-Potentials. Voltage "$_4V$" is called "potential". By analogy, "$_4J$" is the first "sub-potential".

$_4\nabla^{*jsn}\bullet(_4V) = 0$ Lorenz Condition
$_4\nabla^{*jsn}\bullet(_4J) = 0$ Electric Charge Conservation Equation

"$_4J$" relates to "$_4V$" by the square of the gradient operator. By analogy, voltage is related to the first super-potential "$_4U$", "$_4\nabla^{*jsn}\bullet(_4U) = 0$".

$(_4J) = (_4\nabla^{sn}\bullet{_4\nabla^{*jsn}})*(_4V) = (\nabla_{tM}^2 - \nabla_{xM}^2 - \nabla_{yM}^2 - \nabla_{zM}^2)*(_4V)$

$(_4V) = (_4\nabla^{sn}\bullet{_4\nabla^{*jsn}})*(_4U) = (\nabla_{tM}^2 - \nabla_{xM}^2 - \nabla_{yM}^2 - \nabla_{zM}^2)*(_4U)$

Super-potentials and sub-potentials extend indefinitely.
"$(_4\nabla^{sn}\bullet{_4\nabla^{*jsn}}) = (\nabla_{tM}^2 - \nabla_{xM}^2 - \nabla_{yM}^2 - \nabla_{zM}^2)$" is called the "harmonic operator" because, if "$_4J = 0$", then "$_4V$" is a summation of sine waves. The physical example is light waves.

Fields. Between each pair of adjacent potentials is a six-component field. "$_6E$" is between "$_4J$" and "$_4V$".

$_4\nabla^{*jsn}*(_4V) = -_6E$; $_4\nabla^{sn}*(-_6E) = {_4J}$

The first sub-field "$_6G$" has "$_4J$" as its potential.

$_4\nabla^{*jsn}*(_4J) = {_4\nabla^{*jsn}}\mathbf{x}(_4J) = -_6G$

$_4\nabla^{*jsn}\mathbf{x}(_4\nabla^{sn}\blacklozenge(_4\nabla^{*jsn}\mathbf{x}(_4V))) = {_4\nabla^{*jsn}}\mathbf{x}(_4\nabla^{sn}\blacklozenge(-_6E)) = {_4\nabla^{*jsn}}\mathbf{x}(_4J) = -_6G$

Time-Component Field Gauge. A potential is not unique to the field. An electromagnetic field has a potential "$_4V'$" ("'" is "prime") from which alternative "$_4V$" are found using a mathematically complex field "$_2P$".

$_4V = {_4V'} + {_4\nabla^{sn}}*{_2P}$

SPECIAL ALGEBRA FOR SPECIAL RELATIVITY

$$_3\nabla^{*jsn}\textbf{x}(_3\nabla^{sn}*_2P) \equiv 0$$
$$_1\nabla^{*jsn}*(_3\nabla^{sn}*_2P) + _3\nabla^{*jsn}*(_1\nabla^{sn}*_2P) \equiv 0$$

$$_4\nabla^{*jsn}\textbf{x}(_4\nabla^{sn}*_2P) \equiv 0$$

$$\begin{aligned}-_6E &= {_4\nabla^{*jsn}}\textbf{x}(_4V) = {_4\nabla^{*jsn}}\textbf{x}(_4V' + {_4\nabla^{sn}*_2P})\\ &= {_4\nabla^{*jsn}}\textbf{x}(_4V') + {_4\nabla^{*jsn}}\textbf{x}(_4\nabla^{sn}*_2P)\\ &= {_4\nabla^{*jsn}}\textbf{x}(_4V')\end{aligned}$$

"$_6E$" is not affected by "$_2P$". "$_2P$" affects the Lorenz Condition because "$_2P$" adds a harmonic oscillator term "$(_4\nabla^{*jsn}\bullet_4\nabla^{sn})*_2P$". To keep "$_4V$" real, "$_2P$" must be real.

$$\begin{aligned}&= {_4\nabla^{*jsn}}\bullet(_4V' + {_4\nabla^{sn}*_2P})\\ 0 = {_4\nabla^{*jsn}}\bullet(_4V) &= {_4\nabla^{*jsn}}\bullet(_4V') + {_4\nabla^{*jsn}}\bullet(_4\nabla^{sn}*_2P)\\ &= {_4\nabla^{*jsn}}\bullet(_4V') + (_4\nabla^{*jsn}\bullet_4\nabla^{sn})*_2P\end{aligned}$$

Complex number field "$_2P$" is called the field gauge.

The simplest version of "$_4\nabla^{sn}*_2P$" is a constant, and a change to that constant is a change to the ground of the voltage time component.

Space-Component Field Gauge

does not affect the Lorenz Condition, but, rather, adds a term to the electromagnetic field "$_6E$".

$$_3\nabla^{*jsn}\textbf{x}(_3\nabla^{sn}\textbf{x}_6Q) \equiv (_3\nabla^{*jsn}\bullet_3\nabla^{sn})*_6Q - {_3\nabla^{*jsn}}*(_3\nabla^{sn}\bullet_6Q)$$

$$\begin{aligned}_4\nabla^{*jsn}\bullet(_4\nabla^{sn}*(-_6Q)) &= {_4\nabla^{*jsn}}\bullet(_4\nabla^{sn}\blacksquare(-_6Q)) + {_4\nabla^{*jsn}}\bullet(_4\nabla^{sn}\blacklozenge(-_6Q))\\ &= {_1\nabla^{*jsn}}*(_3\nabla^{sn}\bullet(-_3Q_i)) + {_3\nabla^{*jsn}}\bullet(_1\nabla^{sn}\blacklozenge(-_3Q_i) + {_3\nabla^{sn}}\textbf{x}(-_3Q_r))\\ &\quad + {_1\nabla^{*jsn}}*(_3\nabla^{sn}\bullet(-_3Q_r)) + {_3\nabla^{*jsn}}\bullet(_1\nabla^{sn}\blacklozenge(-_3Q_r) + {_3\nabla^{sn}}\textbf{x}(-_3Q_i))\\ &\equiv 0\end{aligned}$$

$$_4V = {_4V'} + {_4\nabla^{sn}}*(-_6Q)$$

$$\begin{aligned}_4\nabla^{*jsn}\bullet(_4V) &= {_4\nabla^{*jsn}}\bullet(_4V') + {_4\nabla^{*jsn}}\bullet(_4\nabla^{sn}*(-_6Q))\\ &= {_4\nabla^{*jsn}}\bullet(_4V') = 0\end{aligned}$$

$$\begin{aligned}-_6E &= {_4\nabla^{*jsn}}\textbf{x}_4V\\ &= {_4\nabla^{*jsn}}\textbf{x}_4V' + {_4\nabla^{*jsn}}\textbf{x}(_4\nabla^{sn}*(-_6Q))\\ &= {_4\nabla^{*jsn}}\textbf{x}_4V' + {_4\nabla^{*jsn}}\textbf{x}(_4\nabla\blacksquare(-_6Q)) + {_4\nabla^{*jsn}}\textbf{x}(_4\nabla\blacklozenge(-_6Q))\end{aligned}$$

CHAPTER 3 – FIELDS

$$= {}_4\nabla^{*jsn}\mathbf{x}_4V' + {}_4\nabla^{*jsn}\mathbf{x}({}_3\nabla^{sn}\bullet(-_3Q_i) + {}_1\nabla^{sn}*(-_3Q_i) + {}_3\nabla^{sn}\mathbf{x}(-_3Q_r))$$
$$+ {}_4\nabla^{*jsn}\mathbf{x}({}_3\nabla^{sn}\bullet(-_3Q_r) + {}_1\nabla^{sn}*(-_3Q_r) + {}_3\nabla^{sn}\mathbf{x}(-_3Q_i))$$

$$= {}_4\nabla^{*jsn}\mathbf{x}_4V' + {}_3\nabla^{*jsn}*({}_3\nabla\bullet(-_3Q_i)) + {}_1\nabla^{*jsn}*({}_1\nabla^{sn}*(-_3Q_i))$$
$$+ {}_3\nabla^{sn}\mathbf{x}(-_3Q_r)) + {}_3\nabla^{*jsn}\mathbf{x}({}_1\nabla^{sn}*(-_3Q_i) + {}_3\nabla^{sn}\mathbf{x}(-_3Q_r))$$
$$+ {}_3\nabla^{*jsn}*({}_3\nabla^{sn}\bullet(-_3Q_r)) + {}_1\nabla^{*jsn}*({}_1\nabla^{sn}*(-_3Q_r) + {}_3\nabla^{sn}\mathbf{x}(-_3Q_i))$$
$$+ {}_3\nabla^{*jsn}\mathbf{x}({}_1\nabla^{sn}*(-_3Q_r) + {}_3\nabla^{sn}\mathbf{x}(-_3Q_i))$$

$$= {}_4\nabla^{*jsn}\mathbf{x}_4V' + {}_3\nabla^{*jsn}*({}_3\nabla^{sn}\bullet(-_3Q_i))$$
$$+ {}_1\nabla^{*jsn}*({}_1\nabla^{sn}*(-_3Q_i)) + {}_3\nabla^{*jsn}\mathbf{x}({}_3\nabla^{sn}\mathbf{x}(-_3Q_r))$$
$$+ {}_3\nabla^{*jsn}*({}_3\nabla^{sn}\bullet(-_3Q_r)) + {}_1\nabla^{*jsn}*({}_1\nabla^{sn}*(-_3Q_r)) + {}_3\nabla^{*jsn}\mathbf{x}({}_3\nabla^{sn}\mathbf{x}(-_3Q_i))$$

$$= {}_4\nabla^{*jsn}\mathbf{x}_4V' + {}_3\nabla^{*jsn}*({}_3\nabla^{sn}\bullet(-_6Q)) + {}_1\nabla^{*jsn}*({}_1\nabla^{sn}*(-_6Q)) + {}_3\nabla^{*jsn}\mathbf{x}({}_3\nabla^{sn}\mathbf{x}(-_6Q))$$
$$= {}_4\nabla^{*jsn}\mathbf{x}_4V' + ({}_3\nabla^{*jsn}\bullet{}_3\nabla^{sn})*(-_6Q) + ({}_1\nabla^{*jsn}*{}_1\nabla^{sn})*(-_6Q)$$
$$= {}_4\nabla^{*jsn}\mathbf{x}_4V' + ({}_4\nabla^{*jsn}\bullet{}_4\nabla^{sn})*(-_6Q)$$

"$(-_6E) = {}_4\nabla^{*jsn}\mathbf{x}_4V' + ({}_4\nabla^{*jsn}\bullet{}_4\nabla^{sn})*(-_6Q)$" and "${}_4\nabla^{*jsn}\bullet(_4V) = {}_4\nabla^{*jsn}\bullet(_4V') + {}_4\nabla^{*jsn}\bullet({}_4\nabla^{sn}*(-_6Q)) = {}_4\nabla^{*jsn}\bullet(_4V') = 0$" define different electromagnetic fields for one Lorentz Condition.

"$_2P$" is a classic feature in electromagnetic theory. "$_6Q$" was not mentioned in reference material available to the author.

Electromagnetic field harmonic term "$({}_4\nabla^{*jsn}\bullet{}_4\nabla^{sn})*(-_6Q)$" is similar in structure to the super-field equation given below.

$$(-_6E) = ({}_4\nabla^{*jsn}\bullet{}_4\nabla^{sn})*(-_6H) = (\nabla_{tM}^2 - \nabla_{xM}^2 - \nabla_{yM}^2 - \nabla_{zM}^2)*(-_6H)$$

Concluding Statement. Super-potentials and gauge fields place Maxwell's Equations into a larger (and mathematically beautiful) structure.

Identities are easily identified in an all-number algebra but not in the geometric algebra used when Maxwell's Equations were first discovered.

3.4 Lorentz Transformation

The Lorentz Transformation transforms components and compound-label-numbers from "M" (moving because seated on the bus) to "S" (stationary because standing on the roadside).

SPECIAL ALGEBRA FOR SPECIAL RELATIVITY

The general form has two "1" factors, left and right.

The technique is checked: An invariant formed as a product of two other invariants must have the same Lorentz Transformation result if the invariant is transformed directly or if the invariant is formed by multiplying two transformed invariants.

Four-Component Vector Lorentz Transformation. General form:

$$_4V = 1_M * V_{tM} + q_{xM} * V_{xM} + q_{yM} * V_{yM} + q_{zM} * V_{zM}$$
$$= \exp(-\kappa * \varsigma/2) * (1 * V_{tM} + q_x * V_{xM} + q_y * V_{yM} + q_z * V_{zM}) * \exp(-\kappa * \varsigma/2)$$
$$= \exp(-\kappa * \varsigma/2) * 1 * (1 * V_{tM} + q_x * V_{xM} + q_y * V_{yM} + q_z * V_{zM}) * 1 * \exp(-\kappa * \varsigma/2)$$

$$= \exp(-\kappa * \varsigma/2) * \exp(-q_x * \alpha_{S/M}/2)$$
$$* \exp(q_x * \alpha_{S/M}/2) * (1_M * V_{tM} + q_{xM} * V_{xM} + q_{yM} * V_{yM} + q_{zM} * V_{zM}) * \exp(q_x * \alpha_{S/M}/2)$$
$$* \exp(-q_x * \alpha_{S/M}/2) * \exp(-\kappa * \varsigma/2)$$

Components:

$$1 * V_{tS} + q_x * V_{xS} + q_y * V_{yS} + q_z * V_{zS}$$
$$= \exp(q_x * \alpha_{S/M}/2) * (1 * V_{tM} + q_x * V_{xM} + q_y * V_{yM} + q_z * V_{zM}) * \exp(q_x * \alpha_{S/M}/2)$$

$$V_{tS} = V_{tM} * \cosh\alpha_{S/M} + V_{xM} * \sinh\alpha_{S/M}$$
$$V_{xS} = V_{tM} * \sinh\alpha_{S/M} + V_{xM} * \cosh\alpha_{S/M}$$
$$V_{yS} = V_{yM} \quad ; \quad V_{zS} = V_{zM}$$

Same mathematics applies to other four-component invariants.

$$J_{tS} = J_{tM} * \cosh\alpha_{S/M} + J_{xM} * \sinh\alpha_{S/M}$$
$$J_{xS} = J_{tM} * \sinh\alpha_{S/M} + J_{xM} * \cosh\alpha_{S/M}$$
$$J_{yS} = J_{yM} \quad ; \quad J_{zS} = J_{zM}$$

Matrix equation form:

$$\begin{vmatrix} V_{tS} \\ V_{xS} \end{vmatrix} = \begin{vmatrix} \cosh\alpha_{S/M} & \sinh\alpha_{S/M} \\ \sinh\alpha_{S/M} & \cosh\alpha_{S/M} \end{vmatrix} * \begin{vmatrix} V_{tM} \\ V_{xM} \end{vmatrix}$$

Lorentz Transformation of the compound-label-numbers:

CHAPTER 3 – FIELDS

$1_S = \exp(-\kappa^*\varsigma/2)^*\exp(-q_x^*\alpha_{S/M}/2)^*\exp(-q_x^*\alpha_{S/M}/2)^*\exp(-\kappa^*\varsigma/2)$
$= \exp(-\kappa^*\varsigma/2)^*\exp(-q_x^*\alpha_{S/M})^*\exp(-\kappa^*\varsigma/2)$

$q_{xS} = \exp(-\kappa^*\varsigma/2)^*\exp(-q_x^*\alpha_{S/M}/2)^*q_x^*\exp(-q_x^*\alpha_{S/M}/2)^*\exp(-\kappa^*\varsigma/2)$
$= \exp(-\kappa^*\varsigma/2)^*q_x^*\exp(-q_x^*\alpha_{S/M})^*\exp(-\kappa^*\varsigma/2)$

$q_{yS} = \exp(-\kappa^*\varsigma/2)^*\exp(-q_x^*\alpha_{S/M}/2)^*q_y^*\exp(-q_x^*\alpha_{S/M}/2)^*\exp(-\kappa^*\varsigma/2)$
$= \exp(-\kappa^*\varsigma/2)^*\exp(-q_x^*\alpha_{S/M}/2)^*\exp(q_x^*\alpha_{S/M}/2)^*q_y^*\exp(-\kappa^*\varsigma/2)$
$= \exp(-\kappa^*\varsigma/2)^*q_y^*\exp(-\kappa^*\varsigma/2) = q_{yM}$

$q_{zS} = \exp(-\kappa^*\varsigma/2)^*\exp(-q_x^*\alpha_{S/M}/2)^*q_z^*\exp(-q_x^*\alpha_{S/M}/2)^*\exp(-\kappa^*\varsigma/2)$
$= \exp(-\kappa^*\varsigma/2)^*\exp(-q_x^*\alpha_{S/M}/2)^*\exp(q_x^*\alpha_{S/M}/2)^*q_z^*\exp(-\kappa^*\varsigma/2)$
$= \exp(-\kappa^*\varsigma/2)^*q_z^*\exp(-\kappa^*\varsigma/2) = q_{zM}$

$1_S^{*j} = (\exp(-\kappa^*\varsigma/2)^*\exp(-q_x^*\alpha_{S/M})^*\exp(-\kappa^*\varsigma/2))^{*j}$
$= (\exp(\kappa^*\varsigma/2)^*\exp(q_x^*\alpha_{S/M})^*\exp(\kappa^*\varsigma/2))$

$q_{xS}^{*j} = (\exp(-\kappa^*\varsigma/2)^*q_x^*\exp(-q_x^*\alpha_{S/M})^*\exp(-\kappa^*\varsigma/2))^{*j}$
$= \exp(\kappa^*\varsigma/2)^*\exp(q_x^*\alpha_{S/M})^*-q_x^*\exp(\kappa^*\varsigma/2)$

$q_{yS}^{*j} = q_{yM}^{*j}$; $q_{zS}^{*j} = q_{zM}^{*j}$

"$_4V$" in "S" equals "$_4V$" in "M", proven using mathematics analogous for the same activity in two-dimensional time-space.

$$_4V = 1_S^*V_{tS} + q_{xS}^*V_{xS} + q_{yS}^*V_{yS} + q_{zS}^*V_{zS}$$
$$= 1_M^*V_{tM} + q_{xM}^*V_{xM} + q_{yM}^*V_{yM} + q_{zM}^*V_{zM} = {_4V}$$

Dot-product of two four-component invariants is an invariant.

$$_4k^{*j} \bullet {_4r} = ((\omega_S/c)^*1_S^{*j} + k_{xS}^*q_{xS}^{*j} + k_{yS}^*q_{yS}^{*j} + k_{zS}^*q_{zS}^{*j})$$
$$\bullet (1_S^*c^*t_S + q_{xS}^*x_S + q_{yS}^*y_S + q_{zS}^*z_S)$$

$$= ((\omega_S/c)^*1_M^{*j}*1_S^*c^*t_S + k_{xS}^*q_{xS}^{*j}*q_{xS}^*x_S$$
$$+ k_{yS}^*q_{yS}^{*j}*q_{yS}^*y_S + k_{zS}^*q_{zS}^{*j}*q_{zS}^*z_S)$$

$$= (\omega_S/c)^*c^*t_S - k_{xS}^*x_S - k_{yS}^*y_S - k_{zS}^*z_S$$

SPECIAL ALGEBRA FOR SPECIAL RELATIVITY

$= ((\omega_M/c)*\cosh\alpha_{S/M} + k_{xM}*\sinh\alpha_{S/M})*(c*t_M*\cosh\alpha_{S/M} + x_M*\sinh\alpha_{S/M})$
$- ((\omega_M/c)*\sinh\alpha_{S/M} + k_{xM}*\cosh\alpha_{S/M})*(c*t_M*\sinh\alpha_{S/M} + x_M*\cosh\alpha_{S/M})$
$- k_{yM}*y_M - k_{zM}*z_M$

$= ((\omega_M/c)*c*t_M*(\cosh^2\alpha_{S/M} - \sinh^2\alpha_{S/M}) + (k_{xM}*x_M*(\sinh^2\alpha_{S/M} - \cosh^2\alpha_{S/M})$
$+ (\omega_M/c)*x_M*(-\sinh\alpha_{S/M}*\cosh\alpha_{S/M} + \cosh\alpha_{S/M}*\sinh\alpha_{S/M})$
$+ (k_{xM}*c*t_M*(-\cosh\alpha_{S/M}*\sinh\alpha_{S/M} + \sinh\alpha_{S/M}*\cosh\alpha_{S/M}) - k_{yM}*y_M - k_{zM}*z_M$

$= (\omega_M/c)*c*t_M - k_{xM}*x_M - k_{yM}*y_M - k_{zM}*z_M = {}_4k^{*j}\bullet_4r$

Space-Negative Lorentz Transformation. The inverted matrix for the space-negative Lorentz Transformation is complemented by an inverted matrix for the space-negative compound-label-numbers.

$\nabla_{tS} = \nabla_{tM}*\cosh\alpha_{S/M} - \nabla_{xM}*\sinh\alpha_{S/M}$
$\nabla_{xS} = -\nabla_{tM}*\sinh\alpha_{S/M} + \nabla_{xM}*\cosh\alpha_{S/M}$
$\nabla_{yS} = \nabla_{yM} \quad ; \quad \nabla_{zS} = \nabla_{zM}$

$1_S^{sn} = (1_M*\cosh\alpha_{S/M} - q_{xM}*\sinh\alpha_{S/M})^{sn}$
$= 1_M^{sn}*\cosh\alpha_{S/M} + q_{xM}^{sn}*\sinh\alpha_{S/M}$

$q_{xS}^{sn} = (-1_M*\sinh\alpha_{S/M} + q_{xM}*\cosh\alpha_{S/M})^{sn}$
$= 1_M^{sn}*\sinh\alpha_{S/M} + q_{xM}^{sn}*\cosh\alpha_{S/M}$

$_4\nabla^{sn} = 1_S^{sn}*\nabla_{tS} + q_{xS}^{sn}*\nabla_{xS} + q_{yS}^{sn}*\nabla_{yS} + q_{zS}^{sn}*\nabla_{zS}$

$= (1_M*\cosh\alpha_{S/M} + q_{xM}*\sinh\alpha_{S/M})^{sn}*(\nabla_{tM}*\cosh\alpha_{S/M} - \nabla_{xM}*\sinh\alpha_{S/M})$
$+ (1_M*\sinh\alpha_{S/M} + q_{xM}*\cosh\alpha_{S/M})^{sn}*(-\nabla_{tM}*\sinh\alpha_{S/M} + \nabla_{xM}*\cosh\alpha_{S/M})$
$+ q_{yM}^{sn}*\nabla_{yM} + q_{zM}^{sn}*\nabla_{zM}$

$= (1_M^{sn}*\cosh\alpha_{S/M} + q_{xM}^{sn}*\sinh\alpha_{S/M})*(\nabla_{tM}*\cosh\alpha_{S/M} - \nabla_{xM}*\sinh\alpha_{S/M})$
$+ (+1_M^{sn}*\sinh\alpha_{S/M} + q_{xM}^{sn}*\cosh\alpha_{S/M})*(-\nabla_{tM}*\sinh\alpha_{S/M} + \nabla_{xM}*\cosh\alpha_{S/M})$
$+ q_{yM}^{sn}*\nabla_{yM} + q_{zM}^{sn}*\nabla_{zM}$

$= (1_M^{sn}*\nabla_{tM} + q_{xM}^{sn}*\nabla_{xM})*(\cosh^2\alpha_{S/M} - \sinh^2\alpha_{S/M})$
$+ (q_{xM}^{sn}*\nabla_{tM} + 1_M^{sn}*\nabla_{xM})*(\cosh\alpha_{S/M}*\sinh\alpha_{S/M} - \cosh\alpha_{S/M}*\sinh\alpha_{S/M})$
$+ q_{yM}^{sn}*\nabla_{yM} + q_{zM}^{sn}*\nabla_{zM}$

$= 1_M^{sn}*\nabla_{tM} + q_{xM}^{sn}*\nabla_{xM} + q_{yM}^{sn}*\nabla_{yM} + q_{zM}^{sn}*\nabla_{zM} = {}_4\nabla^{sn}$

CHAPTER 3 – FIELDS

"$_4\nabla^{*jsn} \bullet (_4V) = 0$" and "$_4\nabla^{*jsn} \bullet (_4J) = 0$" are valid after the Lorentz Transformation.

$$_4\nabla^{*jsn} \bullet (_4J) = (\nabla_{tS}*1_S^{*jsn} + \nabla_{xS}*q_{xS}^{*jsn} + \nabla_{yS}*q_{yS}^{*jsn} + \nabla_{zS}*q_{zS}^{*jsn})$$
$$\bullet (1_S*J_{tS} + q_{xS}*J_{xS} + q_{yS}*J_{yS} + q_{zS}*J_{zS})$$

$$= \nabla_{tS}*1_M^{*jsn}*1_S*J_{tS} + \nabla_{xS}*q_{xS}^{*jsn*}(q_{xS})*J_{xS}$$
$$+ \nabla_{yS}*q_{yS}^{*jsn*}(q_{yS})*J_{yS} + \nabla_{zS}*q_{zS}^{*jsn*}(q_{zS})*J_{zS}$$

$$= \nabla_{tS}*J_{tS} + \nabla_{xS}*J_{xS} + \nabla_{yS}*J_{yS} + \nabla_{zS}*J_{zS}$$

$$= (\nabla_{tM}*\cosh\alpha_{S/M} - \nabla_{xM}*\sinh\alpha_{S/M})*(J_{tM}*\cosh\alpha_{S/M} + J_{xM}*\sinh\alpha_{S/M})$$
$$+ (-\nabla_{tM}/c*\sinh\alpha_{S/M} + \nabla_{xM}*\cosh\alpha_{S/M})*(J_{tM}*\sinh\alpha_{S/M} + J_{xM}*\cosh\alpha_{S/M})$$
$$+ \nabla_{yM}*J_{yM} + \nabla_{zM}*J_{zM}$$

$$= \nabla_{tM}*J_{tM}*(\cosh^2\alpha_{S/M} - \sinh^2\alpha_{S/M}) + \nabla_{xM}*J_{xM}*(-\sinh^2\alpha_{S/M} + \cosh^2\alpha_{S/M})$$
$$+ \nabla_{tM}*J_{xM}*(\sinh\alpha_{S/M}*\cosh\alpha_{S/M} - \cosh\alpha_{S/M}*\sinh\alpha_{S/M})$$
$$+ \nabla_{xM}*J_{tM}*(-\cosh\alpha_{S/M}*\sinh\alpha_{S/M} + \sinh\alpha_{S/M}*\cosh\alpha_{S/M}) + \nabla_{yM}*J_{yM} + \nabla_{zM}*J_{zM}$$

$$= \nabla_{tM}*J_{tM} + \nabla_{xM}*J_{xM} + \nabla_{yM}*J_{yM} + \nabla_{zM}*J_{zM}$$
$$= 0$$

Electromagnetic Field Lorentz Transformation. On the right is "$\exp(q_x*\alpha_{S/M}/2)$" from the voltage invariant. On the left is the reciprocal "$\exp(-q_x*\alpha_{S/M}/2)$" from the gradient invariant, but with the argument changed to positive by the conjugate operation, and then back to negative by the space-negative.

Begin with the general form of the Lorentz Transformation.

$$_6E = p_{xM}*E_{xM} + p_{yM}*E_{yM} + p_{zM}*E_{zM} + k_{xM}*K_{xM} + k_{yM}*K_{yM} + k_{zM}*K_{zM}$$

$$= \exp(\kappa*\varsigma/2)*1*(q_x*E_{xM} + q_y*E_{yM} + q_z*E_{zM}$$
$$+ j_x*K_{xM} + j_y*K_{yM} + j_z*K_{zM})*1*\exp(-\kappa*\varsigma/2)$$

$$= \exp(\kappa*\varsigma/2)*\exp(q_x*\alpha_{S/M}/2)$$
$$*\exp(-q_x*\alpha_{S/M}/2)*(q_x*E_{xM} + q_y*E_{yM} + q_z*E_{zM} + j_x*K_{xM} + j_y*K_{yM} + j_z*K_{zM})$$
$$*\exp(q_x*\alpha_{S/M}/2)$$
$$*\exp(-q_x*\alpha_{S/M}/2)*\exp(-\kappa*\varsigma/2)$$

SPECIAL ALGEBRA FOR SPECIAL RELATIVITY

The equation to find "S" component values for "$_6$E" is below.

$q_x*(E_{xS} + i*K_{xS}) + q_y*(E_{yS} + i*K_{yS}) + q_z*(E_{zS} + i*K_{zS})$
  $= \exp(-q_x*\alpha_{S/M}/2)*(q_x*(E_{xM} + i*K_{xM}) + q_y*(E_{yM} + i*K_{yM})$
    $+ q_z*(E_{zM} + i*K_{zM}))*\exp(q_x*\alpha_{S/M}/2)$

Derivations:

$q_y*E_{yS} = \cosh(\alpha_{S/M}/2)*(q_y*E_{yM})*\cosh(\alpha_{S/M}/2)$
  $+ -q_x*\sinh(\alpha_{S/M}/2)*(q_y*E_{yM})*q_x*\sinh(\alpha_{S/M}/2)$
  $+ \cosh(\alpha_{S/M}/2)*(q_z*i*K_{zM})*q_x*\sinh(\alpha_{S/M}/2)$
  $+ -q_x*\sinh(\alpha_{S/M}/2)*(q_z*i*K_{zM})*\cosh(\alpha_{S/M}/2)$
  $= \cosh(\alpha_{S/M}/2)*\cosh(\alpha_{S/M}/2)*(q_y*E_{yM})$
  $+ q_x^2*\sinh(\alpha_{S/M}/2)*\sinh(\alpha_{S/M}/2)*(q_y*E_{yM})$
  $+ -2*q_x*\sinh(\alpha_{S/M}/2)*\cosh(\alpha_{S/M}/2)*(q_z*i*K_{zM})$
  $= \cosh\alpha_{S/M}*(q_y*E_{yM}) - q_x*\sinh\alpha_{S/M}*(q_z*i*K_{zM})$
  $= q_y*\cosh\alpha_{S/M}*E_{yM} - q_x*q_z*i*\sinh\alpha_{S/M}*K_{zM}$
  $= q_y*(\cosh\alpha_{S/M}*E_{yM} + \sinh\alpha_{S/M}*K_{zM})$

$-q_x*q_z*i = -((j_x/i)*(j_z/i)*i) = j_x*j_z*i = -j_y*i = q_y$

$q_y*i*K_{yS} = \cosh(\alpha_{S/M}/2)*(q_y*i*K_{yM})*\cosh(\alpha_{S/M}/2)$
  $+ -q_x*\sinh(\alpha_{S/M}/2)*(q_y*i*K_{yM})*q_x*\sinh(\alpha_{S/M}/2)$
  $+ \cosh(\alpha_{S/M}/2)*(q_z*E_{zM})*q_x*\sinh(\alpha_{S/M}/2)$
  $+ -q_x*\sinh(\alpha_{S/M}/2)*(q_z*E_{zM})*\cosh(\alpha_{S/M}/2)$
  $= \cosh(\alpha_{S/M}/2)*\cosh(\alpha_{S/M}/2)*(q_y*i*K_{yM})$
  $+ q_x^2*\sinh(\alpha_{S/M}/2)*\sinh(\alpha_{S/M}/2)*(q_y*i*K_{yM})$
  $- 2*q_x*\sinh(\alpha_{S/M}/2)*\cosh(\alpha_{S/M}/2)*(q_z*E_{zM})$
  $= \cosh\alpha_{S/M}*(q_y*i*K_{yM}) - q_x*\sinh\alpha_{S/M}*(q_z*E_{zM})$
  $= q_y*i*\cosh\alpha_{S/M}*K_{yM} - q_x*q_z*\sinh\alpha_{S/M}*E_{zM}$
  $= q_y*i*(\cosh\alpha_{S/M}*K_{yM} - \sinh\alpha_{S/M}*E_{zM})$

$-q_x*q_z = -(j_x/i)*(j_z/i) = -j_y = -q_y*i$

$q_z*E_{zS} = \cosh(\alpha_{S/M}/2)*(q_z*E_{zM})*\cosh(\alpha_{S/M}/2)$
  $+ -q_x*\sinh(\alpha_{S/M}/2)*(q_z*E_{zM})*q_x*\sinh(\alpha_{S/M}/2)$
  $+ \cosh(\alpha_{S/M}/2)*(q_y*i*K_{yM})*q_x*\sinh(\alpha_{S/M}/2)$
  $+ -q_x*\sinh(\alpha_{S/M}/2)*(q_y*i*K_{yM})*\cosh(\alpha_{S/M}/2)$
  $= \cosh(\alpha_{S/M}/2)*\cosh(\alpha_{S/M}/2)*(q_z*E_{zM})$
  $+ q_x^2*\sinh(\alpha_{S/M}/2)*\sinh(\alpha_{S/M}/2)*(q_z*E_{zM})$
  $+ -2*q_x*\sinh(\alpha_{S/M}/2)*\cosh(\alpha_{S/M}/2)*(q_y*i*K_{yM})$

CHAPTER 3 – FIELDS

$\phantom{q_z^*i^*K_{zS}} = \cosh\alpha_{S/M}*(q_z*E_{zM}) - q_x*\sinh\alpha_{S/M}*(q_y*i*K_{yM})$
$\phantom{q_z^*i^*K_{zS}} = q_z*\cosh\alpha_{S/M}*E_{zM} - q_x*q_y*i*\sinh\alpha_{S/M}*K_{yM}$
$\phantom{q_z^*i^*K_{zS}} = q_z*(\cosh\alpha_{S/M}*E_{zM} - \sinh\alpha_{S/M}*K_{yM})$

$-q_x*q_y*i = -((j_x/i)*(j_y/i)*i) = j_x*j_y*i = j_z*i = -q_z$

$q_z*i*K_{zS} = \cosh(\alpha_{S/M}/2)*(q_z*i*K_{zM})*\cosh(\alpha_{S/M}/2)$
$\phantom{q_z*i*K_{zS} =} + -q_x*\sinh(\alpha_{S/M}/2)*(q_z*i*K_{zM})*q_x*\sinh(\alpha_{S/M}/2)$
$\phantom{q_z*i*K_{zS} =} + \cosh(\alpha_{S/M}/2)*(q_y*E_{yM})*q_x*\sinh(\alpha_{S/M}/2)$
$\phantom{q_z*i*K_{zS} =} + -q_x*\sinh(\alpha_{S/M}/2)*(q_y*E_{yM})*\cosh(\alpha_{S/M}/2)$
$\phantom{q_z*i*K_{zS}} = \cosh(\alpha_{S/M}/2)*\cosh(\alpha_{S/M}/2)*(q_z*i*K_{zM})$
$\phantom{q_z*i*K_{zS} =} + q_x^2*\sinh(\alpha_{S/M}/2)*\sinh(\alpha_{S/M}/2)*(q_z*i*K_{zM})$
$\phantom{q_z*i*K_{zS} =} + -2*q_x*\sinh(\alpha_{S/M}/2)*\cosh(\alpha_{S/M}/2)*(q_y*E_{yM})$
$\phantom{q_z*i*K_{zS}} = \cosh\alpha_{S/M}*(q_z*i*K_{zM}) - q_x*\sinh\alpha_{S/M}*(q_y*E_{yM})$
$\phantom{q_z*i*K_{zS}} = q_z*i*\cosh\alpha_{S/M}*K_{zM} - q_x*q_y*\sinh\alpha_{S/M}*E_{yM}$
$\phantom{q_z*i*K_{zS}} = q_z*i*(\cosh\alpha_{S/M}*K_{zM} + \sinh\alpha_{S/M}*E_{yM})$

$-q_x*q_y = -(j_x/i)*(j_y/i) = j_z = q_z*i$

$E_{xS} = E_{xM}$
$K_{xS} = K_{xM}$
$E_{yS} = E_{yM}*\cosh\alpha_{S/M} + K_{zM}*\sinh\alpha_{S/M}$
$K_{yS} = -E_{zM}*\sinh\alpha_{S/M} + K_{yM}*\cosh\alpha_{S/M}$
$E_{zS} = E_{zM}*\cosh\alpha_{S/M} - K_{yM}*\sinh\alpha_{S/M}$
$K_{zS} = E_{yM}*\sinh\alpha_{S/M} + K_{zM}*\cosh\alpha_{S/M}$

$$\begin{array}{c} E_{yS} + i*K_{yS} \\ \\ E_{zS} + i*K_{zS} \end{array} = \begin{array}{cc} \cosh\alpha_{S/M} & -i*\sinh\alpha_{S/M} \\ \\ i*\sinh\alpha_{S/M} & \cosh\alpha_{S/M} \end{array} * \begin{array}{c} E_{yM} + i*K_{yM} \\ \\ E_{zM} + i*K_{zM} \end{array}$$

The equation to find "S" compound-label-numbers for "$_6E = {}_4\nabla^{*jsn}*(_4V)$" uses the left and right denominators.

$p_{xS} = p_{xM}$

$p_{yS} = \exp(\kappa*\varsigma/2)*\exp(q_x*\alpha_{S/M}/2)*q_y*\exp(-q_x*\alpha_{S/M}/2)*\exp(-\kappa*\varsigma/2)$
$\phantom{p_{yS}} = \exp(\kappa*\varsigma/2)*q_y*\exp(q_x*\alpha_{S/M})*\exp(-\kappa*\varsigma/2)$

$p_{zS} = \exp(\kappa*\varsigma/2)*\exp(q_x*\alpha_{S/M}/2)*q_z*\exp(-q_x*\alpha_{S/M}/2)*\exp(-\kappa*\varsigma/2)$
$\phantom{p_{zS}} = \exp(\kappa*\varsigma/2)*q_z*\exp(q_x*\alpha_{S/M})*\exp(-\kappa*\varsigma/2)$

SPECIAL ALGEBRA FOR SPECIAL RELATIVITY

$k_{xS} = k_{xM}$
$k_{yS} = \exp(\kappa^*\varsigma/2)^* j_y^* \exp(q_x^* \alpha_{S/M})^* \exp(-\kappa^*\varsigma/2)$
$k_{zS} = \exp(\kappa^*\varsigma/2)^* j_z^* \exp(q_x^* \alpha_{S/M})^* \exp(-\kappa^*\varsigma/2)$

A check ensured "$_6E = {_6E}$": To check that the theory is correct, "$_M$" Lorentz Transformation components and label-numbers for "$_4\nabla^{*/sn}$" and "$_4V$" substitute into the "$_S$" version of "$-_6E = {_4\nabla^{*/sn}}\mathbf{x}(_4V)$", so that the subscripts are all "$_M$" and not "$_S$". Components and compound-label-numbers are grouped to "$-_6E$" expressed in "$_M$".

Also, the component transformations are confirmed by reference to Page 71 of *Introduction to Modern Physics* by Richtmyer, Kennard and Lauritsen, McGraw-Hill Book Company, Inc., 1955 and by reference to Page 210 of *Methods of Theoretical Physics Part I* by Morse and Feshbach, McGraw-Hill Book Company, Inc., 1953.

A third check ensures "$_4\nabla^{sn}\blacklozenge(-_6E) = {_4}J$" (calculated in "S" from components in "M") matches current density (calculated from "$_4\nabla^{sn}$" and "$_6E$" in "S").

$1_S*J_{tS} + q_{xS}*J_{xS} + q_{yS}*J_{yS} + q_{zS}*J_{zS}$
$= (1_S{}^{sn}*\nabla_{tS} + q_{xS}{}^{sn}*\nabla_{xS} + q_{yS}{}^{sn}*\nabla_{yS} + q_{zS}{}^{sn}*\nabla_{zS})$
$\blacklozenge(p_{xS}*E_{xS} + p_{yS}*E_{yS} + p_{zS}*E_{zS} + k_{xS}*K_{xS} + k_{yS}*K_{yS} + k_{zS}*K_{zS})*(-1)$

$1_S*J_{tS} = -\nabla_{xS}*q_{xS}{}^{sn}*p_{xS}*E_{xS} - \nabla_{yS}*q_{yS}{}^{sn}*p_{yS}*E_{yS} - \nabla_{zS}*q_{zS}{}^{sn}*p_{zS}*E_{zS}$
$q_{xS}*J_{xS} = -\nabla_{tS}*1_S{}^{sn}*p_{xS}*E_{xS} - \nabla_{yS}*q_{yS}{}^{sn}*k_{zS}*K_{zS} - \nabla_{zS}*q_{zS}{}^{sn}*k_{yS}*K_{yS}$
$q_{yS}*J_{yS} = -\nabla_{tS}*1_S{}^{sn}*p_{yS}*E_{yS} - \nabla_{zS}*q_{zS}{}^{sn}*k_{xS}*K_{xS} - \nabla_{xS}*q_{xS}{}^{sn}*k_{zS}*K_{zS}$
$q_{zS}*J_{zS} = -\nabla_{tS}*1_S{}^{sn}*p_{zS}*E_{zS} - \nabla_{xS}*q_{xS}{}^{sn}*k_{yS}*K_{yS} - \nabla_{yS}*q_{yS}{}^{sn}*k_{xS}*K_{xS}$

Compound-label-numbers multiply as

$1_S = -q_{xS}{}^{sn}*p_{xS} = -q_{yS}{}^{sn}*p_{yS} = -q_{zS}{}^{sn}*p_{zS}$
$q_{xS} = -1_S{}^{sn}*p_{xS}*(-1) = -q_{yS}{}^{sn}*k_{zS} = -q_{zS}{}^{sn}*k_{yS}*(-1)$
$q_{yS} = -1_S{}^{sn}*p_{yS}*(-1) = -q_{zS}{}^{sn}*k_{xS} = -q_{xS}{}^{sn}*k_{zS}*(-1)$
$q_{zS} = -1_S{}^{sn}*p_{zS}*(-1) = -q_{xS}{}^{sn}*k_{yS} = -q_{yS}{}^{sn}*k_{xS}*(-1)$

$J_{tS} = \nabla_{xS}*E_{xS} + \nabla_{yS}*E_{yS} + \nabla_{zS}*E_{zS}$
$J_{xS} = -\nabla_{tS}*E_{xS} + \nabla_{yS}*K_{zS} - \nabla_{zS}*K_{yS}$
$J_{yS} = -\nabla_{tS}*E_{yS} + \nabla_{zS}*K_{xS} - \nabla_{xS}*K_{zS}$
$J_{zS} = -\nabla_{tS}*E_{zS} + \nabla_{xS}*K_{yS} - \nabla_{yS}*K_{xS}$

CHAPTER 3 – FIELDS

To prove each of the four component equations is correct, "M" component expressions substitute for "S" component expressions. After manipulation of the equations, previously stated component Lorentz Transformation equations for current density "$_4J$" are found to apply.

Time "t" component: $\qquad J_{tS} = \nabla_{xS}*E_{xS} + \nabla_{yS}*E_{yS} + \nabla_{zS}*E_{zS}$

$(J_{tM}*\cosh\alpha_{S/M} + J_{xM}*\sinh\alpha_{S/M}) = (-\nabla_{tM}*\sinh\alpha_{S/M} + \nabla_{xM}*\cosh\alpha_{S/M})*(E_{xS})$
$\qquad\qquad + (\nabla_{yM})*(E_{yM}*\cosh\alpha_{S/M} + K_{zM}*\sinh\alpha_{S/M})$
$\qquad\qquad + (\nabla_{zM})*(E_{zM}*\cosh\alpha_{S/M} - K_{yM}*\sinh\alpha_{S/M})$

$\qquad J_{tM} = \nabla_{xM}*E_{xS} + \nabla_{yM}*E_{yM} + \nabla_{zM}*E_{zM}$
$\qquad J_{xM} = -\nabla_{tM}*E_{xS} + \nabla_{yM}*K_{zM} - \nabla_{zM}*K_{yM}$

"x" component: $\qquad J_{xS} = -\nabla_{tS}*E_{xS} + \nabla_{yS}*K_{zS} - \nabla_{zS}*K_{yS}$

$(J_{tM}*\sinh\alpha_{S/M} + J_{xM}*\cosh\alpha_{S/M}) = -(\nabla_{tM}*\cosh\alpha_{S/M} - \nabla_{xM}*\sinh\alpha_{S/M})*(E_{xS})$
$\qquad\qquad + (\nabla_{yM})*(E_{yM}*\sinh\alpha_{S/M} + K_{zM}*\cosh\alpha_{S/M})$
$\qquad\qquad - (\nabla_{zM})*(-E_{zM}*\sinh\alpha_{S/M} + K_{yM}*\cosh\alpha_{S/M})$

$\qquad J_{tM} = \nabla_{xM}*E_{xS} + \nabla_{yM}*E_{yM} + \nabla_{zM}*E_{zM}$
$\qquad J_{xM} = -\nabla_{tM}*E_{xS} + \nabla_{yM}*K_{zM} - \nabla_{zM}*K_{yM}$

"y" component: $\qquad J_{yS} = -\nabla_{tS}*E_{yS} + \nabla_{zS}*K_{xS} - \nabla_{xS}*K_{zS}$

$J_{yM} = -(\nabla_{tM}*\cosh\alpha_{S/M} - \nabla_{xM}*\sinh\alpha_{S/M})*(E_{yM}*\cosh\alpha_{S/M} + K_{zM}*\sinh\alpha_{S/M}) + \nabla_{zM}*K_{xM} - (-\nabla_{tM}*\sinh\alpha_{S/M} + \nabla_{xM}*\cosh\alpha_{S/M})*(E_{yM}*\sinh\alpha_{S/M} + K_{zM}*\cosh\alpha_{S/M})$

$J_{yM} = (-\nabla_{tM}*E_{yM} - \nabla_{xM}*K_{zM})*(\cosh^2\alpha_{S/M} - \sinh^2\alpha_{S/M}) + \nabla_{zM}*K_{xM}$
$\qquad + (-\nabla_{tM}*K_{zM} - \nabla_{xM}*E_{yM})*(2*\sinh\alpha_{S/M}*\cosh\alpha_{S/M} - 2*\sinh\alpha_{S/M}*\cosh\alpha_{S/M})$

$\qquad J_{yM} = -\nabla_{tM}*E_{yM} - \nabla_{xM}*K_{zM} + \nabla_{zM}*K_{xM}$

"z" component: $\qquad J_{zS} = -\nabla_{tS}*E_{zS} + \nabla_{xS}*K_{yS} - \nabla_{yS}*K_{xS}$

$J_{zM} = -(\nabla_{tM}*\cosh\alpha_{S/M} - \nabla_{xM}*\sinh\alpha_{S/M})*(E_{zM}*\cosh\alpha_{S/M} - K_{yM}*\sinh\alpha_{S/M}) - \nabla_{yM}*K_{xM} + (-\nabla_{tM}*\sinh\alpha_{S/M} + \nabla_{xM}*\cosh\alpha_{S/M})*(-E_{zM}*\sinh\alpha_{S/M} + K_{yM}*\cosh\alpha_{S/M})$

SPECIAL ALGEBRA FOR SPECIAL RELATIVITY

$J_{zM} = (-\nabla_{tM}*E_{zM} + \nabla_{xM}*K_{yM})*(\cosh^2\alpha_{S/M} - \sinh^2\alpha_{S/M}) - \nabla_{yM}*K_{xM}$
$+ (\nabla_{tM}*K_{yM} + \nabla_{xM}*E_{zM})*(2*\sinh\alpha_{S/M}*\cosh\alpha_{S/M} - 2*\sinh\alpha_{S/M}*\cosh\alpha_{S/M})$

$J_{zM} = -\nabla_{tM}*E_{zM} + \nabla_{xM}*K_{yM} - \nabla_{yM}*K_{xM}$

3.5 Biot-Savart Law

The Biot-Savart Law for the electromagnetic field of a moving particle is derived from the electric field of a stationary particle.

$$_6E = (q/(4*\pi*\mathfrak{z}*r_M^2))*((x_M/r_M)*p_{xM} + (y_M/r_M)*p_{yM} + (z_M/r_M)*p_{zM})$$

$E_{xM} = (q/(4*\pi*\mathfrak{z}*r_M^2))*(x_M/r_M)$; $K_{xM} = 0$
$E_{yM} = (q/(4*\pi*\mathfrak{z}*r_M^2))*(y_M/r_M)$; $K_{yM} = 0$
$E_{zM} = (q/(4*\pi*\mathfrak{z}*r_M^2))*(z_M/r_M)$; $K_{zM} = 0$
$E_{xS} = E_{xM} = (q/(4*\pi*\mathfrak{z}*r_M^3))*x_M$

$$\begin{vmatrix} E_{yS} + i*K_{yS} \\ E_{zS} + i*K_{zS} \end{vmatrix} = \begin{vmatrix} \cosh\alpha_{S/M} & -i*\sinh\alpha_{S/M} \\ i*\sinh\alpha_{S/M} & \cosh\alpha_{S/M} \end{vmatrix} * \begin{vmatrix} (q/(4*\pi*\mathfrak{z}*r_M^2))*(y_M/r_M) \\ (q/(4*\pi*\mathfrak{z}*r_M^2))*(z_M/r_M) \end{vmatrix}$$

$E_{yS} = (q/(4*\pi*\mathfrak{z}*r_M^2))*(y_M/r_M)*(\cosh\alpha_{S/M})$
$K_{yS} = -(q/(4*\pi*\mathfrak{z}*r_M^2))*(z_M/r_M)*(\sinh\alpha_{S/M})$
$E_{zS} = (q/(4*\pi*\mathfrak{z}*r_M^2))*(z_M/r_M)*(\cosh\alpha_{S/M})$
$K_{zS} = (q/(4*\pi*\mathfrak{z}*r_M^2))*(y_M/r_M)*(\sinh\alpha_{S/M})$

"r_M^2" must be independently Lorentz Transformed into "S" from "M". In "M" there is no time "t_M" difference between the electrically charged particle and point (x_M, y_M, z_M) where the electromagnetic field is being measured.

$$location_M^2 = x_M^2 + y_M^2 + z_M^2 - c^2*t_M^2 = r_M^2 - c^2*t_M^2$$

Equations in "S" have that same assumption of no time discrepancy, but in "S" not "M". And points in "S" are at different times with respect to "M".

$x_S^2 + y_S^2 + z_S^2 - c^2*t_S^2 = x_M^2 + y_M^2 + z_M^2 - c^2*t_M^2$
$r_S^2 - c^2*t_S^2 = r_M^2 - c^2*t_M^2$
$r_S^2 = r_M^2 - c^2*t_M^2$; $r_M^2 = r_S^2 + c^2*t_M^2$

CHAPTER 3 – FIELDS

$$\begin{bmatrix} c*t_S \\ x_S \end{bmatrix} = \begin{bmatrix} \cosh\alpha_{S/M} & \sinh\alpha_{S/M} \\ \sinh\alpha_{S/M} & \cosh\alpha_{S/M} \end{bmatrix} * \begin{bmatrix} c*t_M \\ x_M \end{bmatrix}$$

$$\begin{bmatrix} c*t_M \\ x_M \end{bmatrix} = \begin{bmatrix} \cosh\alpha_{S/M} & -\sinh\alpha_{S/M} \\ -\sinh\alpha_{S/M} & \cosh\alpha_{S/M} \end{bmatrix} * \begin{bmatrix} c*t_S \\ x_S \end{bmatrix}$$

$c*t_M = c*t_S*\cosh\alpha_{S/M} - x_S*\sinh\alpha_{S/M}$

$c^2*t_M^2 = c^2*t_S^2*\cosh^2\alpha_{S/M} + x_S^2*\sinh^2\alpha_{S/M} - c^2*t_S*x_S*\sinh(2*\alpha_{S/M})$
$\quad\quad = x_S^2*\sinh^2\alpha_{S/M}$

$r_M^2 = r_S^2 + c^2*t_M^2 = r_S^2 + x_S^2*\sinh^2\alpha_{S/M}$

$x_M = x_S*\cosh\alpha_{S/M} - c*t_S*\sinh\alpha_{S/M} = x_S*\cosh\alpha_{S/M}$

$y_S = y_M \quad ; \quad z_S = z_M$

$E_{xS} = (q/(4*\pi*\text{ə}*(r_S^2 + x_S^2*\sinh^2\alpha_{S/M})^{3/2}))*x_S*\cosh\alpha_{S/M} \quad ; \quad K_{xS} = 0$
$E_{yS} = (q/(4*\pi*\text{ə}*(r_S^2 + x_S^2*\sinh^2\alpha_{S/M})^{3/2}))*y_S*\cosh\alpha_{S/M}$
$K_{yS} = -(q/(4*\pi*\text{ə}*(r_S^2 + x_S^2*\sinh^2\alpha_{S/M})^{3/2}))*z_S*\sinh\alpha_{S/M}$
$E_{zS} = (q/(4*\pi*\text{ə}*(r_S^2 + x_S^2*\sinh^2\alpha_{S/M})^{3/2}))*z_S*\cosh\alpha_{S/M}$
$K_{zS} = (q/(4*\pi*\text{ə}*(r_S^2 + x_S^2*\sinh^2\alpha_{S/M})^{3/2}))*y_S*\sinh\alpha_{S/M}$

$_6E = p_{xS}*(E_{xS} + i*K_{xS}) + p_{yS}*(E_{yS} + i*K_{yS}) + p_{zS}*(E_{zS} + i*K_{zS})$

(The author hasn't seen the above component equations in a book or paper. That statement is common in this book, but with the expectation there is a book or paper not yet seen.)

From Wikipedia ("Biot-Savart Law") attributed to Oliver Heaviside in year 1888 (before the discovery of Special Relativity in 1905):

$E_{xSW} = (q/(4*\pi*\text{ə}))*((1 - v_{S/M}^2/c^2)/(1 - \sin^2\theta*(v_{S/M}^2/c^2))^{3/2})*(x_S/r_S^3)$
$K_{xSW} = 0$
$E_{ySW} = (q/(4*\pi*\text{ə}))*((1 - v_{S/M}^2/c^2)/(1 - \sin^2\theta*(v_{S/M}^2/c^2))^{3/2})*(y_S/r_S^3)$
$K_{ySW} = -(v_{S/M}/c)*E_{zS}$
$E_{zSW} = (q/(4*\pi*\text{ə}))*((1 - v_{S/M}^2/c^2)/(1 - \sin^2\theta*(v_{S/M}^2/c^2))^{3/2})*(z_S/r_S^3)$
$K_{zSW} = (v_{S/M}/c)*E_{yS}$

SPECIAL ALGEBRA FOR SPECIAL RELATIVITY

$$\cos\theta = x_S/r_S \; ; \quad \sin^2\theta = (y_S^2 + z_S^2)/r_S^2 \; ; \quad r_S^2 = x_S^2 + y_S^2 + z_S^2$$

The two sets of electromagnetic component expressions in "S" appear very different. Component values deviate with slightly higher values for Heaviside's expressions (with the "w" subscript) for relativistic speeds and for "x_S" locations far from the origin. For example, at "$\alpha_{S/M} = 6$" and "$x_S = y_S = 1$" with "$z_S = 0$" there is "$E_{xS} = E_{yS} = 0.0000246$" and "$E_{xSW} = E_{ySW} = 0.0000548$", for "$q/(4*\pi*\mathfrak{d}) = 1$". The deviation between "E_{yS}" and "E_{ySW}" may possibly be attributed to a non-relativistic approximation in Heaviside's derivation (for example "$\sinh^2\alpha_{S/M} \approx v_{S/M}^2/c^2$"). That guess has not been confirmed.

<u>Translation from All-Number Algebra to Geometry</u>. We could use measurements to resolve the discrepancy. Translate all-number algebra into geometry per the third step. "i_{xS}", "i_{yS}" and "i_{zS}" substitute for "p_{xS}", "p_{yS}" and "p_{zS}" and for "k_{xS}", "k_{yS}" and "k_{zS}".

3.6 Electric Energy-Momentum of an Electron

Energy-momentum invariant "$_4p$" is calculated from electro-magnetic field invariant "$_6E$" using empirically derived relationships.

Bus "M" moves at speed "$v_{S/M}$" relative to roadside "S". "$_6E$" components are used to find energy-momentum invariant "$_4p$" components. As a check, energy-momentum invariant ("$_4p$") components in "S" found from "$_6E$" must also be found from energy-momentum invariant ("$_4p$") components in "M".

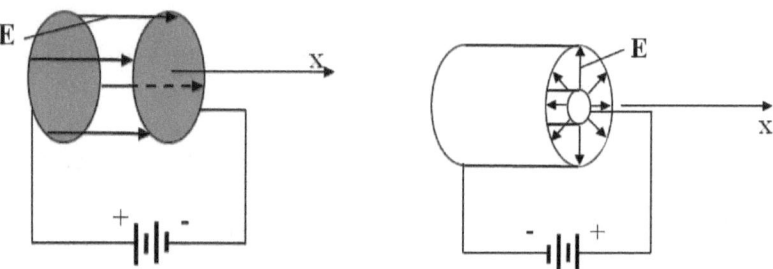

Figure 32. Left: Two parallel capacitor plates.
            Right: Two concentric capacitor plates.

CHAPTER 3 – FIELDS

Select the simplest geometry possible.

- For a field parallel with motion, the two capacitor plates are perpendicular to motion.

- For a field perpendicular to motion, the two capacitor plates are cylindrical concentric circular surfaces with the axis parallel to motion.

Electric charge may be placed onto capacitor plates using a battery. Alternatively, the plates are close together and then electrically charged, and pulled apart: Energy equals force times distance.
Bus mass converts to electric field energy and back again. Energy and momentum (and therefore speed) remain constant during the mass to electric field energy conversion.

Electromagnetic Field Perpendicular to Motion.

$$E_{yM} = A*y/(y^2 + z^2) = A*y/r^2 \quad ; \quad E_{zM} = A*z/(y^2 + z^2) = A*z/r^2$$

The outer cylinder has an excess of negatively charged electrons because the electric field points outward from positive to negative. (Electric field is negative of gradient of voltage, and, therefore, the inner cylinder has higher voltage.)
Energy density "Σ_{tM}" (sigma) was discovered empirically through mathematical modeling of measured energy in experiments.

$$\Sigma_{tM} = \vartheta*(E_{xM}*E_{xM} + E_{yM}*E_{yM} + E_{zM}*E_{zM}$$
$$+ K_{xM}*K_{xM} + K_{yM}*K_{yM} + K_{zM}*K_{zM})/2$$

Symbol "$\perp$" applies to fields perpendicular to motion.

$$\Sigma_{\perp tM} = \vartheta*(E_{yM}*E_{yM} + E_{zM}*E_{zM})/2$$
$$= \vartheta*((A*y/r^2)*(A*y/r^2) + (A*z/r^2)*(A*z/r^2))/2$$
$$= \vartheta*(A^2*(y^2 + z^2))/((r^2)^2)/2$$
$$= \vartheta*(A^2/r^2)/2$$

$$\begin{matrix} E_{yS} + i*K_{yS} \\ \\ E_{zS} + i*K_{zS} \end{matrix} = \begin{matrix} \cosh\alpha_{S/M} & -i*\sinh\alpha_{S/M} \\ \\ i*\sinh\alpha_{S/M} & \cosh\alpha_{S/M} \end{matrix} * \begin{matrix} A*y/r^2 \\ \\ A*z/r^2 \end{matrix}$$

SPECIAL ALGEBRA FOR SPECIAL RELATIVITY

$E_{yS} = \cosh(\alpha_{S/M})*A*y/r^2$; $K_{yS} = -\sinh(\alpha_{S/M})*A*z/r^2$
$E_{zS} = \cosh(\alpha_{S/M})*A*z/r^2$; $K_{zS} = \sinh(\alpha_{S/M})*A*y/r^2$

$\Sigma_{\perp tS} = \partial*(E_{yS}*E_{yS} + E_{zS}*E_{zS} + K_{yS}*K_{yS} + K_{zS}*K_{zS})/2$
$= \partial*((y^2 + z^2)*(\cosh^2(\alpha_{S/M}) + \sinh^2(\alpha_{S/M}))*A^2/((r^2)^2)/2$
$= \partial*(\cosh(2*\alpha_{S/M})*A^2/r^2)/2$

To complement energy density "$\Sigma_{\perp tS}$", there is energy per area per time Poynting Vector "$\Sigma_{\perp xS}*\mathbf{i}_{xS} + \Sigma_{\perp yS}*\mathbf{i}_{yS} + \Sigma_{\perp zS}*\mathbf{i}_{zS}$", empirically found equal to the cross product of electric and magnetic fields, "$\partial*(_3\mathbf{E}x_3\mathbf{B})$".

$\Sigma_{\perp xS} = \partial*(E_{yS}*K_{zS} - E_{zS}*K_{yS})$
$\Sigma_{\perp yS} = \partial*(E_{zS}*K_{xS} - E_{xS}*K_{zS})$
$\Sigma_{\perp zS} = \partial*(E_{xS}*K_{yS} - E_{yS}*K_{xS})$

$\Sigma_{\perp xS} = \partial*(E_{yS}*K_{zS} - E_{zS}*K_{yS})$
$= \partial*(y^2 + z^2)*\cosh(\alpha_{S/M})*\sinh(\alpha_{S/M})*A^2/(r^2)^2$
$= \partial*(\sinh(2*\alpha_{S/M})*A^2/(y^2 + z^2))/2$
$= \partial*(\sinh(2*\alpha_{S/M})*A^2/r^2)/2$

Energy density "$\Sigma_{\perp tS}$" is positive because of squares. And, energy per area per time "$\Sigma_{\perp xS}$" is positive for positive "$\alpha_{S/M}$", so that energy moves in the direction of motion. Because they were empirically derived, "$\Sigma_{\perp tS}$" and "$\Sigma_{\perp xS}$" are combined without a compound-label-number.

$\Sigma_{\perp tS} + q_x*\Sigma_{\perp xS} = \partial*((A^2/r^2)/2)*(\cosh(2*\alpha_{S/M}) + q_x*\sinh(2*\alpha_{S/M}))$
$= \partial*((A^2/r^2)/2)*\exp(q_x*2*\alpha_{S/M})$
$= \partial*((A^2/r^2)/2)*\exp^2(q_x*\alpha_{S/M})$

$\exp(q_x*2*\alpha_{S/M}) = \cosh(2*\alpha_{S/M}) + q_x*\sinh(2*\alpha_{S/M})$
$= \cosh^2(\alpha_{S/M}) + \sinh^2(\alpha_{S/M}) + q_x*2*\sinh(\alpha_{S/M})*\cosh(\alpha_{S/M})$
$= \cosh^2(\alpha_{S/M}) + q_x*q_x*\sinh^2(\alpha_{S/M}) + q_x*2*\sinh(\alpha_{S/M})*\cosh(\alpha_{S/M})$
$= (\cosh(\alpha_{S/M}) + q_x*\sinh(\alpha_{S/M}))^2$
$= \exp^2(q_x*\alpha_{S/M})$

To remove the square operation in "$\exp^2(q_x*\alpha_{S/M})$", "$\Sigma_{\perp tS} + q_x*\Sigma_{\perp xS}$" is multiplied by volume between plates. Multiplication by volume must be an integral because of radius "r" dependency in "$\Sigma_{\perp tS}$" and "$\Sigma_{\perp xS}$". We continue the

CHAPTER 3 – FIELDS

empirical analysis using space-negative of volume, and by making the volume invariant mathematically real rather than imaginary.

Differential volume element "$d(_4Vol^{sn})$" consists of an "L_{xM}" factor in the x-direction, a factor in the radial direction, and a factor in the tangential direction, "$d(_4Vol^{sn}) = 1_S*(L_{xM})*(2*\pi*r)*dr*exp(-q_x*\alpha_{S/M})$". The integration was from "$r_{innerplate}$" to "$r_{outerplate}$".

$$d_4p = (\Sigma_{\perp tS} + q_x*\Sigma_{\perp xS})*d(_4Vol^{sn})/c$$
$$= \eth*((A^2/r^2)/2)*exp^2(q_x*\alpha_{S/M})*(1_S*dVol_B*exp(-q_x*\alpha_{S/M}))/c$$
$$= \eth*((A^2/r^2)/2)*exp^2(q_x*\alpha_{S/M})*(1_S*(L_{xM})*(2*\pi*r)*dr*exp(-q_x*\alpha_{S/M}))/c$$
$$= 1_S*(\eth/c)*(A^2/2)*exp(q_x*\alpha_{S/M})*L_{xM}*(2*\pi)*(dr/r)$$

$$_4p = 1_S*(\eth/c)*(A^2/2)*L_{xM}*(2*\pi)*\ln(r_{outerplate}/r_{innerplate})*exp(q_x*\alpha_{S/M})$$
$$= 1_S*m_B*c*exp(q_x*\alpha_{S/M})$$

The above expression is the energy-momentum invariant for an electromagnetic field perpendicular to motion and formed from two cylindrical concentric circular capacitor plates. Rest mass "m_B*c":

$$m_B*c = (\eth/c)*(A^2/2)*L_{xM}*(2*\pi)*\ln(r_{outerplate}/r_{innerplate})$$

We conclude, in general, any electromagnetic field perpendicular to the direction of motion may be replaced with an equivalent rest mass.

An Electromagnetic Field Parallel to Motion.

$$E_{xS} = E_{xM} \quad ; \quad K_{xS} = K_{xM}$$

Positive "E_{xM}" points right, in the direction of motion, because of a deficit of electrons on the left plate, and excess electrons on the right. (Left plate has the higher voltage.) Per "$E_{xS} = E_{xM}$", a positive test charge "q_{Test}" stationary with roadside "S" experiences the same force "$q_{Test}*E_{xS}$" regardless of speed "$v_{S/M}$" of the bus. But note that energy must be applied to the bus to keep its speed constant because the test charge is repulsed by the left capacitor plate.

If the test charge is stationary with respect to the bus, there is force "$F_{xM} = q_{Test}*E_{xM}$" observed with both time and space components from the roadside because "$x_M = x_S*\cosh(\alpha_{S/M}) - c*t_S*\sinh(\alpha_{S/M})$". To satisfy energy and momentum conservation, use the same basic form of energy density equations given for the case of the perpendicular electric field. (The below equations weren't found in a book or paper.)

SPECIAL ALGEBRA FOR SPECIAL RELATIVITY

$$\Sigma_{\|tS} = \vartheta*(E_{xS}*E_{xS} + K_{xS}*K_{xS})*\cosh(2*\alpha_{S/M})/2$$
$$= \vartheta*(E_{xM}*E_{xM} + K_{xM}*K_{xM})*\cosh(2*\alpha_{S/M})/2$$

$$\Sigma_{\|xS} = \vartheta*(E_{xS}*E_{xS} + K_{xS}*K_{xS})*\sinh(2*\alpha_{S/M})/2$$
$$= \vartheta*(E_{xM}*E_{xM} + K_{xM}*K_{xM})*\sinh(2*\alpha_{S/M})/2$$

"$\Sigma_{\|tS}$" is positive because of squares and the "cosh" function. Energy per area per time "$\Sigma_{\|xS}$" (that passes a location "x_S") is positive for positive "$\alpha_{S/M}$", so that energy moves in the direction of motion.

$$\Sigma_{\|tS} + q_x*\Sigma_{\|xS} = \vartheta*(E_{xM}^2/2)*(\cosh(2*\alpha_{S/M}) + q_x*\sinh(2*\alpha_{S/M}))$$
$$= \vartheta*(E_{xM}^2/2)*\exp(q_x*2*\alpha_{S/M}) = \vartheta*(E_{xM}^2/2)*\exp^2(q_x*\alpha_{S/M})$$

$$Vol_M = L_{xM}*L_{yM}*L_{zM}$$

$$_4Vol^{sn} = 1_S^{sn}*Vol_M*\exp(q_x*\alpha_{S/M}) = 1_S*Vol_M*\exp(-q_x*\alpha_{S/M})$$

$$_4p = (\Sigma_{\|tS} + q_x*\Sigma_{\|xS})*_4Vol^{sn}/c$$
$$= \vartheta*(E_{xM}^2/2)*\exp^2(q_x*\alpha_{S/M})*(1_S*Vol_M*\exp(-q_x*\alpha_{S/M}))/c$$
$$= 1_S*\vartheta*(E_{xM}^2/2)*(Vol_M)*\exp(q_x*\alpha_{S/M})/c$$

$$_4p = 1_S*m_B*c*\exp(q_x*\alpha_{S/M}) \quad ; \quad m_B*c = (\vartheta/c)*(E_{xM}^2/2)*(Vol_M)$$

Electric Field of a Stationary Electron. Radius of an electron is calculated from rest mass and electric charge using a macroscopic model.

Assume negative electric charge is evenly distributed on the surface of a spherical electron. Electric field lines of force extend radially inward to the surface. Electric field magnitude decreases inversely with the square of the distance from the center.

$$E_{xM} = A*(x_M/r_M^3) \quad ; \quad E_{yM} = A*(y_M/r_M^3) \quad ; \quad E_{zM} = A*(z_M/r_M^3)$$

$$A = q/(4*\pi*\vartheta) \quad ; \quad r_M^2 = x_M^2 + y_M^2 + z_M^2$$

$$\Sigma_{tM} = \vartheta*(E_{xM}*E_{xM} + E_{yM}*E_{yM} + E_{zM}*E_{zM}$$
$$+ K_{xM}*K_{xM} + K_{yM}*K_{yM} + K_{zM}*K_{zM})/2$$

$$= \vartheta*(E_{xM}*E_{xM} + E_{yM}*E_{yM} + E_{zM}*E_{zM})/2$$

CHAPTER 3 – FIELDS

$$= \vartheta*(A^2*(x_M^2 + y_M^2 + z_M^2)/r_M^6)/2 = \vartheta*(A^2/r_M^4)/2$$

Energy density "$\Sigma_{tM} = \vartheta*(A^2/r_M^4)/2$" is integrated from infinite radius inward (following electric field lines) to the assumed classical radius "r_e" of the electron. "dVol = $4*\pi*r_M^2*dr_M$".

$$\begin{aligned}d(E_B/c) &= (\vartheta*(A^2/r_M^4)/2)*dVol/c \\ &= ((\vartheta/c)*(A^2/r_M^4)/2)*4*\pi*r_M^2*dr_M \\ &= (1/2)*((\vartheta/c)*(q/(4*\pi*\vartheta))^2/r_M^4)*4*\pi*r_M^2*dr_M \\ &= (1/2)*(q^2/(4*\pi*\vartheta*c))*(dr_M/r_M^2)\end{aligned}$$

Integrand "$(1/2)*(q^2/(4*\pi*\vartheta*c))*(dr_M/r_M^2)$" is integrated to the below rest energy (divided by speed-of-light) of an electron.

$$m_B*c = (1/2)*(q^2/(4*\pi*\vartheta*c))*(1/r_e)$$

Substitute-in measured rest mass "$m_B = 9.11*10\text{^}-31$ kg" and measured electric charge "$q = -1.60*10\text{^}-19$ C" to calculate the classical radius "r_e".

$$r_e = (1/2)*(q^2/(4*\pi*\vartheta*c))*(1/(m_B*c)) = 1.409\ldots *10^{-15} \text{ meters}$$

If an electron had a classical radius less than "r_e", then more electric field would cause more rest mass than what is measured.

- Inside the classical radius is the electric charge. If we count electrons, then we are counting a quantity of electric charge.

- Outside the classical radius is the electric field. Inertia of the energy-momentum invariant is, per this model, due to electromagnetic field induction.

The classical radius macroscopic model of the electron fails when extended to the magnetic field caused by rotation, to suggest the classical radius is not a physically real surface.

Mu and tau particles differ from the electron by having more rest mass and by being unstable. Rather than calculate a smaller classical radius for these particles, the extra mass can be attributed to whatever field causes the instability. The whatever field behaves like mass inertia, as does the electromagnetic field, and so probably creates an induced field when in motion. A similar statement applies to protons, etc.

Energy-Momentum Density for the Biot-Savart Law. $A = q/(4*\pi*ə)$

$$E_{xS} = (A/(r_S^2 + x_S^2*\sinh^2\alpha_{S/M})^{3/2})*x_S*\cosh\alpha_{S/M}$$
$$K_{xS} = 0$$
$$E_{yS} = (A/(r_S^2 + x_S^2*\sinh^2\alpha_{S/M})^{3/2})*y_S*\cosh\alpha_{S/M}$$
$$K_{yS} = -(A/(r_S^2 + x_S^2*\sinh^2\alpha_{S/M})^{3/2})*z_S*\sinh\alpha_{S/M}$$
$$E_{zS} = (A/(r_S^2 + x_S^2*\sinh^2\alpha_{S/M})^{3/2})*z_S*\cosh\alpha_{S/M}$$
$$K_{zS} = (A/(r_S^2 + x_S^2*\sinh^2\alpha_{S/M})^{3/2})*y_S*\sinh\alpha_{S/M}$$

$$\Sigma_{\perp tS} = ə*(E_{yS}*E_{yS} + E_{zS}*E_{zS} + K_{yS}*K_{yS} + K_{zS}*K_{zS})/2$$
$$= ə*A^2*(y_S^2 + z_S^2)*(\cosh^2\alpha_{S/M} + \sinh^2\alpha_{S/M})*(1/(r_S^2 + x_S^2*\sinh^2\alpha_{S/M})^3)/2$$
$$= ə*A^2*(y_S^2 + z_S^2)*\cosh(2*\alpha_{S/M})*(1/(r_S^2 + x_S^2*\sinh^2\alpha_{S/M})^3)/2$$

$$\Sigma_{\perp xS} = ə*(E_{yS}*K_{zS} - E_{zS}*K_{yS})$$
$$= ə*A^2*(y_S^2 + z_S^2)*(\cosh\alpha_{S/M}*\sinh\alpha_{S/M})*(1/(r_S^2 + x_S^2*\sinh^2\alpha_{S/M})^3)$$
$$= ə*A^2*(y_S^2 + z_S^2)*\sinh(2*\alpha_{S/M})*(1/(r_S^2 + x_S^2*\sinh^2\alpha_{S/M})^3)/2$$

$$\Sigma_{\|tS} = ə*(E_{xS}*E_{xS} + K_{xS}*K_{xS})*\cosh(2*\alpha_{S/M})/2$$
$$= ə*A^2*(x_S^2)*\cosh^2\alpha_{S/M}*\cosh(2*\alpha_{S/M})*(1/(r_S^2 + x_S^2*\sinh^2\alpha_{S/M})^3)/2$$

$$\Sigma_{\|xS} = ə*(E_{xS}*E_{xS} + K_{xS}*K_{xS})*\sinh(2*\alpha_{S/M})/2$$
$$= ə*A^2*(x_S^2)*\cosh^2\alpha_{S/M}*\sinh(2*\alpha_{S/M})*(1/(r_S^2 + x_S^2*\sinh^2\alpha_{S/M})^3)/2$$

$$\Sigma_{tS} = \Sigma_{\perp tS} + \Sigma_{\|tS}$$
$$= ə*A^2*(x_S^2*\cosh^2\alpha_{S/M} + y_S^2 + z_S^2)*\cosh(2*\alpha_{S/M})*(1/(r_S^2 + x_S^2*\sinh^2\alpha_{S/M})^3)/2$$

$$\Sigma_{xS} = \Sigma_{\perp xS} + \Sigma_{\|xS}$$
$$= ə*A^2*(x_S^2*\cosh^2\alpha_{S/M} + y_S^2 + z_S^2)*\sinh(2*\alpha_{S/M})*(1/(r_S^2 + x_S^2*\sinh^2\alpha_{S/M})^3)/2$$

$$\Sigma_{tS} + q_x*\Sigma_{xS} =$$
$$ə*A^2*(x_S^2*\cosh^2\alpha_{S/M} + y_S^2 + z_S^2)*\exp(2*q_x*\alpha_{S/M})*(1/(r_S^2 + x_S^2*\sinh^2\alpha_{S/M})^3)/2$$
$$= ə*A^2*(x_S^2*\cosh^2\alpha_{S/M} + y_S^2 + z_S^2)*\exp^2(q_x*\alpha_{S/M})*(1/(r_S^2 + x_S^2*\sinh^2\alpha_{S/M})^3)/2$$
$$= ə*A^2*(x_M^2 + y_M^2 + z_M^2)*\exp^2(q_x*\alpha_{S/M})*(1/(r_S^2 + x_S^2*\sinh^2\alpha_{S/M})^3)/2$$
$$= ə*A^2*r_M^2*\exp^2(q_x*\alpha_{S/M})*(1/(r_S^2 + (c*t_M)^2)^3)/2$$
$$= ə*A^2*r_M^2*\exp^2(q_x*\alpha_{S/M})*(1/(r_M^2)^3)/2$$
$$= ə*((A^2/r_M^4)/2)*\exp^2(q_x*\alpha_{S/M})$$

Hyperbolic-radius "$ə*(A^2/r_M^4)/2$" is the same as energy density of the stationary electron, "$\Sigma_{tM} = ə*(A^2/r_M^4)/2$".

CHAPTER 3 – FIELDS

3.7 Maxwell's Wave Equation

Maxwell's Wave Equation is given in three forms below.

$$_4\nabla^2 *(-_6E) = 0 \quad ; \quad (_4\nabla^{*jsn} \bullet {_4\nabla^{sn}})*(-_6E) = 0$$
$$(\nabla_{tM}^2 - (\nabla_{xM}^2 + \nabla_{yM}^2 + \nabla_{zM}^2))*(-_6E) = 0$$

"$_4\nabla^2$" is called the "harmonic" operator because it applies to functions that are sine waves.

$$(\nabla_{tM}^2 - \nabla_{xM}^2)*\sin(k_x*(x - c*t)) = 0$$

$$\partial(\sin(k_x*(x - c*t)))/\partial x = k_x*\cos(k_x*(x - c*t))$$
$$\partial(k_x*\cos(k_x*(x - c*t)))/\partial x = k_x^2*\sin(k_x*(x - c*t))$$

$$\partial(\sin(k_x*(x - c*t)))/\partial ct = k_x*\cos(k_x*(x - c*t))$$
$$\partial(k_x*\cos(k_x*(x - c*t)))/\partial ct = k_x^2*\sin(k_x*(x - c*t))$$

An electromagnetic field "$_6E$" that satisfies Maxwell's Wave Equation "$_4\nabla^2*(-_6E) = 0$" must be a summation of sine waves. These include cosine waves (because ninety degrees can be added into the argument) and any motion at the speed of light (because its wave form can be a summation using different wavenumbers per a Fourier transform).

Maxwell's Wave Equation is derived by taking the gradient of both sides of Maxwell's Equations "$_4\nabla^{sn}*(-_6E) = {_4J}$".

$$_4\nabla^{*jsn} * {_4\nabla^{sn}}*(-_6E) = {_4\nabla^{*jsn}}*(_4J)$$

Apply identity "$(_4\nabla^{*jsn} \mathbf{x} {_4\nabla^{sn}})*(-_6E) \equiv 0$" to the left side and apply the electric charge conservation equation "$_4\nabla^{*jsn} \bullet (_4J) = 0$" to the right side.

$$_4\nabla^{*jsn} * {_4\nabla^{sn}}*(-_6E) = (_4\nabla^{*jsn} \bullet {_4\nabla^{sn}})*(-_6E) \quad ; \quad {_4\nabla^{*jsn}}*(_4J) = {_4\nabla^{*jsn}} \mathbf{x}(_4J)$$

Set both sides equal to zero. Typically, "$_4J = 0$".

$$(_4\nabla^{*jsn} \bullet {_4\nabla^{sn}})*(-_6E) = 0 \quad ; \quad {_4\nabla^{*jsn}} \mathbf{x}(_4J) = 0$$

SPECIAL ALGEBRA FOR SPECIAL RELATIVITY

<u>Spiral Waves</u>. Each spiral wave is the sum of two perpendicular waves.

First Spiral Wave

$(-_6E_{first}) = E_{amp}*(k_{yM} + p_{zM})*\exp(i*(k_{xM}*(x_M + c*t_M)))$
$-(E_{yM} + i*K_{yM})_{first} = i*E_{amp}*\exp(i*(k_{xM}*(x_M + c*t_M)))$
$-(E_{zM} + i*K_{zM})_{first} = E_{amp}*\exp(i*(k_{xM}*(x_M + c*t_M)))$
$E_{yMfirst} = E_{amp}*\sin(k_{xM}*(x_M + c*t_M))$
$K_{yMfirst} = -E_{amp}*\cos(k_{xM}*(x_M + c*t_M))$
$E_{zMfirst} = -E_{amp}*\cos(k_{xM}*(x_M + c*t_M))$
$K_{zMfirst} = -E_{amp}*\sin(k_{xM}*(x_M + c*t_M))$

Second Spiral Wave

$(-_6E_{second}) = E_{amp}*(k_{yM} - p_{zM})*\exp(i*(k_{xM}*(x_M - c*t_M)))$
$-(E_{yM} + i*K_{yM})_{second} = i*E_{amp}*\exp(i*(k_{xM}*(x_M - c*t_M)))$
$-(E_{zM} + i*K_{zM})_{second} = -E_{amp}*\exp(i*(k_{xM}*(x_M - c*t_M)))$
$E_{yMsecond} = E_{amp}*\sin(k_{xM}*(x_M - c*t_M))$
$K_{yMsecond} = -E_{amp}*\cos(k_{xM}*(x_M - c*t_M))$
$E_{zMsecond} = E_{amp}*\cos(k_{xM}*(x_M - c*t_M))$
$K_{zMsecond} = E_{amp}*\sin(k_{xM}*(x_M - c*t_M))$

Third Spiral Wave

$(-_6E_{third}) = E_{amp}*(k_{yM} + p_{zM})*\exp(-i*(k_{xM}*(x_M + c*t_M)))$
$-(E_{yM} + i*K_{yM})_{third} = i*E_{amp}*\exp(-i*(k_{xM}*(x_M + c*t_M)))$
$-(E_{zM} + i*K_{zM})_{third} = E_{amp}*\exp(-i*(k_{xM}*(x_M + c*t_M)))$
$E_{yMthird} = -E_{amp}*\sin(k_{xM}*(x_M + c*t_M))$
$K_{yMthird} = -E_{amp}*\cos(k_{xM}*(x_M + c*t_M))$
$E_{zMthird} = -E_{amp}*\cos(k_{xM}*(x_M + c*t_M))$
$K_{zMthird} = E_{amp}*\sin(k_{xM}*(x_M + c*t_M))$

Fourth Spiral Wave

$(-_6E_{fourth}) = E_{amp}*(k_{yM} - p_{zM})*\exp(-i*(k_{xM}*(x_M - c*t_M)))$
$-(E_{yM} + i*K_{yM})_{fourth} = i*E_{amp}*\exp(-i*(k_{xM}*(x_M - c*t_M)))$
$-(E_{zM} + i*K_{zM})_{fourth} = -E_{amp}*\exp(-i*(k_{xM}*(x_M - c*t_M)))$
$E_{yMfourth} = -E_{amp}*\sin(k_{xM}*(x_M - c*t_M))$
$K_{yMfourth} = -E_{amp}*\cos(k_{xM}*(x_M - c*t_M))$
$E_{zMfourth} = E_{amp}*\cos(k_{xM}*(x_M - c*t_M))$
$K_{zMfourth} = -E_{amp}*\sin(k_{xM}*(x_M - c*t_M))$

To check that the Fourth Spiral Wave is valid, first check it against Maxwell's Wave Equation by taking derivatives:

CHAPTER 3 – FIELDS

$$\partial^2(-_6E_{fourth})/\partial(ct_M)^2 = -k_{xM}^2*(-_6E_{fourth})$$
$$\partial^2(-_6E_{fourth})/\partial(x_M)^2 = -k_{xM}^2*(-_6E_{fourth})$$

$$\partial^2(-_6E_{fourth})/\partial(ct_M)^2 - \partial^2(-_6E_{fourth})/\partial(x_M)^2 = 0$$

The above check only confirmed the Fourth Spiral Wave equation was constructed of sine waves that move at the speed-of-light. A second check confirms electric and magnetic fields induce each other, to ensure there is not a misplaced negative: "$_1\nabla^{sn}*(-_6E) + _3\nabla^{sn}\mathbf{x}(-_6E) = 0$"

$$_1\nabla^{sn}*(-_6E_{fourth}) = (1_M^{sn})*\partial(-_6E_{fourth})/\partial(ct_M)$$
$$= (1_M)*(k_{yM} - p_{zM})*E_{amp}*\partial(\exp(-i*(k_{xM}*(x_M - c*t_M))))/\partial(ct_M)$$
$$= (i*k_{xM})*E_{amp}*(j_{yM} - q_{zM})*\exp(-i*(k_{xM}*(x_M - c*t_M)))$$

$$_3\nabla^{sn}\mathbf{x}(-_6E_{fourth}) = (q_{xM}^{sn})*\partial(-_6E_{fourth})/\partial(x_M)$$
$$= E_{amp}*(q_{xM}^{sn}*k_{yM} - q_{xM}^{sn}*p_{zM})*\partial(\exp(-i*(k_{xM}*(x_M - c*t_M))))/\partial(x_M)$$
$$= (-i*k_{xM})*E_{amp}*(-q_{zM} + j_{yM})*\exp(-i*(k_{xM}*(x_M - c*t_M)))$$

$$q_{xM}^{sn}*k_{yM} = -q_{xM}*k_{yM} = -q_{zM}$$
$$-q_{xM}^{sn}*p_{zM} = q_{xM}*k_{zM}/i = -q_{zM}*k_{xM}/i = -q_{yM}*(-i) = j_{yM}$$

Alternatively, as a quick check on the math, the Poynting Vector gives the direction of energy travel and must point in the direction of motion. The Poynting Vector is the cross product of the electric field by the magnetic field. For example, for the First Spiral Wave there are "$E_{yMfirst} = E_{amp}*\sin(k_{xM}*(x_M + c*t_M))$" and "$K_{zMfirst} = -E_{amp}*\sin(k_{xM}*(x_M + c*t_M))$" as two components of a plane wave that induce each other. For a slightly positive "x_M" at "$t_M = 0$" there is "$E_{yMfirst} > 0$" and "$K_{zMfirst} < 0$". Right-hand fingers pass through positive "y_M" and then through negative "z_M" and the thumb points in negative "x_M", which is correct for the First Spiral Wave. Therefore, induction is correctly modeled.

<u>Right-Handed and Left-Handed Spiral Waves</u>. Point the thumb in the direction of wave travel ("+x" for "$x_M - c*t_M$" and "-x" for "$x_M + c*t_M$"). The maximum of the helix of the wave spiral is similar to threads on a screw. Curl fingers to ride the maximum as the thumb moves in the direction of wave travel.

The coordinate system is a right-hand coordinate system with "x" the front table edge going right, with "y" the left table edge going away, and with "z" up.

The example is the Fourth Spiral Wave.

SPECIAL ALGEBRA FOR SPECIAL RELATIVITY

$$_3E_{fourth} = E_{amp}*(p_{zM}*\cos(k_{xM}*(x_M - c*t_M)) - p_{yM}*\sin(k_{xM}*(x_M - c*t_M)))$$

Set "$t_M = 0$". Fingers pass positive "E_z", then at "$k_{xM}*x_M = \pi/2$", negative "E_y" using a right hand. The Fourth Spiral Wave is right-handed.

The Third Spiral Wave is left-handed because fingers pass negative "E_z" at "$x_M = 0$", and then positive "E_y" at "$k_{xM}*x_M = -\pi/2$".

$$_3E_{third} = E_{amp}*(-p_{zM}*\cos(k_{xM}*(x_M + c*t_M)) - p_{yM}*\sin(k_{xM}*(x_M + c*t_M)))$$

Second Spiral Wave is left-handed.

$$_3E_{second} = E_{amp}*(p_{zM}*\cos(k_{xM}*(x_M - c*t_M)) + p_{yM}*\sin(k_{xM}*(x_M - c*t_M)))$$

First Spiral Wave is right-handed.

$$_3E_{first} = E_{amp}*(-p_{zM}*\cos(k_{xM}*(x_M + c*t_M)) + p_{yM}*\sin(k_{xM}*(x_M + c*t_M)))$$

A (non-spinning) source disintegrates into two photons with the same handedness, like a nut that separates from a bolt that depart in opposite directions with the same handedness of spin.

Right-Handed Photon Pair (first and fourth):

$$_6E_{+fourth\text{-}first} = -E_{amp}*(k_{yM} - \pm p_{zM})*\exp(-\pm i*(k_{xM}*(x_M - \pm c*t_M)))$$

Left-Handed Photon Pair (third and second):

$$_6E_{+second\text{-}third} = -E_{amp}*(k_{yM} - \pm p_{zM})*\exp(\pm i*(k_{xM}*(x_M - \pm c*t_M)))$$

Quantum Effect on Spin. The four spiral waves each represent an idealized hypothetical photon's wave. They are idealized because direction is known, as is frequency, and spin axis is known to be parallel or else anti-parallel to direction of motion. Actual photon waves, per quantum effects, have spin in the direction of motion only if the measurement of the axis of spin is in the direction of motion, and when spin axis is not being measured, it includes a component perpendicular to direction of motion, and same with an electron.

CHAPTER 3 – FIELDS

Matter and Anti-Matter for Electromagnetic Waves. Because a photon has zero length due to the extreme of length contraction, its back side (taillights) is confounded with its front side (headlights), so there is no distinguishing between matter and anti-matter. Regardless, they are paired with direction of motion: First with third and second with fourth.

Lorentz Transformation of a Spiral Wave. The example below has wave fronts that move in the positive "x" direction because of "$x_M - c*t_M$".

$$(-_6E_{fourth}) = E_{amp}*(k_{yM} - p_{zM})*\exp(-i*(k_{xM}*(x_M - c*t_M)))$$

$$-(E_{yM} + i*K_{yM}) = i*E_{amp}*\exp(-i*(k_{xM}*(x_M - c*t_M)))$$

$$-(E_{zM} + i*K_{zM}) = -E_{amp}*\exp(-i*(k_{xM}*(x_M - c*t_M)))$$

$$\begin{array}{c} -(E_{yS} + i*K_{yS}) \\ -(E_{zS} + i*K_{zS}) \end{array} = \begin{array}{cc} \cosh\alpha_{S/M} & -i*\sinh\alpha_{S/M} \\ i*\sinh\alpha_{S/M} & \cosh\alpha_{S/M} \end{array} * \begin{array}{c} -(E_{yM} + i*K_{yM}) \\ -(E_{zM} + i*K_{zM}) \end{array}$$

$$-(E_{yS} + i*K_{yS}) = E_{amp}*\exp(-i*(k_{xM}*(x_M - c*t_M)))*(i*\cosh\alpha_{S/M} - -i*\sinh\alpha_{S/M})$$
$$= i*E_{amp}*\exp(-i*(k_{xM}*(x_M - c*t_M)))*\exp(\alpha_{S/M})$$
$$= i*E_{amp}*\exp(-i*(k_{xS}*(x_S - c*t_S)))*\exp(\alpha_{S/M})$$

$$-(E_{zS} + i*K_{zS}) = E_{amp}*\exp(-i*(k_{xM}*(x_M - c*t_M)))*(i*i*\sinh\alpha_{S/M} - \cosh\alpha_{S/M})$$
$$= -E_{amp}*\exp(-i*(k_{xM}*(x_M - c*t_M)))*\exp(\alpha_{S/M})$$
$$= -E_{amp}*\exp(-i*(k_{xS}*(x_S - c*t_S)))*\exp(\alpha_{S/M})$$

Observer "S" sees amplitude increased by factor "$\exp(\alpha_{S/M})$" compared to "M", accompanied by an increase in frequency "$k_{xS} > k_{xM}$".
"$k_{xS}*(x_S - c*t_S)$" equals "$k_{xM}*(x_M - c*t_M)$" because it is an invariant.

Geometric-Vector Notation. Induction equations "$_3\nabla x_3E = -\partial(_3B)/\partial t$" and "$_3\nabla x_3B = (\partial(_3E)/\partial t)/c^2$" are satisfied for the fourth wave.

$_3E = E_{xM}*i_{xM} + E_{yM}*i_{yM} + E_{zM}*i_{zM}$; $_3B = B_{xM}*i_{xM} + B_{yM}*i_{yM} + B_{zM}*i_{zM}$
$\quad = -E_{amp}*\sin(k_{xM}*(x_M - c*t_M))*i_{yM} \quad\quad = -(E_{amp}/c)*\cos(k_{xM}*(x_M - c*t_M))*i_{yM}$
$\quad + E_{amp}*\cos(k_{xM}*(x_M - c*t_M))*i_{zM} \quad\quad + -(E_{amp}/c)*\sin(k_{xM}*(x_M - c*t_M))*i_{zM}$

$$_3\nabla \mathbf{x}_3\mathbf{E} = (\partial/\partial x_M * \mathbf{i}_{xM})\mathbf{x}(E_{amp}*\cos(k_{xM}*(x_M - c*t_M))*\mathbf{i}_{zM})$$
$$+ (\partial/\partial x_M * \mathbf{i}_{xM})\mathbf{x}(-E_{amp}*\sin(k_{xM}*(x_M - c*t_M))*\mathbf{i}_{yM})$$

$$= -k_{xM}*E_{amp}*\sin(k_{xM}*(x_M - c*t_M))*(-\mathbf{i}_{yM})$$
$$+ -k_{xM}*E_{amp}*\cos(k_{xM}*(x_M - c*t_M))*(+\mathbf{i}_{zM})$$

$$= k_{xM}*E_{amp}*\sin(k_{xM}*(x_M - c*t_M))*(\mathbf{i}_{yM})$$
$$+ -k_{xM}*E_{amp}*\cos(k_{xM}*(x_M - c*t_M))*(\mathbf{i}_{zM})$$

$$\partial(_3\mathbf{B})/\partial t = (\partial/\partial t_M)(-(E_{amp}/c)*\cos(k_{xM}*(x_M - c*t_M))*\mathbf{i}_{yM})$$
$$+ (\partial/\partial t_M)(-(E_{amp}/c)*\sin(k_{xM}*(x_M - c*t_M))*\mathbf{i}_{zM})$$

$$= -k_{xM}*-E_{amp}*-\sin(k_{xM}*(x_M - c*t_M))*(\mathbf{i}_{yM})$$
$$+ -k_{xM}*-E_{amp}*\cos(k_{xM}*(x_M - c*t_M))*(\mathbf{i}_{zM})$$

$$= -k_{xM}*E_{amp}*\sin(k_{xM}*(x_M - c*t_M))*(\mathbf{i}_{yM})$$
$$+ k_{xM}*E_{amp}*\cos(k_{xM}*(x_M - c*t_M))*(\mathbf{i}_{zM})$$

$$_3\nabla \mathbf{x}_3\mathbf{B} = (\partial/\partial x_M * \mathbf{i}_{xM})\mathbf{x}((-E_{amp}/c)*\sin(k_{xM}*(x_M - c*t_M))*\mathbf{i}_{zM})$$
$$+ (\partial/\partial x_M * \mathbf{i}_{xM})\mathbf{x}((-E_{amp}/c)*\cos(k_{xM}*(x_M - c*t_M))*\mathbf{i}_{yM})$$

$$= k_{xM}*(-E_{amp}/c)*\cos(k_{xM}*(x_M - c*t_M))*(-\mathbf{i}_{yM})$$
$$+ k_{xM}*(-E_{amp}/c)*-\sin(k_{xM}*(x_M - c*t_M))*(+\mathbf{i}_{zM})$$

$$= k_{xM}*(E_{amp}/c)*\cos(k_{xM}*(x_M - c*t_M))*(\mathbf{i}_{yM})$$
$$+ k_{xM}*(E_{amp}/c)*\sin(k_{xM}*(x_M - c*t_M))*(\mathbf{i}_{zM})$$

$$(\partial(_3\mathbf{E})/\partial t)/c^2 = (\partial/\partial t_M)((-E_{amp})*\sin(k_{xM}*(x_M - c*t_M))*\mathbf{i}_{yM})/c^2$$
$$+ (\partial/\partial t_M)((E_{amp})*\cos(k_{xM}*(x_M - c*t_M))*\mathbf{i}_{zM})/c^2$$
$$= -k_{xM}*c*-E_{amp}*\cos(k_{xM}*(x_M - c*t_M))*(\mathbf{i}_{yM})/c^2$$
$$+ -k_{xM}*c*E_{amp}*-\sin(k_{xM}*(x_M - c*t_M))*(\mathbf{i}_{zM})/c^2$$
$$= k_{xM}*(E_{amp}/c)*\cos(k_{xM}*(x_M - c*t_M))*(\mathbf{i}_{yM})$$
$$+ k_{xM}*(E_{amp}/c)*\sin(k_{xM}*(x_M - c*t_M))*(\mathbf{i}_{zM})$$

3.8 Forces Using Geometric-Vector Notation

Evidence the electromagnetic field "$_6E$" and electric charge density "$_4J$" are physically real comes from measurements of force. Force causes a reaction force or else accelerates test electric charge "q".

CHAPTER 3 – FIELDS

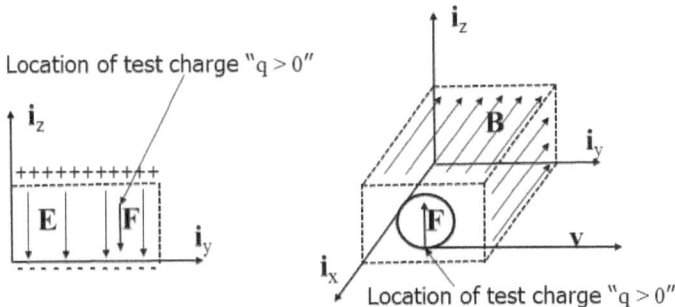

Figure 33a. "E_z" is negative because "$E_z = -\partial V_t/\partial z < 0$". "$q > 0$" imposes a force "$F_z = q*E_z < 0$" on whatever is holding it stationary.

Figure 33b. Magnetic field "$B_x < 0$" surrounds electric charge "$q > 0$" that moves at speed "$v_y > 0$" to create force "$F_z = -q*v_y*B_x > 0$", which accelerates the particle so that it travels in a circle. More mass "m" means less curvature.

Imaginary Force. There is no measurable force due to a stationary electric charge in a magnetic field, or a moving charge in an electric field. But, "$q*_3B$" and "$(q*_3v) \times _3E$" do not simply equal zero.

$_3F_r = q*_3E + (q*_3v) \times _3B$; $_3F_i = q*c*_3B - (q*_3v/c) \times _3E$

$F_{rx} = q*E_x + q*v_y*B_z - q*v_z*B_y$; $F_{ix} = q*c*B_x - ((q*v_y/c)*E_z - (q*v_z/c)*E_y)$
$F_{ry} = q*E_y + q*v_z*B_x - q*v_x*B_z$; $F_{iy} = q*c*B_y - ((q*v_z/c)*E_x - (q*v_x/c)*E_z)$
$F_{rz} = q*E_z + q*v_x*B_y - q*v_y*B_x$; $F_{iz} = q*c*B_z - ((q*v_x/c)*E_y - (q*v_y/c)*E_x)$

$_3F = {_3F_r} \pm i*_3F_i$

Energy Rate. A current of positive charge flows from a high voltage "+" battery terminal to the "-" at a lower value of "x", "$q*v_x < 0$". "$E_x < 0$" because "$E_x = -\partial V_t/\partial x$". Energy transfers out from the electrical system into heat when both "v_x" and "E_x" are the same sign (both are negative) because "$F_{tr} = q*(_3v/c) \bullet _3E > 0$".

SPECIAL ALGEBRA FOR SPECIAL RELATIVITY

An imaginary rate of energy change "F_{ti}" that cannot be measured occurs when current flows parallel to a magnetic field.

$$F_t = F_{tr} \pm i{*}F_{ti} = q{*}(_3v/c)\bullet_3E \pm i{*}q{*}_3v\bullet_3B$$

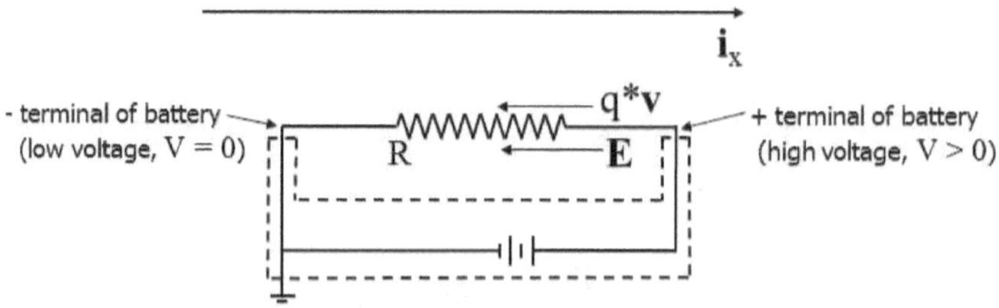

Figure 34. Energy is lost in a resistor so that "$F_{tr} = q{*}(_3v/c)\bullet_3E > 0$".

3.9 Force Density Invariant

"$_3F$" and "F_t" are to be expressed using all-number algebra. First guess:

$$_4g = -(_4J){*}(-_6E) = -(_4J)\blacksquare(-_6E) - (_4J)\blacklozenge(-_6E)$$
$$= -((_3J)\bullet(-_3K) + (_3J)\mathbf{x}(-_3E) + (_1J){*}(-_3K)) - ((_3J)\bullet(-_3E) + (_3J)\mathbf{x}(-_3K) + (_1J){*}(-_3E))$$

Two terms of "$_4g$" are expanded out.

$$1_M{*}g_{tM} = (J_{xM}{*}K_{xM}{*}q_{xM}{*}k_{xM} + J_{yM}{*}K_{yM}{*}q_{yM}{*}k_{yM} + J_{zM}{*}K_{zM}{*}q_{zM}{*}k_{zM})$$
$$+ (J_{xM}{*}E_{xM}{*}q_{xM}{*}p_{xM} + J_{yM}{*}E_{yM}{*}q_{yM}{*}p_{yM} + J_{zM}{*}E_{zM}{*}q_{zM}{*}p_{zM})$$

$$= 1_M{*}((J_{xM}{*}K_{xM} + J_{yM}{*}K_{yM} + J_{zM}{*}K_{zM}){*}i$$
$$+ (J_{xM}{*}E_{xM} + J_{yM}{*}E_{yM} + J_{zM}{*}E_{zM}))$$

$$q_{xM}{*}k_{xM} = j_{xM}{*}k_{xM}{*}(-i) = (-1_M){*}(-i) = 1_M{*}i$$
$$q_{xM}{*}p_{xM} = j_{xM}{*}k_{xM}{*}(-1) = (-1_M){*}(-1) = 1_M$$

$$q_{xM}{*}g_{xM} = ((J_{yM}{*}E_{zM}{*}q_{yM}{*}p_{zM} + J_{zM}{*}E_{yM}{*}q_{zM}{*}p_{yM}) + J_{tM}{*}K_{xM}{*}1_M{*}k_{xM})$$
$$+ ((J_{yM}{*}K_{zM}{*}q_{yM}{*}k_{zM} + J_{zM}{*}K_{yM}{*}q_{zM}{*}k_{yM}) + J_{tM}{*}E_{xM}{*}1_M{*}p_{xM})$$

$$= q_{xM}{*}(-(J_{yM}{*}E_{zM} - J_{zM}{*}E_{yM}) + J_{tM}{*}K_{xM}){*}i$$
$$+ ((J_{yM}{*}K_{zM} - J_{zM}{*}K_{yM}) + J_{tM}{*}E_{xM}))$$

CHAPTER 3 – FIELDS

$q_{yM}*p_{zM} = -j_{yM}*k_{zM} = -j_{xM} = q_{xM}*(-i)$
$q_{yM}*k_{zM} = -j_{yM}*k_{zM}*i = -j_{xM}*i = q_{xM}$

"$_4g$" has the correct arrangement of plus and minus signs.

$g_{tr} = J_x*E_x + J_y*E_y + J_z*E_z$; $F_{rt} = q*(v_x*E_x + v_y*B_y + v_z*E_z)/c$
$g_{ti} = J_x*K_x + J_y*K_y + J_z*K_z$; $F_{it} = q*(v_x*B_x + v_y*E_y + v_z*B_z)$

$g_{xr} = (J_y*K_z - J_z*K_y) + J_t*E_x$; $F_{rx} = (q*v_y*B_z - q*v_z*B_y) + q*E_x$
$g_{xi} = -(J_y*E_z - J_z*E_y) + J_t*K_x$; $F_{ix} = -(q*(v_y/c)*E_z - q*(v_z/c)*E_y) + q*c*B_x$

Applying Maxwell's Equations. If the field from test charge "$_4J$" were included in the externally applied electromagnetic field "$_6E$", then the force would be there, regardless. Substitute "$_4\nabla^{sn}*(-_6E) = (_4J)$" into "$_4g$".

$_4g = -(_4\nabla^{sn}*(-_6E))*(-_6E) = -(_4\nabla^{sn}*_6E)*_6E$
  $= -((_4\nabla^{sn}*_6E)\blacklozenge_6E + (_4\nabla^{sn}*_6E)\blacksquare_6E)$
  $= -((_4\nabla^{sn}\blacklozenge_6E)\blacklozenge_6E + (_4\nabla^{sn}\blacksquare_6E)\blacklozenge_6E + (_4\nabla^{sn}\blacklozenge_6E)\blacksquare_6E + (_4\nabla^{sn}\blacksquare_6E)\blacksquare_6E)$

$= -((_3\nabla^{sn}\bullet_3E)*_3E + (_1\nabla^{sn}*_3E + _3\nabla^{sn}\mathbf{x}_3K)\mathbf{x}_3K + (_1\nabla^{sn}*_3E + _3\nabla^{sn}\mathbf{x}_3K)\bullet_3E$
$+ (_3\nabla^{sn}\bullet_3K)*_3E + (_1\nabla^{sn}*_3K + _3\nabla^{sn}\mathbf{x}_3E)\mathbf{x}_3K + (_1\nabla^{sn}*_3K + _3\nabla^{sn}\mathbf{x}_3E)\bullet_3E$
$+ (_3\nabla^{sn}\bullet_3E)*_3K + (_1\nabla^{sn}*_3E + _3\nabla^{sn}\mathbf{x}_3K)\mathbf{x}_3E + (_1\nabla^{sn}*_3E + _3\nabla^{sn}\mathbf{x}_3K)\bullet_3K$
$+ (_3\nabla^{sn}\bullet_3K)*_3K + (_1\nabla^{sn}*_3K + _3\nabla^{sn}\mathbf{x}_3E)\mathbf{x}_3E + (_1\nabla^{sn}*_3K + _3\nabla^{sn}\mathbf{x}_3E)\bullet_3K)$

First and fourth rows have mathematically real components. Second and Third have imaginary components.

The first row sums to the real portion of "$_4g$". The third row sums to the imaginary portion of "$_4g$". The second and fourth rows each sum to zero but individual terms in the second and fourth rows are not zero, for example, "$(_1\nabla^{sn}*_3K)\bullet_3K \neq 0$".

Time component terms of the first row are given below. Gradient operator "∇" only applies to the component factor immediately behind it, and not to both field component factors, to simplify what is written.

SPECIAL ALGEBRA FOR SPECIAL RELATIVITY

Portion of $(1_M*g_{trM}) = (_1\nabla^{sn}*_3E + _3\nabla^{sn}\mathbf{x}_3K) \bullet_3 E$

$= -(\nabla_{tM}*E_{xM}*E_{xM}*1_M{}^{sn}*p_{xM}*p_{xM}$
$+ \nabla_{tM}*E_{yM}*E_{yM}*1_M{}^{sn}*p_{yM}*p_{yM} + \nabla_{tM}*E_{zM}*E_{zM}*1_M{}^{sn}*p_{zM}*p_{zM})$
$+ -(\nabla_{yM}*K_{zM}*E_{xM}*q_{yM}{}^{sn}*k_{zM}*p_{xM} + \nabla_{zM}*K_{yM}*E_{xM}*q_{zM}{}^{sn}*k_{yM}*p_{xM}$
$+ \nabla_{zM}*K_{xM}*E_{yM}*q_{zM}{}^{sn}*k_{xM}*p_{yM} + \nabla_{xM}*K_{zM}*E_{yM}*q_{xM}{}^{sn}*k_{zM}*p_{yM}$
$+ \nabla_{xM}*K_{yM}*E_{zM}*q_{xM}{}^{sn}*k_{yM}*p_{zM} + \nabla_{yM}*K_{xM}*E_{zM}*q_{yM}{}^{sn}*k_{xM}*p_{zM})$

$= -1_M*((\nabla_{tM}*E_{xM}*E_{xM} + \nabla_{tM}*E_{yM}*E_{yM} + \nabla_{tM}*E_{zM}*E_{zM})$
$+ (-\nabla_{yM}*K_{zM}*E_{xM} + \nabla_{zM}*K_{yM}*E_{xM})$
$+ (-\nabla_{zM}*K_{xM}*E_{yM} + \nabla_{xM}*K_{zM}*E_{yM})$
$+ (-\nabla_{xM}*K_{yM}*E_{zM} + \nabla_{yM}*K_{xM}*E_{zM}))$

$q_{yM}{}^{sn}*k_{zM}*p_{xM} = j_{yM}*k_{zM}*k_{xM} = j_{xM}*k_{xM} = -1_M$

A Negative is Needed. "g_{trM}", like "F_{rt}", models the loss of energy from the system, per the example of the battery and resistor, and this is properly modelled by the "$-(_1\nabla^{sn}*_3E)\bullet_3E$" term in the first row with the example "$-(\nabla_{tM}*E_{xM})*E_{xM} = -\nabla_{tM}*(E_{xM}*E_{xM})/2$". A positive value is a loss of energy, and the term is positive because energy "$E_{xM}*E_{xM}$" is positive. The gradient is negative because it is a loss of energy, and the "-1" makes "g_{tM}" positive. The math is as expected. Good.

"$-(_1\nabla^{sn}*_3K)\bullet_3K$" of the fourth row has "$-(\nabla_{tM}*K_{xM})*K_{xM}*i^2 = \nabla_{tM}*(K_{xM}*K_{xM})/2$". "$i^2$" was introduced by label-numbers because "$_3K$" is mathematically imaginary if "$_3E$" is mathematically real. The extra negative is bad because "$_3K\bullet_3K = -K^2$" is subtracted from "$_3E\bullet_3E = +E^2$" in "$_3E\bullet_3E + _3K\bullet_3K = E^2 - K^2$", in contradiction to experiments for which total energy density is "$E^2 + K^2$". To fix this, a negative is needed for "$_3K$".

To help specify how to insert the negative, second row term "$_3E\bullet_3K$" plus third row term "$_3K\bullet_3E$" should subtract to cancel. And first row "$_3E\mathbf{x}_3K$" needs to subtract from its negative "$_3K\mathbf{x}_3E$" of the fourth row for conformance to the experimentally derived Poynting Vector.

The conclusion is that the fourth row needs a negative relative to the first row, and that negative is acceptable because the fourth row sums to zero. The second row also sums to zero and it needs a negative relative to the third row. Also, because the imaginary force of the third row is not measurable, the second and third rows can be either plus or minus relative to the first row.

Place "±1" with the first "■" and "-±1" with the second "■".

CHAPTER 3 – FIELDS

$_4f = -((_4\nabla^{sn}\blacklozenge_6E)\blacklozenge_6E \pm (_4\nabla^{sn}\blacksquare_6E)\blacklozenge_6E \pm -(_4\nabla^{sn}\blacklozenge_6E)\blacksquare_6E - (_4\nabla^{sn}\blacksquare_6E)\blacksquare_6E)$

The components of "$_4f$" separate into models of physics with two examples being electromagnetic field energy density and the Poynting vector. Note that "$_4f_r = {_4g_r}$". Also, "$_4f = {_4f_r} \pm i*{_4f_i}$".

Mathematically Ugly. Because of the inserted negatives, "$_4f$" cannot be expressed using "*" operator gradients, and that means "$_4f$" is ugly.

By analogy, the dot product in the exponential function, "$_4k^{*j} \bullet _4r$", is ugly because cross-product "$_4k^{*j} \mathbf{x} _4r$" is not added to it. This ugliness is explained by equating "$_4k^{*j} \bullet _4r$" to the product of hyperbolic radii.

The ugliness of "$_4f$", too, goes away. "$_4f$" becomes beautiful using the proposed algebra in this book's last chapter.

Terms of the Force Density Invariant.

$-_4f = ((_3\nabla^{sn}\bullet_3E)*_3E + (_1\nabla^{sn}*_3E + _3\nabla^{sn}\mathbf{x}_3K)\mathbf{x}_3K + (_1\nabla^{sn}*_3E + _3\nabla^{sn}\mathbf{x}_3K)\bullet_3E)$
$\pm ((_3\nabla^{sn}\bullet_3K)*_3E + (_1\nabla^{sn}*_3K + _3\nabla^{sn}\mathbf{x}_3E)\mathbf{x}_3K + (_1\nabla^{sn}*_3K + _3\nabla^{sn}\mathbf{x}_3E)\bullet_3E)$
$\pm -((_3\nabla^{sn}\bullet_3E)*_3K + (_1\nabla^{sn}*_3E + _3\nabla^{sn}\mathbf{x}_3K)\mathbf{x}_3E + (_1\nabla^{sn}*_3E + _3\nabla^{sn}\mathbf{x}_3K)\bullet_3K)$
$- ((_3\nabla^{sn}\bullet_3K)*_3K + (_1\nabla^{sn}*_3K + _3\nabla^{sn}\mathbf{x}_3E)\mathbf{x}_3E + (_1\nabla^{sn}*_3K + _3\nabla^{sn}\mathbf{x}_3E)\bullet_3K)$

$= ((-_1J)*_3E + (-_3J)\mathbf{x}_3K + (-_3J)\bullet_3E)$
$\pm ((_10)*_3E + (_30)\mathbf{x}_3K + (_30)\bullet_3E)$
$\pm -((-_1J)*_3K + (-_3J)\mathbf{x}_3E + (-_3J)\bullet_3K)$
$- ((_10)*_3K + (_30)\mathbf{x}_3E + (_30)\bullet_3K)$

$-_1f = (-_3J)\bullet_3E \pm (_30)\bullet_3E \pm -(-_3J)\bullet_3K - (_30)\bullet_3K$
$= ((_1\nabla^{sn}*_3E + _3\nabla^{sn}\mathbf{x}_3K)\bullet_3E) \pm ((_1\nabla^{sn}*_3K + _3\nabla^{sn}\mathbf{x}_3E)\bullet_3E)$
$\pm -((_1\nabla^{sn}*_3E + _3\nabla^{sn}\mathbf{x}_3K)\bullet_3K) - ((_1\nabla^{sn}*_3K + _3\nabla^{sn}\mathbf{x}_3E)\bullet_3K)$

$-f_{tr} = -(J_x*E_x + J_y*E_y + J_z*E_z)$
$= ((\partial E_x/\partial ct)*E_x + (\partial E_y/\partial ct)*E_y + (\partial E_z/\partial ct)*E_z)$
$- (-(\partial K_x/\partial ct)*K_x - (\partial K_y/\partial ct)*K_y - (\partial K_z/\partial ct)*K_z)$
$+ ((-\partial K_z/\partial y + \partial K_y/\partial z)*E_x + (-\partial K_x/\partial z + \partial K_z/\partial x)*E_y + (-\partial K_y/\partial x + \partial K_x/\partial y)*E_z)$
$- ((\partial E_y/\partial z - \partial E_z/\partial y)*K_x + (\partial E_z/\partial x - \partial E_x/\partial z)*K_y + (\partial E_x/\partial y - \partial E_y/\partial x)*K_z)$

SPECIAL ALGEBRA FOR SPECIAL RELATIVITY

$\nabla_y * K_z * E_x * (-q_{yM}) * k_{zM} * p_{xM} = \nabla_y * K_z * E_x * (-q_{yM}) * p_{yM} = -\nabla_y * K_z * E_x * 1_M$
$-\nabla_z * E_y * K_x * (-q_{zM}) * p_{yM} * k_{xM} = -\nabla_z * E_y * K_x * (-q_{zM}) * (-p_{zM}) = -\nabla_z * E_y * K_x * 1_M$
$\nabla_t * E_x * E_x * (1_M) * p_{xM} * p_{xM} = \nabla_t * E_x * E_x * (1_M) * (1) = \nabla_t * E_x * E_x * 1_M$
$-\nabla_t * K_x * K_x * (1_M) * k_{xM} * k_{xM} = -\nabla_t * K_x * K_x * (1_M) * (-1) = \nabla_t * K_x * K_x * 1_M$
$-J_x * E_x * q_{xM} * p_{xM} = -J_x * E_x * 1_M$

$-f_{ti} = \pm (J_x * K_x + J_y * K_y + J_z * K_z)$
$= \pm ((\partial K_x/\partial ct) * E_x + (\partial K_y/\partial ct) * E_y + (\partial K_z/\partial ct) * E_z)$
$\pm (-(\partial E_x/\partial ct) * K_x - (\partial E_y/\partial ct) * K_y - (\partial E_z/\partial ct) * K_z)$
$\pm ((\partial K_z/\partial y - \partial K_y/\partial z) * K_x + (\partial K_x/\partial z - \partial K_z/\partial x) * K_y + (\partial K_y/\partial x - \partial K_x/\partial y) * K_z)$
$\pm -((\partial E_y/\partial z - \partial E_z/\partial y) * E_x + (\partial E_z/\partial x - \partial E_x/\partial z) * E_y + (\partial E_x/\partial y - \partial E_y/\partial x) * E_z)$

$\pm -\nabla_y * K_z * K_x * (-q_{yM}) * k_{zM} * k_{xM} = \pm \nabla_y * K_z * K_x * (q_{yM}) * (k_{yM}) = \pm i * \nabla_y * K_z * K_x * 1_M$
$\pm \nabla_z * E_y * E_x * (-q_{zM}) * p_{yM} * p_{xM} = \pm -\nabla_z * E_y * E_x * (q_{zM}) * (k_{zM}) = \pm -i * \nabla_z * E_y * E_x * 1_M$
$\pm \nabla_t * K_x * E_x * (1_M) * k_{xM} * p_{xM} = \pm \nabla_t * K_x * E_x * (1_M) * (i) = \pm i * \nabla_t * K_x * E_x * 1_M$
$\pm -\nabla_t * E_x * K_x * (1_M) * p_{xM} * k_{xM} = \pm -\nabla_t * K_x * K_x * (1_M) * (i) = \pm -i * \nabla_t * E_x * K_x * 1_M$
$\pm J_x * K_x * q_{xM} * k_{xM} = \pm i * J_x * K_x * 1_M$

$-_3f = ((-_1J) *_3E + (-_3J) \mathbf{x}_3K) \pm ((0) *_3E + (_30) \mathbf{x}_3K)$
$\pm -((-_1J) *_3K + (-_3J) \mathbf{x}_3E) - ((0) *_3K + (_30) \mathbf{x}_3E)$

$= ((_3\nabla^{sn} \bullet_3E) *_3E + (_1\nabla^{sn} *_3E +_3\nabla^{sn} \mathbf{x}_3K) \mathbf{x}_3K)$
$\pm ((_3\nabla^{sn} \bullet_3K) *_3E + (_1\nabla^{sn} *_3K +_3\nabla^{sn} \mathbf{x}_3E) \mathbf{x}_3K)$
$\pm -((_3\nabla^{sn} \bullet_3E) *_3K + (_1\nabla^{sn} *_3E +_3\nabla^{sn} \mathbf{x}_3K) \mathbf{x}_3E)$
$- ((_3\nabla^{sn} \bullet_3K) *_3K + (_1\nabla^{sn} *_3K +_3\nabla^{sn} \mathbf{x}_3E) \mathbf{x}_3E)$

$-f_{zr} = -J_t * E_z + (J_y * K_x - J_x * K_y)$
$= ((\partial K_y/\partial ct) * E_x - (\partial K_x/\partial ct) * E_y) + ((\partial E_x/\partial ct) * K_y - (\partial E_y/\partial ct) * K_x)$
$+ ((-\partial K_z/\partial y + \partial K_y/\partial z) * K_y) - ((\partial K_z/\partial x - \partial K_x/\partial z) * K_x)$
$+ ((-\partial E_z/\partial y + \partial E_y/\partial z) * E_y) - ((\partial E_z/\partial x - \partial E_x/\partial z) * E_x)$
$+ (-(\partial E_x/\partial x + \partial E_y/\partial y + \partial E_z/\partial z) * E_z) - ((\partial K_x/\partial x + \partial K_y/\partial y + \partial K_z/\partial z) * K_z)$

$-\nabla_t * K_y * E_x * (1_M) * k_{yM} * p_{xM} = -\nabla_t * K_y * E_x * (1_M) * (-p_{zM}) = \nabla_t * K_y * E_x * q_{zM}$
$\nabla_t * E_x * K_y * (1_M) * p_{xM} * k_{yM} = \nabla_t * E_x * K_y * (1_M) * (p_{zM}) = \nabla_t * E_x * K_y * q_{zM}$
$-\nabla_y * E_z * E_y * (-q_{yM}) * p_{zM} * p_{yM} = -\nabla_y * E_z * E_y * (-q_{yM}) * k_{xM} = -\nabla_y * E_z * E_y * q_{zM}$
$\nabla_y * K_z * K_y * (-q_{yM}) * k_{zM} * k_{yM} = \nabla_y * K_z * K_y * (-q_{yM}) * (-k_{xM}) = -\nabla_y * K_z * K_y * q_{zM}$
$\nabla_x * E_x * E_z * (-q_{xM}) * p_{xM} * p_{zM} = \nabla_x * E_x * E_z * (-q_{xM}) * k_{yM} = -\nabla_x * E_x * E_z * q_{zM}$
$-\nabla_x * K_x * K_z * (-q_{xM}) * k_{xM} * k_{zM} = -\nabla_x * K_x * K_z * (-q_{xM}) * (-k_{yM}) = -\nabla_x * K_x * K_z * q_{zM}$

CHAPTER 3 – FIELDS

$-J_x*K_y*q_{xM}*k_{yM} = -J_x*K_y*q_{zM}$
$-J_t*E_z*1_M*p_{zM} = -J_t*E_z*q_{zM}$

$-f_{zi} = \pm J_t*K_z \pm (J_y*E_x - J_x*E_y)$
$= \pm(-(\partial K_y/\partial ct)*K_x + (\partial K_x/\partial ct)*K_y) \pm ((\partial E_x/\partial ct)*E_y - (\partial E_y/\partial ct)*E_x)$
$\pm (((-\partial K_z/\partial y + \partial K_y/\partial z)*E_y) - ((\partial K_z/\partial x - \partial K_x/\partial z)*E_x))$
$\pm (((\partial E_z/\partial y - \partial E_y/\partial z)*K_y) + ((\partial E_z/\partial x - \partial E_x/\partial z)*K_x))$
$\pm ((\partial E_x/\partial x + \partial E_y/\partial y + \partial E_z/\partial z)*K_z) \pm (-(\partial K_x/\partial x + \partial K_y/\partial y + \partial K_z/\partial z)*E_z)$

$\pm \nabla_t*K_y*K_x*(1_M)*k_{yM}*k_{xM} = \pm \nabla_t*K_y*K_x*(1_M)*(-k_{zM}) = \pm -\nabla_t*K_y*K_x*j_{zM}$
$\pm -\nabla_t*E_x*E_y*(1_M)*p_{xM}*p_{yM} = \pm -\nabla_t*E_x*E_y*(1_M)*(-k_{zM}) = \pm \nabla_t*E_x*E_y*j_{zM}$
$\pm \nabla_y*E_z*K_y*(-q_{yM})*p_{zM}*k_{yM} = \pm \nabla_y*E_z*K_y*q_{yM}*p_{xM} = \pm \nabla_y*E_z*K_y*j_{zM}$
$\pm -\nabla_y*K_z*E_y*(-q_{yM})*k_{zM}*p_{yM} = \pm -\nabla_y*K_z*E_y*q_{yM}*p_{xM} = \pm -\nabla_y*K_z*E_y*j_{zM}$
$\pm -\nabla_x*E_x*K_z*(-q_{xM})*p_{xM}*k_{zM} = \pm -\nabla_x*E_x*K_z*(-1_M)*k_{zM} = \nabla_x*E_x*K_z*j_{zM}$
$\pm \nabla_x*K_x*E_z*(-q_{xM})*k_{xM}*p_{zM} = \pm \nabla_x*K_x*E_z*(-i*1_M)*(p_{zM}) = \pm -\nabla_x*K_x*E_z*j_{zM}$
$\pm J_x*E_y*q_{xM}*p_{yM} = \pm -J_x*E_y*j_{zM}$
$\pm J_t*K_z*1_M*k_{zM} = \pm J_t*K_z*j_{zM}$

As a check on compound-label-number products, component force equations have as factors Maxwell's Equations.
Component Maxwell's Equations:

$-\partial K_x/\partial ct - \partial E_z/\partial y + \partial E_y/\partial z = 0$; $-\partial E_x/\partial ct + \partial K_z/\partial y - \partial K_y/\partial z = J_x$
$-\partial K_y/\partial ct - \partial E_x/\partial z + \partial E_z/\partial x = 0$; $-\partial E_y/\partial ct + \partial K_x/\partial z - \partial K_z/\partial x = J_y$
$-\partial K_z/\partial ct - \partial E_y/\partial x + \partial E_x/\partial y = 0$; $-\partial E_z/\partial ct + \partial K_y/\partial x - \partial K_x/\partial y = J_z$
$\partial K_x/\partial x + \partial K_y/\partial y + \partial K_z/\partial z = 0$; $\partial E_x/\partial x + \partial E_y/\partial y + \partial E_z/\partial z = J_t$

In the component equation for "-f_{zt}" are six of Maxwell's Equations.

- "$-J_t*E_z$" equates to "$-(\partial E_x/\partial x + \partial E_y/\partial y + \partial E_z/\partial z)*E_z$"
- "J_y*K_x" equates to "$-(\partial E_y/\partial ct)*K_x - (\partial K_z/\partial x - \partial K_x/\partial z)*K_x$"
- "$-J_x*K_y$" equates to "$(\partial E_x/\partial ct)*K_y + (-\partial K_z/\partial y + \partial K_y/\partial z)*K_y$"
- "0" equates to "$-(\partial K_x/\partial x + \partial K_y/\partial y + \partial K_z/\partial z)*K_z$"
- "0" equates to "$(\partial K_y/\partial ct)*E_x - (\partial E_z/\partial x - \partial E_x/\partial z)*E_x$"
- "0" equates to "$-(\partial K_x/\partial ct)*E_y + (-\partial E_z/\partial y + \partial E_y/\partial z)*E_y$"

SPECIAL ALGEBRA FOR SPECIAL RELATIVITY

In the component equation for "-f_{zi}" are these other six Equations that also conform to Maxwell's Equations.

- "$\pm J_t*K_z$" equates to "$\pm(\partial E_x/\partial x + \partial E_y/\partial y + \partial E_z/\partial z)*K_z$"
- "$\pm J_y*E_x$" equates to "$\pm$-$(\partial E_y/\partial ct)*E_x \pm$ -$(\partial K_z/\partial x - \partial K_x/\partial z)*E_x$"
- "$\pm$ -J_x*E_y" equates to "$\pm(\partial E_x/\partial ct)*E_y \pm (-\partial K_z/\partial y + \partial K_y/\partial z)*E_y$"
- "0" equates to "$\pm$-$(\partial K_x/\partial x + \partial K_y/\partial y + \partial K_z/\partial z)*E_z$"
- "0" equates to "$\pm$-$(\partial K_y/\partial ct)*K_x \pm (\partial E_z/\partial x - \partial E_x/\partial z)*K_x$"
- "0" equates to "$\pm(\partial K_x/\partial ct)*K_y \pm (\partial E_z/\partial y - \partial E_y/\partial z)*K_y$"

<u>Example Use of First Case Force Density</u>. Gradient components of the negative of the Third Spiral Wave Solution are given below.

Third Spiral Wave
$(-_6E_{third}) = E_{amp}*(k_{yM} + p_{zM})*\exp(-i*(k_{xM}*(x_M + c*t_M)))$
$-(E_{yM} + i*K_{yM})_{third} = i*E_{amp}*\exp(-i*(k_{xM}*(x_M + c*t_M)))$
$-(E_{zM} + i*K_{zM})_{third} = E_{amp}*\exp(-i*(k_{xM}*(x_M + c*t_M)))$
$E_{yMthird} = -E_{amp}*\sin(k_{xM}*(x_M + c*t_M))$
$K_{yMthird} = -E_{amp}*\cos(k_{xM}*(x_M + c*t_M))$
$E_{zMthird} = -E_{amp}*\cos(k_{xM}*(x_M + c*t_M))$
$K_{zMthird} = E_{amp}*\sin(k_{xM}*(x_M + c*t_M))$

$_3E = E_{amplitude}*(p_{yM}*\sin(k_{xM}*(x_M + c*t_M)) + p_{zM}*\cos(k_{xM}*(x_M + c*t_M)))$
$_3K = E_{amplitude}*(k_{yM}*\cos(k_{xM}*(x_M + c*t_M)) - k_{zM}*\sin(k_{xM}*(x_M + c*t_M)))$
$_6E = {_3E} + {_3K}$

$_1\nabla^{sn}*{_3E}$
$= 1_M*\partial/\partial ct*{_3E}$
$= E_{amplitude}*k_{xM}*(q_{yM}*\cos(k_{xM}*(x_M + c*t_M)) - q_{zM}*\sin(k_{xM}*(x_M + c*t_M)))$
Using: $1_M*p_{yM} = q_{yM}$ $1_M*p_{zM} = q_{zM}$

$_3\nabla^{sn}\mathbf{x}{_3E}$
$= (-q_{xM}*\partial/\partial x - q_{yM}*\partial/\partial y - q_{zM}*\partial/\partial z)\mathbf{x}{_3E}$
$= E_{amplitude}*k_{xM}*(j_{zM}*\cos(k_{xM}*(x_M + c*t_M)) + j_{yM}*\sin(k_{xM}*(x_M + c*t_M)))$
Using: $-q_{xM}*p_{yM} = j_{xM}*k_{yM} = j_{zM}$ $-q_{xM}*p_{zM} = j_{xM}*k_{zM} = -j_{yM}$

$_3\nabla^{sn}\bullet{_3E}$
$= (-q_{xM}*\partial/\partial x - q_{yM}*\partial/\partial y - q_{zM}*\partial/\partial z)\bullet{_3E} = 0$

CHAPTER 3 – FIELDS

$_1\nabla^{sn}{}^*{}_3K$
$= 1_M{}^*\partial/\partial ct{}^*{}_3K$
$= E_{amplitude}{}^*k_{xM}{}^*(-j_{yM}{}^*\sin(k_{xM}{}^*(x_M + c{}^*t_M)) - j_{zM}{}^*\cos(k_{xM}{}^*(x_M + c{}^*t_M)))$
Using: $\qquad 1_M{}^*k_{yM} = j_{yM} \qquad 1_M{}^*{-}k_{zM} = -j_{zM}$

$_3\nabla^{sn}\mathbf{x}_3K$
$= (-q_{xM}{}^*\partial/\partial x - q_{yM}{}^*\partial/\partial y - q_{zM}{}^*\partial/\partial z)\mathbf{x}_3K$
$= E_{amplitude}{}^*k_{xM}{}^*(q_{zM}{}^*\sin(k_{xM}{}^*(x_M + c{}^*t_M)) - q_{yM}{}^*\cos(k_{xM}{}^*(x_M + c{}^*t_M)))$
Using: $-q_{xM}{}^*k_{yM} = -j_{xM}{}^*k_{yM}/i = -q_{zM} \qquad -q_{xM}{}^*{-}k_{zM} = j_{xM}{}^*k_{zM}/i = -q_{yM}$

$_3\nabla^{sn}\bullet_3K$
$= (-q_{xM}{}^*\partial/\partial x - q_{yM}{}^*\partial/\partial y - q_{zM}{}^*\partial/\partial z)\bullet_3K = 0$

$(_1\nabla^{sn}{}^*{}_3E)\bullet_3E$
$= (E_{amplitude}{}^*k_{xM}{}^*(q_{yM}{}^*\cos(k_{xM}{}^*(x_M + c{}^*t_M)) - q_{zM}{}^*\sin(k_{xM}{}^*(x_M + c{}^*t_M))))$
$\bullet (E_{amplitude}{}^*(p_{yM}{}^*\sin(k_{xM}{}^*(x_M + c{}^*t_M)) + p_{zM}{}^*\cos(k_{xM}{}^*(x_M + c{}^*t_M))))$
$= E_{amplitude}{}^2{}^*k_{xM}{}^*(1_M - 1_M){}^*\sin(k_{xM}{}^*(x_M + c{}^*t_M)){}^*\cos(k_{xM}{}^*(x_M + c{}^*t_M)) = 0$
Using: $\qquad q_{yM}{}^*p_{yM} = 1_M \qquad -q_{zM}{}^*p_{zM} = -1_M$

$(_1\nabla^{sn}{}^*{}_3E)\mathbf{x}_3E$
$= (E_{amplitude}{}^*k_{xM}{}^*(q_{yM}{}^*\cos(k_{xM}{}^*(x_M + c{}^*t_M)) - q_{zM}{}^*\sin(k_{xM}{}^*(x_M + c{}^*t_M))))$
$\mathbf{x}(E_{amplitude}{}^*(p_{yM}{}^*\sin(k_{xM}{}^*(x_M + c{}^*t_M)) + p_{zM}{}^*\cos(k_{xM}{}^*(x_M + c{}^*t_M))))$
$= E_{amplitude}{}^2{}^*k_{xM}{}^*(-j_{xM}){}^*(\cos^2(k_{xM}{}^*(x_M + c{}^*t_M)) + \sin^2(k_{xM}{}^*(x_M + c{}^*t_M)))$
$= E_{amplitude}{}^2{}^*k_{xM}{}^*(-j_{xM})$
Using: $\qquad q_{yM}{}^*p_{zM} = -j_{yM}{}^*k_{zM} = -j_{xM} \qquad -q_{zM}{}^*p_{yM} = -j_{yM}{}^*k_{zM} = -j_{xM}$

$(_3\nabla^{sn}\mathbf{x}_3E)\bullet_3E$
$= (E_{amplitude}{}^*k_{xM}{}^*(j_{zM}{}^*\cos(k_{xM}{}^*(x_M + c{}^*t_M)) + j_{yM}{}^*\sin(k_{xM}{}^*(x_M + c{}^*t_M))))$
$\bullet(E_{amplitude}{}^*(p_{yM}{}^*\sin(k_{xM}{}^*(x_M + c{}^*t_M)) + p_{zM}{}^*\cos(k_{xM}{}^*(x_M + c{}^*t_M))))$
$= E_{amplitude}{}^2{}^*k_{xM}{}^*i{}^*1_M{}^*(\cos^2(k_{xM}{}^*(x_M + c{}^*t_M)) + \sin^2(k_{xM}{}^*(x_M + c{}^*t_M)))$
$= E_{amplitude}{}^2{}^*k_{xM}{}^*i{}^*1_M$
Using: $\qquad j_{zM}{}^*p_{zM} = i{}^*1_M \qquad j_{yM}{}^*p_{yM} = i{}^*1_M$

$(_3\nabla^{sn}\mathbf{x}_3E)\mathbf{x}_3E$
$= (E_{amplitude}{}^*k_{xM}{}^*(j_{zM}{}^*\cos(k_{xM}{}^*(x_M + c{}^*t_M)) + j_{yM}{}^*\sin(k_{xM}{}^*(x_M + c{}^*t_M))))$
$\mathbf{x}(E_{amplitude}{}^*(p_{yM}{}^*\sin(k_{xM}{}^*(x_M + c{}^*t_M)) + p_{zM}{}^*\cos(k_{xM}{}^*(x_M + c{}^*t_M))))$
$= E_{amplitude}{}^2{}^*k_{xM}{}^*(q_{xM} - q_{xM}){}^*\sin(k_{xM}{}^*(x_M + c{}^*t_M)){}^*\cos(k_{xM}{}^*(x_M + c{}^*t_M)) = 0$
Using: $\qquad j_{zM}{}^*p_{yM} = -j_{yM}{}^*k_{zM}/i = -q_{xM} \qquad j_{yM}{}^*p_{zM} = j_{yM}{}^*k_{zM}/i = q_{xM}$

SPECIAL ALGEBRA FOR SPECIAL RELATIVITY

$(_1\nabla^{sn}*_3E) \bullet _3K$
$= (E_{amplitude}*k_{xM}*(q_{yM}*\cos(k_{xM}*(x_M + c*t_M)) - q_{zM}*\sin(k_{xM}*(x_M + c*t_M))))$
$\bullet(E_{amplitude}*(k_{yM}*\cos(k_{xM}*(x_M + c*t_M)) - k_{zM}*\sin(k_{xM}*(x_M + c*t_M))))$
$= E_{amplitude}^2*k_{xM}*(i*1_M)*(\cos^2(k_{xM}*(x_M + c*t_M)) + \sin^2(k_{xM}*(x_M + c*t_M)))$
$= E_{amplitude}^2*k_{xM}*(i*1_M)$
Using: $\quad q_{yM}*k_{yM} = i*j_{yM}*-k_{yM} = i*1_M \quad\quad -q_{zM}*-k_{zM} = i*1_M$

$(_1\nabla^{sn}*_3E) \mathbf{x} _3K$
$= (E_{amplitude}*k_{xM}*(q_{yM}*\cos(k_{xM}*(x_M + c*t_M)) - q_{zM}*\sin(k_{xM}*(x_M + c*t_M))))$
$\mathbf{x}(E_{amplitude}*(k_{yM}*\cos(k_{xM}*(x_M + c*t_M)) - k_{zM}*\sin(k_{xM}*(x_M + c*t_M))))$
$= E_{amplitude}^2*k_{xM}*(p_{xM} - p_{xM})*\sin(k_{xM}*(x_M + c*t_M))*\cos(k_{xM}*(x_M + c*t_M)) = 0$
Using: $\quad q_{yM}*-k_{zM} = -p_{xM} \quad\quad -q_{zM}*k_{yM} = p_{xM}$

$(_3\nabla^{sn}\mathbf{x}_3E) \bullet _3K$
$= (E_{amplitude}*k_{xM}*(j_{zM}*\cos(k_{xM}*(x_M + c*t_M)) + j_{yM}*\sin(k_{xM}*(x_M + c*t_M))))$
$\bullet(E_{amplitude}*(k_{yM}*\cos(k_{xM}*(x_M + c*t_M)) - k_{zM}*\sin(k_{xM}*(x_M + c*t_M))))$
$= E_{amplitude}^2*k_{xM}*(1_M - 1_M)*\sin(k_{xM}*(x_M + c*t_M))*\cos(k_{xM}*(x_M + c*t_M)) = 0$
Using: $\quad j_{zM}*-k_{zM} = 1_M \quad\quad j_{yM}*k_{yM} = -1_M$

$(_3\nabla^{sn}\mathbf{x}_3E)\mathbf{x}_3K$
$= (E_{amplitude}*k_{xM}*(j_{zM}*\cos(k_{xM}*(x_M + c*t_M)) + j_{yM}*\sin(k_{xM}*(x_M + c*t_M))))$
$\mathbf{x}(E_{amplitude}*(k_{yM}*\cos(k_{xM}*(x_M + c*t_M)) - k_{zM}*\sin(k_{xM}*(x_M + c*t_M))))$
$= E_{amplitude}^2*k_{xM}*-j_{xM}*(\cos^2(k_{xM}*(x_M + c*t_M)) + \sin^2(k_{xM}*(x_M + c*t_M)))$
$= E_{amplitude}^2*k_{xM}*-j_{xM}$
Using: $\quad j_{zM}*k_{yM} = -j_{yM}*k_{zM} = -j_{xM} \quad\quad j_{yM}*-k_{zM} = -j_{xM}$

$(_1\nabla^{sn}*_3K) \bullet _3E$
$= (E_{amplitude}*k_{xM}*(-j_{yM}*\sin(k_{xM}*(x_M + c*t_M)) - j_{zM}*\cos(k_{xM}*(x_M + c*t_M))))$
$\bullet(E_{amplitude}*(p_{yM}*\sin(k_{xM}*(x_M + c*t_M)) + p_{zM}*\cos(k_{xM}*(x_M + c*t_M))))$
$= E_{amplitude}^2*k_{xM}*(-i*1_M)*(\cos^2(k_{xM}*(x_M + c*t_M)) + \sin^2(k_{xM}*(x_M + c*t_M)))$
$= E_{amplitude}^2*k_{xM}*(-i*1_M)$
Using: $\quad -j_{yM}*p_{yM} = -i*1_M \quad\quad -j_{zM}*p_{zM} = -i*1_M$

$(_1\nabla^{sn}*_3K)\mathbf{x}_3E$
$= (E_{amplitude}*k_{xM}*(-j_{yM}*\sin(k_{xM}*(x_M + c*t_M)) - j_{zM}*\cos(k_{xM}*(x_M + c*t_M))))$
$\mathbf{x}(E_{amplitude}*(p_{yM}*\sin(k_{xM}*(x_M + c*t_M)) + p_{zM}*\cos(k_{xM}*(x_M + c*t_M))))$
$= E_{amplitude}^2*k_{xM}*(q_{xM} - q_{xM})*\sin(k_{xM}*(x_M + c*t_M))*\cos(k_{xM}*(x_M + c*t_M)) = 0$
Using: $\quad -j_{yM}*p_{zM} = -q_{xM} \quad\quad -j_{zM}*p_{yM} = q_{xM}$

CHAPTER 3 – FIELDS

$(_3\nabla^{sn}\mathbf{x}_3K)\bullet_3E$
$= (E_{amplitude}*k_{xM}*(q_{zM}*\sin(k_{xM}*(x_M + c*t_M)) - q_{yM}*\cos(k_{xM}*(x_M + c*t_M))))$
$\bullet(E_{amplitude}*(p_{yM}*\sin(k_{xM}*(x_M + c*t_M)) + p_{zM}*\cos(k_{xM}*(x_M + c*t_M))))$
$= E_{amplitude}^2*k_{xM}*(1_M - 1_M)*\sin(k_{xM}*(x_M + c*t_M))*\cos(k_{xM}*(x_M + c*t_M)) = 0$
Using: $q_{zM}*p_{zM} = 1_M$ $-q_{yM}*p_{yM} = -1_M$

$(_3\nabla^{sn}\mathbf{x}_3K)\mathbf{x}_3E$
$= (E_{amplitude}*k_{xM}*(q_{zM}*\sin(k_{xM}*(x_M + c*t_M)) - q_{yM}*\cos(k_{xM}*(x_M + c*t_M))))$
$\mathbf{x}(E_{amplitude}*(p_{yM}*\sin(k_{xM}*(x_M + c*t_M)) + p_{zM}*\cos(k_{xM}*(x_M + c*t_M))))$
$= E_{amplitude}^2*k_{xM}*j_{xM}*(\cos^2(k_{xM}*(x_M + c*t_M)) + \sin^2(k_{xM}*(x_M + c*t_M)))$
$= E_{amplitude}^2*k_{xM}*j_{xM}$
Using: $q_{zM}*p_{yM} = j_{yM}*k_{zM} = j_{xM}$ $-q_{yM}*p_{zM} = j_{yM}*k_{zM} = j_{xM}$

$(_1\nabla^{sn}*_3K)\bullet_3K$
$= (E_{amplitude}*k_{xM}*(-j_{yM}*\sin(k_{xM}*(x_M + c*t_M)) + -j_{zM}*\cos(k_{xM}*(x_M + c*t_M))))$
$\bullet(E_{amplitude}*(k_{yM}*\cos(k_{xM}*(x_M + c*t_M)) + -k_{zM}*\sin(k_{xM}*(x_M + c*t_M))))$
$= E_{amplitude}^2*k_{xM}*(1_M - 1_M)*\sin(k_{xM}*(x_M + c*t_M))*\cos(k_{xM}*(x_M + c*t_M)) = 0$
Using: $-j_{yM}*k_{yM} = 1_M$ $-j_{zM}*-k_{zM} = -1_M$

$(_1\nabla^{sn}*_3K)\mathbf{x}_3K$
$= (E_{amplitude}*k_{xM}*(-j_{yM}*\sin(k_{xM}*(x_M + c*t_M)) - j_{zM}*\cos(k_{xM}*(x_M + c*t_M))))$
$\mathbf{x}(E_{amplitude}*(k_{yM}*\cos(k_{xM}*(x_M + c*t_M)) - k_{zM}*\sin(k_{xM}*(x_M + c*t_M))))$
$= E_{amplitude}^2*k_{xM}*(j_{xM})*(\cos^2(k_{xM}*(x_M + c*t_M)) + \sin^2(k_{xM}*(x_M + c*t_M)))$
$= E_{amplitude}^2*k_{xM}*(j_{xM})$
Using: $-j_{yM}*-k_{zM} = j_{xM}$ $-j_{zM}*k_{yM} = j_{xM}$

$(_3\nabla^{sn}\mathbf{x}_3K)\bullet_3K$
$= (E_{amplitude}*k_{xM}*(q_{zM}*\sin(k_{xM}*(x_M + c*t_M)) - q_{yM}*\cos(k_{xM}*(x_M + c*t_M))))$
$\bullet((E_{amplitude}*(k_{yM}*\cos(k_{xM}*(x_M + c*t_M)) - k_{zM}*\sin(k_{xM}*(x_M + c*t_M))))$
$= E_{amplitude}^2*k_{xM}*(-i*1_M)*(\cos^2(k_{xM}*(x_M + c*t_M)) + \sin^2(k_{xM}*(x_M + c*t_M)))$
$= E_{amplitude}^2*k_{xM}*(-i*1_M)$
Using: $q_{zM}*-k_{zM} = -i*1_M$ $-q_{yM}*k_{yM} = -i*1_M$

$(_3\nabla^{sn}\mathbf{x}_3K)\mathbf{x}_3K$
$= (E_{amplitude}*k_{xM}*(q_{zM}*\sin(k_{xM}*(x_M + c*t_M)) - q_{yM}*\cos(k_{xM}*(x_M + c*t_M))))$
$\mathbf{x}(E_{amplitude}*(k_{yM}*\cos(k_{xM}*(x_M + c*t_M)) - k_{zM}*\sin(k_{xM}*(x_M + c*t_M))))$
$= E_{amplitude}^2*k_{xM}*(q_{xM} - q_{xM})*\sin(k_{xM}*(x_M + c*t_M))*\cos(k_{xM}*(x_M + c*t_M)) = 0$
Using: $q_{zM}*k_{yM} = -q_{yM}*k_{zM} = -q_{xM}$ $-q_{yM}*-k_{zM} = q_{xM}$

SPECIAL ALGEBRA FOR SPECIAL RELATIVITY

$-_4f = -_1f + -_3f$

$= ((_1\nabla^{sn}*_3E + _3\nabla^{sn}\mathbf{x}_3K) \bullet_3 E) \pm_W ((_1\nabla^{sn}*_3K + _3\nabla^{sn}\mathbf{x}_3E) \bullet_3 E)$
$\pm_W -((_1\nabla^{sn}*_3E + _3\nabla^{sn}\mathbf{x}_3K) \bullet_3 K) - ((_1\nabla^{sn}*_3K + _3\nabla^{sn}\mathbf{x}_3E) \bullet_3 K)$
$+ (((_3\nabla^{sn}\bullet_3E)*_3E + (_1\nabla^{sn}*_3E + _3\nabla^{sn}\mathbf{x}_3K)\mathbf{x}_3K)$
$\pm_W ((_3\nabla^{sn}\bullet_3K)*_3E + (_1\nabla^{sn}*_3K + _3\nabla^{sn}\mathbf{x}_3E)\mathbf{x}_3K)$
$\pm_W -((_3\nabla^{sn}\bullet_3E)*_3K + (_1\nabla^{sn}*_3E + _3\nabla^{sn}\mathbf{x}_3K)\mathbf{x}_3E)$
$- ((_3\nabla^{sn}\bullet_3K)*_3K + (_1\nabla^{sn}*_3K + _3\nabla^{sn}\mathbf{x}_3E)\mathbf{x}_3E))$

$= (_1\nabla^{sn}*_3E)\bullet_3E + (_3\nabla^{sn}\mathbf{x}_3K)\bullet_3E \pm_W (_1\nabla^{sn}*_3K)\bullet_3E \pm_W (_3\nabla^{sn}\mathbf{x}_3E)\bullet_3E$
$\pm_W -(_1\nabla^{sn}*_3E)\bullet_3K \pm_W -(_3\nabla^{sn}\mathbf{x}_3K)\bullet_3K - (_1\nabla^{sn}*_3K)\bullet_3K - (_3\nabla^{sn}\mathbf{x}_3E)\bullet_3K$
$+ (_3\nabla^{sn}\bullet_3E)*_3E + (_1\nabla^{sn}*_3E)\mathbf{x}_3K + (_3\nabla^{sn}\mathbf{x}_3K)\mathbf{x}_3K$
$\pm_W (_3\nabla^{sn}\bullet_3K)*_3E \pm_W (_1\nabla^{sn}*_3K)\mathbf{x}_3K \pm_W (_3\nabla^{sn}\mathbf{x}_3E)\mathbf{x}_3K$
$\pm_W -(_3\nabla^{sn}\bullet_3E)*_3K \pm_W -(_1\nabla^{sn}*_3E)\mathbf{x}_3E \pm_W -(_3\nabla^{sn}\mathbf{x}_3K)\mathbf{x}_3E$
$- (_3\nabla^{sn}\bullet_3K)*_3K - (_1\nabla^{sn}*_3K)\mathbf{x}_3E - (_3\nabla^{sn}\mathbf{x}_3E)\mathbf{x}_3E$

$= 0 + 0 \pm_W E_{amplitude}^2 * k_{xM} * -i * 1_M \pm_W E_{amplitude}^2 * k_{xM} * i * 1_M$
$\pm_W -E_{amplitude}^2 * k_{xM} * i * 1_M \pm_W -E_{amplitude}^2 * k_{xM} * -i * 1_M - 0 - 0 + 0 + 0$
$\pm_W 0 \pm_W E_{amplitude}^2 * k_{xM} * j_{xM} \pm_W E_{amplitude}^2 * k_{xM} * -j_{xM}$
$\pm_W -0 \pm_W -E_{amplitude}^2 * k_{xM} * -j_{xM} \pm_W -E_{amplitude}^2 * k_{xM} * j_{xM} - 0 - 0 - 0 = 0$

"$_4f = 0$" because "$_4J = 0$".

3.10 Area and Volume Differential Operators

- Per nothing
- Gradient (per length) differential operator "$_4\nabla^{sn}$" (del)
- Per area differential operator "$_6\vartheta^{sn}$" (theta)
- Per volume differential operator "$_4\Xi^{sn}$" (xi)
- Per world volume differential operator "$_1\varpi$" (omega)

The count of terms matches a row of Pascal's Triangle: 1, 4, 6, 4, 1. The first "1" is for a One-Component Invariant. It is Lorentz Transformed by Lorentz Transforming components it depends on.

Electric current density "$_4J$" uses the per volume differential operator.

CHAPTER 3 – FIELDS

<u>Finding Wavenumber</u>. "$_4\nabla^{sn}$" can be made equivalent to a wavenumber/frequency invariant "$_4k$" with consideration as to what function "$_1T$" is. The first example is "$_1T = A*\sin(k_{xM}*x_M - \omega_M*t_M)$".

$_4\nabla^{sn}*(A*\sin(k_{xM}*x_M - \omega_M*t_M)) = A*(_4\nabla^{sn}*\sin(k_{xM}*x_M - \omega_M*t_M))$
$= A*(1_M^{sn}*\partial/\partial ct + q_{xM}^{sn}*\partial/\partial x + q_{yM}^{sn}*\partial/\partial y + q_{zM}^{sn}*\partial/\partial z)*\sin(k_{xM}*x_M - \omega_M*t_M)$
$= A*(1_M^{sn}*\partial/\partial ct_M + q_{xM}^{sn}*\partial/\partial x_M)*\sin(k_{xM}*x_M - \omega_M*t_M)$
$= A*(1_M^{sn}*(-\omega_M/c) + q_{xM}^{sn}*(k_{xM}))*\cos(k_{xM}*x_M - \omega_M*t_M)$
$= A*(1_M*(-\omega_M/c) + q_{xM}*(-k_{xM}))*\cos(k_{xM}*x_M - \omega_M*t_M)$
$= A*(-_4k)*\cos(k_{xM}*x_M - \omega_M*t_M)$

"$_1T = A*\exp(i*(k_{xM}*x_M - \omega_M*t_M))$" avoids the change from sine to cosine.

$_4\nabla^{sn}*(A*\exp(i*(k_{xM}*x_M - \omega_M*t_M))) = A*(_4\nabla^{sn}*\exp(i*(k_{xM}*x_M - \omega_M*t_M)))$
$= A*(1_M^{sn}*\partial/\partial ct_M + q_{xM}^{sn}*\partial/\partial x_M)*\exp(i*(k_{xM}*x_M - \omega_M*t_M))$
$= A*(1_M^{sn}*(-i*\omega_M/c) + q_{xM}^{sn}*(i*k_{xM}))*\exp(i*(k_{xM}*x_M - \omega_M*t_M))$
$= A*i*(1_M*(-\omega_M/c) + q_{xM}*(-k_{xM}))*\exp(i*(k_{xM}*x_M - \omega_M*t_M))$
$= i*(-_4k)*A*\exp(i*(k_{xM}*x_M - \omega_M*t_M))$

"$_4\nabla^{sn} = -i*_4k$" (which has terms "$1_M^{sn}*\partial/\partial ct_M = 1_M*(-i*\omega_M/c)$" and "$q_{xM}^{sn}*\partial/\partial x_M = q_{xM}*(-i*k_{xM})$") is an abbreviation of "$_4\nabla^{sn}*(_1T) = -i*_4k*(_1T)$" and is exclusive to "$_1T = A*\exp(i*(k_{xM}*x_M - \omega_M*t_M))$".

"$i*_4\nabla^{sn} = _4k$" is used in the next chapter on waves.

* Note: The remaining portion of this chapter can be bypassed

<u>A Linear Scalar Field's Gradient</u>. A count (one, two, and higher) is equally spaced along the "x" axis. Count is an invariant scalar field represented by the below expression. The compound-label-number in "$_1$Count" is "1".

$_1\text{Count} = b*x_M$

Visualize "$_1$Count" as a long rod on which numbers are written: Zero at the back of the bus and increasing forward ("b > 0").

$_4\nabla^{sn}*(_1\text{Count}) = _4\nabla^{sn}*(b*x_M)$
$= (1_M^{sn}*\partial/\partial ct_M + q_{xM}^{sn}*\partial/\partial x_M + q_{yM}^{sn}*\partial/\partial y_M + q_{zM}^{sn}*\partial/\partial z_M)*(b*x_M)$
$= (q_{xM}^{sn}*\partial/\partial x_M)*(b*x_M)$
$= q_{xM}^{sn}*b$

SPECIAL ALGEBRA FOR SPECIAL RELATIVITY

"$_4\nabla^{sn}*(_1\text{Count}) = q_{xM}{}^{sn}*b$" states "$_1\text{Count}$" varies by gradient "$b$" in the "$q_{xM}{}^{sn}$" direction (that is, positive "i_{xM}" direction).

"$_1\text{Count}$" is Lorentz Transformed from "$_1\text{Count} = b*x_M$" to "$_1\text{Count} = b*(x_S*\cosh(\alpha_{S/M}) - c*t_S*\sinh(\alpha_{S/M}))$", after which a gradient "$_4\nabla^{sn}$" is taken.

$$_4\nabla^{sn}*(b*x_M) = {}_4\nabla^{sn}*(b*(x_S*\cosh(\alpha_{S/M}) - c*t_S*\sinh(\alpha_{S/M})))$$
$$= b*(-1_M{}^{sn}*\partial/\partial ct_M*c*t_S*\sinh(\alpha_{S/M}) + q_{xM}{}^{sn}*\partial/\partial x_M*x_S*\cosh(\alpha_{S/M}))$$
$$= b*(-1_M{}^{sn}*\sinh(\alpha_{S/M}) + q_{xM}{}^{sn}*\cosh(\alpha_{S/M}))$$

Alternatively, "$_4\nabla^{sn}*(_1\text{Count}) = q_{xM}{}^{sn}*b$" is Lorentz Transformed to the result "$_4\nabla^{sn}*(_1\text{Count}) = 1_S{}^{sn}*b*\sinh(\alpha_{S/M}) + q_{xS}{}^{sn}*b*\cosh(\alpha_{S/M})$", per the matrix equation below. "$\partial/\partial ct_M*(_1\text{Count}) = 0$".

$$\begin{array}{c} \partial/\partial ct_S*(_1\text{Count}) \\ \partial/\partial x_S*(_1\text{Count}) \end{array} = \begin{array}{cc} \cosh(\alpha_{S/M}) & -\sinh(\alpha_{S/M}) \\ -\sinh(\alpha_{S/M}) & \cosh(\alpha_{S/M}) \end{array} * \begin{array}{c} \partial/\partial ct_M*(_1\text{Count}) \\ \partial/\partial x_M*(_1\text{Count}) \end{array}$$

$$= \begin{array}{cc} \cosh(\alpha_{S/M}) & -\sinh(\alpha_{S/M}) \\ -\sinh(\alpha_{S/M}) & \cosh(\alpha_{S/M}) \end{array} * \begin{array}{c} 0 \\ b \end{array} = \begin{array}{c} -b*\sinh(\alpha_{S/M}) \\ b*\cosh(\alpha_{S/M}) \end{array}$$

Either method has the same result.

$$_4\nabla^{sn}*(_1\text{Count}) = 1_S{}^{sn}*b*(-\sinh(\alpha_{S/M})) + q_{xS}{}^{sn}*b*\cosh(\alpha_{S/M})$$

Time component "$-b*\sinh(\alpha_{S/M})$" of "$_4\nabla^{sn}*(_1\text{Count})$" is negative because "$_1\text{Count}$" decreases at rate "$-b*\sinh(\alpha_{S/M})$" when observed from one location "x_S" (for "$\alpha_{S/M} > 0$"). If we are standing on roadside "S" looking at numbers, then the numbers decrease.

Rather than care about numbers, we can care about spacing of the numbers and the rate numbers pass us. Spacing of numbers, called wavenumber, is a positive value (if we select it to be positive). The rate at which numbers pass us, called frequency, is positive (per that selection). We continue to use the gradient operator, but we must transition away from the gradient operator invariant with its space-negative to the wavenumber-frequency invariant that has no space-negative (because now "q_{xM}" translates to "i_{xM}").

Remove the space-negative in "$_4\nabla^{sn}*(_1\text{Count}) = q_{xM}{}^{sn}*b$" by replacing "$q_{xM}{}^{sn}$" with "$-q_{xM}$" so that "$_4\nabla^{sn}*(_1\text{Count}) = -q_{xM}*b = -_4k$".

CHAPTER 3 – FIELDS

$$_4\nabla^{sn}*(_1\text{Count}) = -_4k \quad ; \quad _4k = q_{xM}*k_{xM} = q_{xM}*b$$
$$= 1_S*b*\sinh(\alpha_{S/M}) + q_{xS}*b*\cosh(\alpha_{S/M})$$
$$= 1_S*\omega_S/c + q_{xS}*k_{xS}$$

<u>Per-Area Differential Operator</u> "$_6\vartheta^{sn}$" has six terms.

$$_6\vartheta^{sn} = p_{xM}{}^{sn}*(\partial^2/(\partial x_M*\partial ct_M) + i*(\partial^2/(\partial y_M*\partial z_M)))$$
$$+ p_{yM}{}^{sn}*(\partial^2/(\partial y_M*\partial ct_M) + i*(\partial^2/(\partial z_M*\partial x_M)))$$
$$+ p_{zM}{}^{sn}*(\partial^2/(\partial z_M*\partial ct_M) + i*(\partial^2/(\partial x_M*\partial y_M)))$$

$$= p_{xM}{}^{sn}*(\vartheta_{xrM} + i*\vartheta_{xiM})$$
$$+ p_{yM}{}^{sn}*(\vartheta_{yrM} + i*\vartheta_{yiM})$$
$$+ p_{zM}{}^{sn}*(\vartheta_{zrM} + i*\vartheta_{ziM})$$

On the floor inside bus "M" is a rectangular array of particles. On each particle is a sequential number in the "x" direction and a sequential number in the "y" direction. The first particle counted is in the rear of the bus on the right side, "x = 0" and "y = 0". The count of particles is "$_1\text{Count} = (b_x*x_M)*(b_y*y_M)$". ("$b_x > 0$", "$b_y > 0$")

"$_6\vartheta^{sn}$" operates on "$_1\text{Count} = (b_x*x_M)*(b_y*y_M)$" to quantify the change in count relative to area.

$$_6\vartheta^{sn}*(_1\text{Count}) = p_{xM}{}^{sn}*(\partial^2/(\partial x_M*\partial ct_M) + i*(\partial^2/(\partial y_M*\partial z_M)))*_1\text{Count}$$
$$+ p_{yM}{}^{sn}*(\partial^2/(\partial y_M*\partial ct_M) + i*(\partial^2/(\partial z_M*\partial x_M)))*_1\text{Count}$$
$$+ p_{zM}{}^{sn}*(\partial^2/(\partial z_M*\partial ct_M) + i*(\partial^2/(\partial x_M*\partial y_M)))*_1\text{Count}$$

$$= p_{xM}{}^{sn}*(\partial^2/(\partial x_M*\partial ct_M) + i*(\partial^2/(\partial y_M*\partial z_M)))*(b_x*x_M)*(b_y*y_M)$$
$$+ p_{yM}{}^{sn}*(\partial^2/(\partial y_M*\partial ct_M) + i*(\partial^2/(\partial z_M*\partial x_M)))*(b_x*x_M)*(b_y*y_M)$$
$$+ p_{zM}{}^{sn}*(\partial^2/(\partial z_M*\partial ct_M) + i*(\partial^2/(\partial x_M*\partial y_M)))*(b_x*x_M)*(b_y*y_M)$$

$$= p_{zM}{}^{sn}*(i*(\partial^2/(\partial x_M*\partial y_M)))*(b_x*x_M)*(b_y*y_M)$$
$$= p_{zM}{}^{sn}*i*b_x*b_y$$

"$_6\vartheta^{sn}*(_1\text{Count}) = p_{zM}{}^{sn}*i*b_x*b_y$" states the change in count with respect to area equals "b_x*b_y", as observed by a person seated on bus "M". The "i" factor in "$p_{zM}{}^{sn}*i$" means the "$p_{zM}{}^{sn}*i = k_{zM}{}^{sn}$" label number translates to the "x / y" plane and not to the "z" direction.

SPECIAL ALGEBRA FOR SPECIAL RELATIVITY

The result "$_6\vartheta^{sn}*(_1Count)$" is Lorentz Transformed.

$$\begin{pmatrix} (\vartheta_{yrS} + i*\vartheta_{yiS})*(_1Count) \\ (\vartheta_{zrS} + i*\vartheta_{ziS})*(_1Count) \end{pmatrix} = \begin{pmatrix} \cosh(\alpha_{S/M}) & i*\sinh(\alpha_{S/M}) \\ -i*\sinh(\alpha_{S/M}) & \cosh(\alpha_{S/M}) \end{pmatrix} * \begin{pmatrix} (\vartheta_{yrM} + i*\vartheta_{yiM})*(_1Count) \\ (\vartheta_{zrM} + i*\vartheta_{ziM})*(_1Count) \end{pmatrix}$$

Right-side column vector has "$(\vartheta_{yrM} + i*\vartheta_{yiM})*(_1Count) = 0$" and "$(\vartheta_{zrM} + i*\vartheta_{ziM})*(_1Count) = i*b_x*b_y$". The matrix operator is a space-negative. Terms of the left side column vector, "$(\vartheta_{yrS} + i*\vartheta_{yiS})*(_1Count) = (i*b_x*b_y)*(i*\sinh(\alpha_{S/M}))$" and "$(\vartheta_{zrS} + i*\vartheta_{ziS})*(_1Count) = (i*b_x*b_y)*(\cosh(\alpha_{S/M}))$", are in the invariant expression below.

$$\begin{aligned} _6\vartheta^{sn}*(_1Count) &= p_{yS}^{sn}*(\vartheta_{yrS}+i*\vartheta_{yiS})*(_1Count) + p_{zS}^{sn}*(\vartheta_{zrS}+i*\vartheta_{ziS})*(_1Count) \\ &= p_{yS}^{sn}*(-b_x*b_y*\sinh(\alpha_{S/M})) + p_{zS}^{sn}*(i*b_x*b_y*\cosh(\alpha_{S/M})) \end{aligned}$$

"$p_{zS}^{sn}*(i*b_x*b_y*\cosh(\alpha_{S/M}))$" of "$_6\vartheta^{sn}*(_1Count)$" states there are more particles for the same amount of x*y area if we use "x_S" rather than "x_M" for that x*y area, by a factor of "$\cosh(\alpha_{S/M})$".

Term "$p_{yS}^{sn}*(-b_x*b_y*\sinh(\alpha_{S/M}))$" of "$_6\vartheta^{sn}*(_1Count)$" is a measure of per-length and per-time, because it is real and is not imaginary. Per-length is in the "y" direction, perpendicular to the direction of motion. Per-time is associated with the negative. The count as written on each particle decreases relative to a person standing on roadside "S".

The complementary invariant to the per-area differential operator "$_6\vartheta^{sn}$" is wave-number-area invariant "$_6kk$". There is no space-negative on "$_6kk$".

$$\begin{aligned} _6kk &= p_{zM}*i*kk_{ziM} = p_{zM}*i*b_x*b_y \\ &= p_{yS}*b_x*b_y*\sinh(\alpha_{S/M}) + p_{zS}*i*b_x*b_y*\cosh(\alpha_{S/M}) \\ &= p_{yS}*kk_{yrS} + p_{zS}*i*kk_{ziS} \end{aligned}$$

"kk_{ziS}" component is spacing of particle rows in the "x" direction, and "kk_{ziS}" includes factor "b_y" for density along particle rows. "kk_{yrS}" is frequency of particle rows that pass a person standing on roadside "S" (and "kk_{yrS}" includes a factor "b_y" for density of particles along rows). The Lorentz Transformation used the below matrix equation. The matrix operator is not a space-negative, as identified by the different location of the negative on off-diagonal terms.

$$\begin{pmatrix} (kk_{yrS} + i*kk_{yiS}) \\ (kk_{zrS} + i*kk_{ziS}) \end{pmatrix} = \begin{pmatrix} \cosh(\alpha_{S/M}) & -i*\sinh(\alpha_{S/M}) \\ i*\sinh(\alpha_{S/M}) & \cosh(\alpha_{S/M}) \end{pmatrix} * \begin{pmatrix} (kk_{yrM} + i*kk_{yiM}) \\ (kk_{zrM} + i*kk_{ziM}) \end{pmatrix}$$

CHAPTER 3 – FIELDS

The above matrix operator was also used for the Lorentz Transformation for electromagnetic field and area invariants.

$$\begin{bmatrix} E_{yS} + i*K_{yS} \\ E_{zS} + i*K_{zS} \end{bmatrix} = \begin{bmatrix} \cosh\alpha_{S/M} & -i*\sinh\alpha_{S/M} \\ i*\sinh\alpha_{S/M} & \cosh\alpha_{S/M} \end{bmatrix} * \begin{bmatrix} E_{yM} + i*K_{yM} \\ E_{zM} + i*K_{zM} \end{bmatrix}$$

$$\begin{bmatrix} B_{yS} + i*A_{yS} \\ B_{zS} + i*A_{zS} \end{bmatrix} = \begin{bmatrix} \cosh\alpha_{S/M} & -i*\sinh\alpha_{S/M} \\ i*\sinh\alpha_{S/M} & \cosh\alpha_{S/M} \end{bmatrix} * \begin{bmatrix} B_{yM} + i*A_{yM} \\ B_{zM} + i*A_{zM} \end{bmatrix}$$

<u>Per-Volume Differential Operator</u> "$_4\Xi^{sn}$" (xi) has four terms.

$_4\Xi^{sn} = 1_M{}^{sn}*(\partial^3/(\partial x_M*\partial y_M*\partial z_M)) + q_{xM}{}^{sn}*(\partial^3/(\partial ct_M*\partial y_M*\partial z_M))$
$\phantom{_4\Xi^{sn} =} + q_{yM}{}^{sn}*(\partial^3/(\partial ct_M*\partial x_M*\partial z_M)) + q_{zM}{}^{sn}*(\partial^3/(\partial ct_M*\partial x_M*\partial y_M))$

$\phantom{_4\Xi^{sn}} = 1_M{}^{sn}*\Xi_{tM} + q_{xM}{}^{sn}*\Xi_{xM} + q_{yM}{}^{sn}*\Xi_{yM} + q_{zM}{}^{sn}*\Xi_{zM}$

A long rectangular prism array of particles sits on the floor of bus "M". On each are three numbers: "x" (to the front of the bus), "y" (from right to left), and "z" (up). At the rear, right side and on the floor is the particle labelled "x = 0", "y = 0" and "z = 0". The count of particles is given by "$_1\text{Count} = (b_x*x_M)*(b_y*y_M)*(b_z*z_M)$". ("$b_x > 0$", "$b_y > 0$", and "$b_z > 0$")

Per-volume differential operator "$_4\Xi^{sn}$" operates on particle count scalar "$_1\text{Count} = (b_x*x_M)*(b_y*y_M)*(b_z*z_M)$" to quantify the change in count relative to volume.

$_4\Xi^{sn}*(_1\text{Count}) = 1_M{}^{sn}*(\partial^3/(\partial x_M*\partial y_M*\partial z_M))*(_1\text{Count})$
$\phantom{_4\Xi^{sn}*(_1\text{Count}) =} + q_{xM}{}^{sn}*(\partial^3/(\partial ct_M*\partial y_M*\partial z_M))*(_1\text{Count})$
$\phantom{_4\Xi^{sn}*(_1\text{Count}) =} + q_{yM}{}^{sn}*(\partial^3/(\partial ct_M*\partial x_M*\partial z_M))*(_1\text{Count})$
$\phantom{_4\Xi^{sn}*(_1\text{Count}) =} + q_{zM}{}^{sn}*(\partial^3/(\partial ct_M*\partial x_M*\partial y_M))*(_1\text{Count})$

$\phantom{_4\Xi^{sn}*(_1\text{Count})} = 1_M{}^{sn}*(\partial^3/(\partial x_M*\partial y_M*\partial z_M))*(b_x*x_M)*(b_y*y_M)*(b_z*z_M)$
$\phantom{_4\Xi^{sn}*(_1\text{Count}) =} + q_{xM}{}^{sn}*(\partial^3/(\partial ct_M*\partial y_M*\partial z_M))*(b_x*x_M)*(b_y*y_M)*(b_z*z_M)$
$\phantom{_4\Xi^{sn}*(_1\text{Count}) =} + q_{yM}{}^{sn}*(\partial^3/(\partial ct_M*\partial x_M*\partial z_M))*(b_x*x_M)*(b_y*y_M)*(b_z*z_M)$
$\phantom{_4\Xi^{sn}*(_1\text{Count}) =} + q_{zM}{}^{sn}*(\partial^3/(\partial ct_M*\partial x_M*\partial y_M))*(b_x*x_M)*(b_y*y_M)*(b_z*z_M)$

SPECIAL ALGEBRA FOR SPECIAL RELATIVITY

$$= 1_M{}^{sn}*(\partial^3/(\partial x_M*\partial y_M*\partial z_M))*(b_x*x_M)*(b_y*y_M)*(b_z*z_M)$$
$$= 1_M{}^{sn}*b_x*b_y*b_z$$

"$_4\Xi^{sn}*(_1Count) = 1_M{}^{sn}*b_x*b_y*b_z$" states the change in count with respect to volume equals "$b_x*b_y*b_z$", as observed by a person seated on bus "M". "$b_x*b_y*b_z$" is density of counts: Counts per Volume or counted particles per volume.

"$_4\Xi^{sn}*(_1Count)$" is Lorentz Transformed.

$$\begin{array}{c}\Xi_{tS}*(_1Count)\\ \\ \Xi_{xS}*(_1Count)\end{array} = \begin{array}{cc}\cosh(\alpha_{S/M}) & -\sinh(\alpha_{S/M})\\ \\ -\sinh(\alpha_{S/M}) & \cosh(\alpha_{S/M})\end{array} * \begin{array}{c}\Xi_{tM}*(_1Count)\\ \\ \Xi_{xM}*(_1Count)\end{array}$$

Right-side column vector has "$\Xi_{tM}*(_1Count) = b_x*b_y*b_z$" and "$\Xi_{xM}*(_1Count) = 0$". The matrix operator is space-negative. Terms of the left side column vector, "$\Xi_{tS}*(_1Count) = b_x*b_y*b_z*\cosh(\alpha_{S/M})$" and "$\Xi_{xS}*(_1Count) = -b_x*b_y*b_z*\sinh(\alpha_{S/M})$", are in the invariant expression below.

$$_4\Xi^{sn}*(_1Count) = 1_S{}^{sn}*(\Xi_{tS}*(_1Count)) + q_{xS}{}^{sn}*(\Xi_{xS}*(_1Count))$$
$$= 1_S{}^{sn}*b_x*b_y*b_z*\cosh(\alpha_{S/M}) + q_{xS}{}^{sn}*(-b_x*b_y*b_z*\sinh(\alpha_{S/M}))$$

"$1_S{}^{sn}*b_x*b_y*b_z*\cosh(\alpha_{S/M})$" of "$_4\Xi^{sn}*(_1Count)$" states there are more particles for the same amount of $x*y*z$ volume if we use "x_S" rather than "x_M" for that $x*y*z$ volume, by a factor "$\cosh(\alpha_{S/M})$".

"$q_{xS}{}^{sn}*(-b_x*b_y*b_z*\sinh(\alpha_{S/M}))$" of "$_4\Xi^{sn}*(_1Count)$" is a measure of per-area and per-time. Per-area is in the "$y*z$" direction, perpendicular to the direction of motion. Per-time is associated with the negative because the x-direction count as written on each particle decreases relative to a person standing on roadside "S".

The complementary invariant is current-density invariant "$_4\rho$" (rho). There is no space-negative on "$_4\rho$".

$$_4\rho = 1_M*\rho_{tM} = 1_M*b_x*b_y*b_z$$
$$= 1_S*b_x*b_y*b_z*\cosh(\alpha_{S/M}) + q_{xS}*b_x*b_y*b_z*\sinh(\alpha_{S/M})$$
$$= 1_S*\rho_{tS} + q_{xS}*\rho_{xS}$$

"ρ_{tS}" is density. "ρ_{xS}" is flow per area in the "x" direction due to movement of bus "M" relative to roadside "S". There was no space-negative in the Lorentz Transformation matrix operator.

CHAPTER 3 – FIELDS

$_4\rho = 1_S * b_x * b_y * b_z * \cosh(\alpha_{S/M}) + q_{xS} * b_x * b_y * b_z * \sinh(\alpha_{S/M})$

compared to

$_4\Xi^{sn} * (_1\text{Count}) = 1_S{}^{sn} * b_x * b_y * b_z * \cosh(\alpha_{S/M}) + q_{xS}{}^{sn} * (-b_x * b_y * b_z * \sinh(\alpha_{S/M}))$

shows "$_4\rho = {}_4\Xi^{sn} * (_1\text{Count})$". It applies to a homogeneous material. If "$b_x = b_y = b_z$", then it is a solid homogeneous block.

Electric Current Density Invariant. "Q" is electric charge per electron particle, "$_1\text{Charge} = Q * {}_1\text{Count}$". "$_4J = Q * {}_4\rho$" is found using per-volume operator "$_4\Xi^{sn}$" (for a homogeneous material):

$_4J = {}_4\Xi^{sn} * (_1\text{Charge})$

Per-World-Volume Differential Operator "$_1\varpi$" (omega) has one term.

$_1\varpi = 1 * (\partial^4 / (\partial ct_M * \partial x_M * \partial y_M * \partial z_M))$

Inside a long rectangular prism on bus "M" is an array of lights that flash. Four numbers are printed onto each light. The first number is a count of flashes the light has had. The second number is a sequential number in the "x" direction, third in the "y" direction, fourth "z". The initial light flash in the rear of the bus, on the right side and on the floor corresponds to "$c*t = x = y = z = 0$". ("$b_t, b_x, b_y, b_z > 0$") "$_1\text{Count} = (b_t * c * t_M) * (b_x * x_M) * (b_y * y_M) * (b_z * z_M)$".

"$_1\varpi$" operates on "$_1\text{Count} = (b_t * c * t_M) * (b_x * x_M) * (b_y * y_M) * (b_z * z_M)$" to quantify the change in count relative to world-volume.

$_1\varpi * (_1\text{Count}) = 1 * (\partial^4 / (\partial ct_M * \partial x_M * \partial y_M * \partial z_M)) * (_1\text{Count})$
$= 1 * (\partial^4 / (\partial c * t_M * \partial x_M * \partial y_M * \partial z_M)) * ((b_t * c * t_M) * (b_x * x_M) * (b_y * y_M) * (b_z * z_M))$
$= b_t * b_x * b_y * b_z$

"$_1\varpi * (_1\text{Count}) = b_t * b_x * b_y * b_z$" is the density of counts per world-volume. The same result "$_1\varpi * (_1\text{Count})$" is observed by a person standing on roadside "S" because length contraction balances time dilation as "$\cosh(\alpha_{S/M}) / \cosh(\alpha_{S/M}) = 1$". With respect to the roadside, lights do not all flash at the same time, but, rather, appear to move forward as a pulse faster than the speed-of-light. Per-world-volume density "$_1\rho\rho$" is not complementary to, but rather, is equal to "$_1\varpi * (_1\text{Count}) = b_t * b_x * b_y * b_z$".

3.11 Exercises

Text Comprehension Exercises.

1) Prove "$-_6E = {_4}\nabla^{*/sn}\mathbf{x}(_4V)$". Use "$_4\nabla^{*j}$" and "$_4V$" expressed in "S". Substitute in "M" expressions for compound-label-numbers and components of "$_4\nabla^{*j}$" and "$_4V$". Reduce the expression to the result of "$_6E$" expressed in "M".

2) Prove the following identities.

   a. $_1\nabla^{sn}*(_3\nabla^{*/sn}\mathbf{x}(_3V)) + {_3}\nabla^{sn}\mathbf{x}(_1\nabla^{*/sn}*(_3V)) \equiv 0$
   b. $_3\nabla^{sn}\mathbf{x}(_3\nabla^{*/sn}*(_1V)) \equiv 0$
   c. $_3\nabla^{sn}\bullet(_3\nabla^{*/sn}\mathbf{x}(_3V)) \equiv 0$
   d. $_4\nabla^{sn}\blacksquare(_4\nabla^{*/sn}\mathbf{x}(_4V)) \equiv 0$
   e. $_3\nabla^{*/sn}\bullet(_1\nabla^{sn}*(_1\nabla^{*/sn}*(_3V))) + {_1}\nabla^{*/sn}*(_3\nabla^{sn}\bullet(_1\nabla^{*/sn}*(_3V))) \equiv 0$
   f. $_3\nabla^{*/sn}\bullet(_1\nabla^{sn}*(_3\nabla^{*/sn}*(_1V))) + {_1}\nabla^{*/sn}*(_3\nabla^{sn}\bullet(_3\nabla^{*/sn}*(_1V))) \equiv 0$
   g. $_3\nabla^{*/sn}\bullet(_3\nabla^{sn}\mathbf{x}(_3\nabla^{*/sn}\mathbf{x}(_3V))) \equiv 0$
   h. $_4\nabla^{*/sn}\bullet(_4\nabla^{sn}\blacklozenge(_4\nabla^{*/sn}\mathbf{x}(_4V))) \equiv 0$
   i. $(_4\nabla^{*/sn}\mathbf{x}_4\nabla^{sn})*(-_6E) \equiv 0$
   j. $_3\nabla^{*/sn}\mathbf{x}(_3\nabla^{sn}*_2P) \equiv 0$
   k. $_1\nabla^{*/sn}*(_3\nabla^{sn}*_2P) + {_3}\nabla^{*/sn}*(_1\nabla^{sn}*_2P) \equiv 0$
   l. $_4\nabla^{*/sn}\mathbf{x}(_4\nabla^{sn}*_2P) \equiv 0$
   m. $_3\nabla^{*/sn}\mathbf{x}(_3\nabla^{sn}\mathbf{x}_6Q) \equiv (_3\nabla^{*/sn}\bullet_3\nabla^{sn})*_6Q - {_3}\nabla^{*/sn}*(_3\nabla^{sn}\bullet_6Q)$
   n. $_4\nabla^{*/sn}\bullet(_4\nabla^{sn}*(-_6Q)) = {_4}\nabla^{*/sn}\bullet(_4\nabla^{sn}\blacksquare(-_6Q)) + {_4}\nabla^{*/sn}\bullet(_4\nabla^{sn}\blacklozenge(-_6Q))$
   $= {_1}\nabla^{*/sn}*(_3\nabla^{sn}\bullet(-_3Q_i)) + {_3}\nabla^{*/sn}\bullet(_1\nabla^{sn}\blacklozenge(-_3Q_i) + {_3}\nabla^{sn}\mathbf{x}(-_3Q_r))$
   $+ {_1}\nabla^{*/sn}*(_3\nabla^{sn}\bullet(-_3Q_r)) + {_3}\nabla^{*/sn}\bullet(_1\nabla^{sn}\blacklozenge(-_3Q_r) + {_3}\nabla^{sn}\mathbf{x}(-_3Q_i)) \equiv 0$

3) Confirm "$_1\nabla^{sn}*(-_6E_{third}) + {_3}\nabla^{sn}\mathbf{x}(-_6E_{third}) = 0$".

4) For "$_2P = a*(c*t)^2 + b*x^2$", find "$_4\nabla^{sn}*_2P$". Find "$_4\nabla^{*/sn}\bullet(_4\nabla^{sn}*_2P)$", find "$(_4\nabla^{*/sn}\bullet_4\nabla^{sn})*_2P$", and identify the relationship between "a" and "b" for "$(_4\nabla^{*/sn}\bullet_4\nabla^{sn})*_2P = 0$". For any values of "a" and "b" show that "$_4\nabla^{*/sn}\mathbf{x}(_4\nabla^{sn}*_2P) = 0$".

CHAPTER 3 – FIELDS

5) Confirm "$_6E = E_{amp}*(-j_{yM} + q_{zM})*\cos(k_{xM}*(x_M - c*t_M))$" complies with Maxwell's Equations. Find the simplest representation of "$_4V$" for "$_6E$". What is the effect on "$_6E$" and on the Lorenz Condition of adding a "$_4V_{constants}$" (that has components that are each a constant relative to time and to space) to "$_4V$"? Confirm that specific scalar gauge function "$_2P = P_{max}*\exp(i*n*k_M*(x_M - c*t_M))$" satisfies "$(_4\nabla^{*/sn} \bullet {_4\nabla^{sn}})*_2P = 0$". Find "$_4\nabla^{sn}*_2P$".

6) Show "$-f_{tr}$" and "$-f_{ti}$" component equations conform to Maxwell's Equations

Select Exercises Solutions.

1) Not Given
2) Not Given

3) $_1\nabla^{sn}*(-_6E_{third}) = (1_M{}^{sn})*\partial(-_6E_{third})/\partial(ct_M)$
   $= (1_M)*(k_{yM} + p_{zM})*E_{amp}*\partial(\exp(-i*(k_{xM}*(x_M + c*t_M))))/\partial(ct_M)$
   $= (-i*k_{xM})*E_{amp}*(j_{yM} + q_{zM})*\exp(-i*(k_{xM}*(x_M + c*t_M)))$

   $_3\nabla^{sn}\mathbf{x}(-_6E_{third}) = (q_{xM}{}^{sn})*\partial(-_6E_{third})/\partial(x_M)$
   $= E_{amp}*(q_{xM}{}^{sn}*k_{yM} + q_{xM}{}^{sn}*p_{zM})*\partial(\exp(-i*(k_{xM}*(x_M + c*t_M))))/\partial(x_M)$
   $= (-i*k_{xM})*E_{amp}*(-q_{zM} - j_{yM})*\exp(-i*(k_{xM}*(x_M + c*t_M)))$

   $q_{xM}{}^{sn}*k_{yM} = -q_{xM}*k_{yM} = -q_{zM}$
   $q_{xM}{}^{sn}*p_{zM} = -q_{xM}*k_{zM}/i = q_{zM}*k_{xM}/i = q_{yM}*(-i) = -j_{yM}$

   $_1\nabla^{sn}*(-_6E_{third}) + _3\nabla^{sn}\mathbf{x}(-_6E_{third}) = 0$ **OK**

4) $_2P = a*(c*t)^2 + b*x^2$

   $_4\nabla^{sn}*_2P = (1_M*\nabla_t + q_{xM}*\nabla_x + q_{xM}*\nabla_x + q_{xM}*\nabla_x)*(a*(c*t)^2 + b*x^2)$
   $= 1_M*2*a*c*t + q_{xM}*2*a*x$

   $_4\nabla^{*/sn} \bullet (_4\nabla^{sn}*_2P) = (\nabla_t*1_M{}^{*/sn} + \nabla_x*q_{xM}{}^{*/sn} + \nabla_y*q_{yM}{}^{*/sn} + \nabla_z*q_{zM}{}^{*/sn})$
   $\bullet (1_M*2*a*c*t + q_{xM}*2*b*x)$

   $= (\nabla_t*(2*a*c*t))*(1_M{}^{*/sn}*1_M) + (\nabla_x*2*b*x)*(q_{xM}{}^{*/sn}*q_{xM})$
   $= 2*a + 2*b$

SPECIAL ALGEBRA FOR SPECIAL RELATIVITY

$$(_4\nabla^{*jsn} \bullet _4\nabla^{sn}) *_2 P = (\nabla_t^2 - \nabla_x^2 - \nabla_y^2 - \nabla_z^2)*(a*(c*t)^2 + b*x^2)$$
$$= 2*a + 2*b$$

$$a = -b$$

$$_4\nabla^{*jsn} \mathbf{x} (_4\nabla^{sn} *_2 P) = (\nabla_t * 1_M^{*jsn} + \nabla_x * q_{xM}^{*jsn} + \nabla_y * q_{yM}^{*jsn} + \nabla_z * q_{zM}^{*jsn})$$
$$\mathbf{x}(1_M * 2*a*c*t + q_{xM}*2*b*x)$$

$$= (\nabla_t *(2*b*x))*(1_M^{*jsn} * q_{xM}) + (\nabla_x * 2*a*c*t)*(q_{xM}^{*jsn} * 1_M)$$
$$= 0 * p_{xM} + 0 * p_{xM} = 0$$

5) $_6E = E_{amp}*(-j_{yM} + q_{zM})*\cos(k_{xM}*(x_M - c*t_M))$

$K_{yM} = -E_{amp}*\cos(k_{xM}*(x_M - c*t_M))$
$E_{zM} = E_{amp}*\cos(k_{xM}*(x_M - c*t_M))$

$-\nabla_{tM}*K_{yM} + \nabla_{xM}*E_{zM}$
$\quad = -((-1)*(-k_{xM}) + k_{xM})*E_{amp}*-\sin(k_{xM}*(x_M - c*t_M)) = 0$ **OK**

First attempt at "$_4V$":

$V_{xM} = -z*K_{yM} = z*E_{amp}*\cos(k_{xM}*(x_M - c*t_M))$
$V_{tM} = -y*E_{zM} = -y*E_{amp}*\cos(k_{xM}*(x_M - c*t_M))$
$_4V = 1_M*V_{tM} + q_{xM}*V_{xM} = 1_M*V_{tM} + q_{xM}*V_{xM}$

$E_{zM} = -\nabla_{zM}*V_{tM} = E_{amp}*\cos(k_{xM}*(x_M - c*t_M))$ **OK**
$-K_{yM} = -\nabla_{zM}*V_{xM} = E_{amp}*\cos(k_{xM}*(x_M - c*t_M))$ **OK**
$_4\nabla^{*jsn} \bullet _4V = \nabla_{tM}*V_{tM} + \nabla_{xM}*V_{xM}$
$\quad = ((-y)*(-k_{xM}) + z*k_{xM})*E_{amp}*-\sin(k_{xM}*(x_M - c*t_M)) \neq 0$ **not OK**

Second attempt at "$_4V$":

$V_{zM} = K_{yM}/k_{xM} = (1/k_{xM})*-E_{amp}*\cos(k_{xM}*(x_M - c*t_M))$
$V_{yM} = E_{zM}/k_{xM} = (1/k_{xM})*E_{amp}*\cos(k_{xM}*(x_M - c*t_M))$
$_4V = q_{yM}*V_{yM} + q_{zM}*V_{zM}$

$E_{zM} = -\nabla_{tM}*V_{zM} = -(-k_{xM})*(1/k_{xM})*E_{amp}*\cos(k_{xM}*(x_M - c*t_M))$ **OK**
$-K_{yM} = \nabla_{xM}*V_{zM} = (k_{xM})*(1/k_{xM})*E_{amp}*\cos(k_{xM}*(x_M - c*t_M))$ **OK**

CHAPTER 3 – FIELDS

$${}_4\nabla^{*/sn} \bullet {}_4V^{sn} = \nabla_{yM} * V_{yM} + \nabla_{zM} * V_{zM} = 0 + 0 = 0 \quad \text{OK}$$

$${}_4V_{constants} = 1_M * Konst_t + q_{xM} * Konst_x + q_{yM} * Konst_y + q_{zM} * Konst_z$$

"${}_4\nabla^{*/sn} \mathbf{x} \,{}_4V_{constants} = 0$" and "${}_4\nabla^{*/sn} \mathbf{x} \,{}_4V = 0$", therefore there is no effect on "${}_6E$" or on Lorenz Condition if "${}_4V + {}_4V_{constants}$" is substituted for "${}_4V$".

$${}_2P = P_{max} * \exp(i * n * k_M * (x_M - c * t_M))$$

$$({}_4\nabla^{*/sn} \bullet {}_4\nabla^{sn}) * {}_2P = (\nabla_{tM}^2 - \nabla_{xM}^2 - \nabla_{yM}^2 - \nabla_{zM}^2) * {}_2P$$
$$= (-i*n*k_M)^2 * {}_2P - (i*n*k_M)^2 * {}_2P = 0 \quad \text{OK}$$

$${}_4\nabla^{sn} * {}_2P = (1_M^{sn} * \nabla_{tM} + q_{xM}^{sn} * \nabla_{xM} + q_{yM}^{sn} * \nabla_{yM} + q_{zM}^{sn} * \nabla_{zM}) * {}_2P$$
$$= (1_M * (-i*n*k_M) - q_{xM} * (i*n*k_M)) * {}_2P$$
$$= (1_M + q_{xM}) * (-i*n*k_M) * {}_2P$$

$${}_4\nabla^{*/sn} \bullet ({}_4\nabla^{sn} * {}_2P) = (\nabla_{tM} * 1_M^{*/sn} + \nabla_{xM} * q_{xM}^{*/sn}$$
$$+ \nabla_{yM} * q_{yM}^{*/sn} + \nabla_{zM} * q_{zM}^{*/sn}) \bullet ({}_4\nabla^{sn} * {}_2P)$$
$$= ((1_M^{*/sn} * 1_M) * (-i*n*k_M)^2 + (q_{xM}^{*/sn} * q_{xM}) * (i*n*k_M) * (-i*n*k_M)) * {}_2P$$
$$= ((1) * (-i*n*k_M)^2 + (1) * (i*n*k_M) * (-i*n*k_M)) * {}_2P$$
$$= 0 \quad \text{OK}$$

$${}_4\nabla^{*/sn} \mathbf{x} ({}_4\nabla^{sn} * {}_2P) = (\nabla_{tM} * 1_M^{*/sn} + \nabla_{xM} * q_{xM}^{*/sn}$$
$$+ \nabla_{yM} * q_{yM}^{*/sn} + \nabla_{zM} * q_{zM}^{*/sn}) \mathbf{x} ({}_4\nabla^{sn} * {}_2P)$$
$$= (\nabla_{tM} * 1_M^{*/sn} + \nabla_{xM} * q_{xM}^{*/sn}) \mathbf{x} ((1_M + q_{xM}) * (-i*n*k_M) * {}_2P)$$
$$+ (\nabla_{yM} * q_{yM}^{*/sn} + \nabla_{zM} * q_{zM}^{*/sn}) \mathbf{x} ((1_M + q_{xM}) * (-i*n*k_M) * {}_2P)$$
$$= ((1_M^{*/sn} * q_{xM}) * (-i*n*k_M)^2 + (q_{xM}^{*/sn} * 1_M) * (i*n*k_M) * (-i*n*k_M)) * {}_2P$$
$$= ((p_{xM}) * (-i*n*k_M)^2 + (p_{xM}) * (i*n*k_M) * (-i*n*k_M)) * {}_2P$$
$$= 0 \quad \text{OK}$$

6) Solution is almost identical to the text given for "$-f_{zr}$" and "$-f_{zi}$".

Further Thought.

1) "$K_{yS} = -(q/(4*\pi*\mathfrak{e}*(r_S^2 + x_S^2 * \sinh^2\alpha_{S/M})^{3/2})) * z_S * \sinh\alpha_{S/M}$" of the Biot-Savart Law reaches a maximum when the two particles are closest. Should the moving particle be a little past the other?

SPECIAL ALGEBRA FOR SPECIAL RELATIVITY

2) Can we violate the Lorenz Condition and still have a consistent mathematics for electromagnetism?

3) The author could not find an all-number identity that combined the below identities, and the reason is speculated to be that the force density math of the last chapter is needed.

$$_3\mathbf{A} = A_x*\mathbf{i}_x + A_y*\mathbf{i}_y + A_z*\mathbf{i}_z \quad ; \quad _3\mathbf{B} = B_x*\mathbf{i}_x + B_y*\mathbf{i}_y + B_z*\mathbf{i}_z$$

$$_3\nabla = \nabla_x*\mathbf{i}_x + \nabla_y*\mathbf{i}_y + \nabla_z*\mathbf{i}_z = \partial/\partial x*\mathbf{i}_x + \partial/\partial y*\mathbf{i}_y + \partial/\partial z*\mathbf{i}_z \quad ; \quad _1\nabla = \nabla_t = \partial/\partial(c*t)$$

$$_3\nabla\bullet(_3\mathbf{A}\mathbf{x}_3\mathbf{B}) \equiv -(_3\nabla\mathbf{x}_3\mathbf{A})\bullet_3\mathbf{B} + (_3\nabla\mathbf{x}_3\mathbf{B})\bullet_3\mathbf{A}$$
$$_1\nabla*(_3\mathbf{A}\bullet_3\mathbf{B}) \equiv (_1\nabla*_3\mathbf{A})\bullet_3\mathbf{B} + (_1\nabla*_3\mathbf{B})\bullet_3\mathbf{A}$$
$$_1\nabla*(_3\mathbf{A}\mathbf{x}_3\mathbf{B}) \equiv (_1\nabla*_3\mathbf{A})\mathbf{x}_3\mathbf{B} - (_1\nabla*_3\mathbf{B})\mathbf{x}_3\mathbf{A}$$
$$_3\nabla\mathbf{x}(_3\mathbf{A}\mathbf{x}_3\mathbf{B}) \equiv ((_3\mathbf{B}\bullet_3\nabla)*_3\mathbf{A}) - ((_3\mathbf{A}\bullet_3\nabla)*_3\mathbf{B}) + (_3\nabla\bullet_3\mathbf{B})*_3\mathbf{A} - (_3\nabla\bullet_3\mathbf{A})*_3\mathbf{B}$$
$$_3\nabla*(_3\mathbf{A}\bullet_3\mathbf{B}) \equiv ((_3\mathbf{B}\bullet_3\nabla)*_3\mathbf{A}) + ((_3\mathbf{A}\bullet_3\nabla)*_3\mathbf{B}) + (_3\nabla\mathbf{x}_3\mathbf{B})\mathbf{x}_3\mathbf{A} + (_3\nabla\mathbf{x}_3\mathbf{A})\mathbf{x}_3\mathbf{B}$$

4) Proposed invariant "$_4S$" has time component "$U = \partial*(_3\mathbf{E}\bullet_3\mathbf{E} + c^2*_3\mathbf{B}\bullet_3\mathbf{B})/2$" and space components "$_3S = \partial*c*(_3\mathbf{E}\mathbf{x}_3\mathbf{B} - _3\mathbf{B}\mathbf{x}_3\mathbf{E})/2$". Try to apply compound label numbers and find "$(p_{xM})^{*i} \neq -p_{xM}$" for "$*^i$" applied to quaternions. Try to perform a Lorentz Transformation. Explain why "$_4S$" is not an actual invariant. "$_{-1}f = _4\nabla^{sn}\bullet_4S$" with respect to components, but why "$_{-4}f \neq _4\nabla^{sn}*_4S$"? Use "$_6E_{fourth} = -E_{amp}*(k_{yM} - p_{zM})*\exp(-i*(k_{xM}*(x_M - c*t_M)))$" and "$_4S = (_3\mathbf{E} - _3\mathbf{K})*(_3\mathbf{E} + _3\mathbf{K})/2$". Notice there's no mathematically beautiful way to create "$_4S$" from "$_6E$".

5) Try to find a linear combination of two of "q_x", "q_y" or "q_z" for "κ".

6) How can anti-matter be worked into a macroscopic approximation theory that inertial mass we measure is electromagnetic field energy of an electron?

7) Triple-vector-product identity "$_4\nabla^{sn}\blacksquare(_4\nabla^{*/sn}\mathbf{x}(_4V)) \equiv 0$" includes two space-component identities that add together to create Faraday's Law of Induction. Is there significance to "$_1\nabla^{sn}*(-_3\mathbf{K}) + _3\nabla^{sn}\mathbf{x}(-_3\mathbf{E}) = 0$" being formed from two identities?

$$_1\nabla^{sn}*(_3\nabla^{*/sn}\mathbf{x}(_3V)) + _3\nabla^{sn}\mathbf{x}(_1\nabla^{*/sn}*(_3V)) \equiv 0 \quad ; \quad _3\nabla^{sn}\mathbf{x}(_3\nabla^{*/sn}*(_1V)) \equiv 0$$

CHAPTER 3 – FIELDS

$$_1\nabla^{sn}*(_3\nabla^{*jsn}\mathbf{x}(_3V)) + (_3\nabla^{sn}\mathbf{x}(_1\nabla^{*jsn}*(_3V)) + {_3}\nabla^{sn}\mathbf{x}(_3\nabla^{*jsn}*(_1V))) \equiv 0$$

$$_1\nabla^{sn}*(_3\nabla^{*jsn}\mathbf{x}_3V) + {_3}\nabla^{sn}\mathbf{x}(_1\nabla^{*jsn}*_3V + {_3}\nabla^{*jsn}*_1V) \equiv 0$$

$$_1\nabla^{sn}*(-_3K) + {_3}\nabla^{sn}\mathbf{x}(-_3E) = 0$$

8) To better understand space-negative, review how space-negative behaves in the non-relativistic approximation example given below. A temperature gradient is given in "B" and is then Lorentz Transformed to "M" and to "S". The very simple non-relativistic approximation example is given so that space-negative is the only complexity. Is there a more optimal algebraic technique that can replace messy space-negative and that performs the same function?

In "B": $_1T = b*x_B$; $_2\nabla^{sn}*_1T = q_{xB}{}^{sn}*b = q_{xB}{}^{sn}*\partial_1 T/\partial x_B$ $(\partial_1 T/\partial ct_B = 0)$
In "M": $_1T = b*(x_M - v_M*t_M)$; $_2\nabla^{sn}*_1T = 1_M{}^{sn}*(-b*v_M/c) + q_{xM}{}^{sn}*b$
In "S": $_1T = b*(x_S - v_S*t_S)$; $_2\nabla^{sn}*_1T = 1_S{}^{sn}*(-b*v_S/c) + q_{xS}{}^{sn}*b$

Because of the non-relativistic approximation:

- $t_B = t_M = t_S$
- $v_S = v_M + v_{S/M}$
- $v_M * v_{S/M}/c^2 = 0$

$$\begin{array}{c} \partial_1 T/\partial ct_M \\ = \\ \partial_1 T/x_M \end{array} \quad \begin{array}{cc} 1 & -v_M/c \\ -v_M/c & 1 \end{array} * \begin{array}{c} \partial_1 T/\partial ct_B \\ \\ \partial_1 T/\partial x_B \end{array}$$

$$\begin{array}{c} \partial_1 T/\partial ct_S \\ = \\ \partial_1 T/x_S \end{array} \quad \begin{array}{cc} 1 & -v_{S/M}/c \\ -v_{S/M}/c & 1 \end{array} * \begin{array}{c} \partial_1 T/\partial ct_M \\ \\ \partial_1 T/\partial x_M \end{array}$$

$$\begin{array}{c} \partial_1 T/\partial ct_S \\ = \\ \partial_1 T/x_S \end{array} \quad \begin{array}{cc} 1 & -v_{S/M}/c \\ -v_{S/M}/c & 1 \end{array} * \begin{array}{cc} 1 & -v_M/c \\ -v_M/c & 1 \end{array} * \begin{array}{c} \partial_1 T/\partial ct_B \\ \\ \partial_1 T/\partial x_B \end{array} = \begin{array}{cc} 1 & -v_S/c \\ -v_S/c & 1 \end{array} * \begin{array}{c} \partial_1 T/\partial ct_B \\ \\ \partial_1 T/\partial x_B \end{array}$$

SPECIAL ALGEBRA FOR SPECIAL RELATIVITY

WAVES

Chapter 4 – Waves

The Dirac Equation is our relativistic model for dynamics of an electron. An electron is so small Newton's Second Law (force equals mass times acceleration) does not apply.

Newton's Second Law was presented in Chapter 1 by writing its geometric-vector equation and then explaining force, mass, and acceleration. In contrast, the Dirac Equation is a set of four first order differential equations that needs to be developed slowly emphasizing a logical thought process.

4.1 Differential Operator

<u>Mechanical Energy-Momentum.</u>   An electron's mechanical energy and momentum combine in the time-space momentum invariant "$_4p$".

$$_4p = \exp(-\kappa*\varsigma/2)*(E_M/c + q_x*p_{xM} + q_y*p_{yM} + q_z*p_{zM})*\exp(-\kappa*\varsigma/2)$$

"E_M" is mechanical energy. "p_{xM}", "p_{yM}", and "p_{zM}" are mechanical momentum components. "$_4p$" is time-like because energy is modeled using "cosh" and momentum "sinh".

$$_4p = \exp(-\kappa*\varsigma/2)*(m_B*c*\cosh\alpha_M + q*m_B*c*\sinh\alpha_M)*\exp(-\kappa*\varsigma/2)$$
$$= \exp(-\kappa*\varsigma/2)*m_B*c*\exp(q*\alpha_M)*\exp(-\kappa*\varsigma/2)$$

"α_M" relates to electron speed by "$v_M = c*\tanh\alpha_M$". Subscript "$_M$" identifies the inertial reference frame of the observer. "$_B$" is rest frame of the particle so that "m_B" is rest mass. "c" is speed-of-light.

"q" is made general through use of knowable circular-angles "θ".

$$q = q_x*\cos(\theta_{x/yz}) + (q_y*\cos(\theta_{y/z}) + q_z*\sin(\theta_{y/z}))*\sin(\theta_{x/yz}) \qquad q*q = 1$$

"1_M" and "q_M" are compound-label-numbers. "κ" is the unknown and unknowable unspecified simple-label-number. "κ" contrasts with "q" because "q" is knowable.

$$1_M = \exp(-\kappa*\varsigma) \quad ; \quad q_M = \exp(-\kappa*\varsigma/2)*q*\exp(-\kappa*\varsigma/2)$$

$$q_M = q_{xM}*\cos(\theta_{x/yz}) + (q_{yM}*\cos(\theta_{y/z}) + q_{zM}*\sin(\theta_{y/z}))*\sin(\theta_{x/yz}) \qquad q_M^{*j}*q_M = -1$$

Special Algebra for Special Relativity

$_4p = 1_M * m_B * c * \cosh\alpha_M + q_M * m_B * c * \sinh\alpha_M$

Electrical Energy-Momentum. The electrical energy-momentum invariant is "$_4q = Q_B *_4V$". "Q_B" is electron electric charge. "$_4V$" is external voltage.

The analogy for energy component "$Q_B *_1V$" is potential energy of a car at the top of a hill.

To visualize electrical momentum component "$Q_B *_3V$" think about what happens when a wire with direct current is cut. The magnetic field around the wire provides inertia to maintain the electric current, typically by ionizing air to make air conductive.

Total Energy-Momentum. Per de Broglie relations, total energy is proportional to frequency "ω" (or "$_1k$") and total momentum is proportional to wavenumber "k" (or "$_3k$") with the constant of proportionality Planck's constant "$\hbar$" (h-bar) ("$\hbar = h/(2*\pi)$").

$\hbar = 1.054571800(13) * 10^{-34}$ Joule*seconds   (angular momentum)

Total "$\hbar *_4k$" equals mechanical plus electrical.

$\hbar *_4k = {_4p} + Q_B *_4V$

The next task is to associate an actual wave to the particle.

Gradient Substitution for Frequency/Wave-Number. Total energy-momentum invariant "$\hbar *_4k$" is replaced with a differential gradient operator.

$\hbar *_4k = i*\tau*\hbar *_4\nabla^{sn}$ and $\hbar *_4k^{sn} = i*\tau*\hbar *(_4\nabla^{sn})^{sn} = i*\tau*\hbar *_4\nabla$

Replacement of frequency "$_4k$" by gradient operator "$i*\tau *_4\nabla^{sn}$" is justified by use of an invariant wave function "$_1T$" of the form below. Assume "τ" equals "+1".

$_1T = \exp(i*\tau*(\pm k_{xM}*x_M - \omega_M*t_M))$

$-_4k^{*/} \bullet {_4r} = k_{xM}*x_M - \omega_M*t_M$ (written for "$k_{yM} = k_{zM} = 0$")
$-_4k^{*/sn} \bullet {_4r} = -k_{xM}*x_M - \omega_M*t_M$ (written for "$k_{yM} = k_{zM} = 0$")

WAVES

"$_4k = i*\tau*_4\nabla^{sn}$" (and "$_4k^{sn} = i*\tau*_4\nabla$")

$1_M*\omega_M/c$ is replaced by $i*\tau*_1\nabla^{sn}$; $1_M^{sn}*\omega_M/c$ is replaced by $i*\tau*_1\nabla$
$_3k$ is replaced by $i*\tau*_3\nabla^{sn}$; $_3k^{sn}$ is replaced by $i*\tau*_3\nabla$

Per the above equations, the space-negative alternative is redundant and so will be dropped along with "±".

"$i*\tau*_1\nabla^{sn}$" is substituted for the total energy-momentum invariant.

$i*\tau*\hbar*(_4\nabla^{sn}) = {_4}p + Q_B*_4V$
$\qquad\qquad\qquad = 1_M*m_B*c*\cosh\alpha_M + q_M*m_B*c*\sinh\alpha_M + Q_B*(_1V + {_3}V)$

$i*\tau*\hbar*(_4\nabla^{sn}) = i*\tau*\hbar*(_1\nabla) - i*\tau*\hbar*(_3\nabla)$

$1_M*m_B*c*\cosh\alpha_M = i*\tau*\hbar*(_1\nabla) - Q_B*_1V$

$q_M*m_B*c*\sinh\alpha_M = -i*\tau*\hbar*(_3\nabla) - Q_B*_3V$

To make the above two equations component equations, divide them left and right by "$\exp(-\kappa*\varsigma/2)$".

$1*m_B*c*\cosh\alpha_M = i*\hbar*\tau*\nabla_{tM} - Q_B*V_{tM}$

$q*m_B*c*\sinh\alpha_M = q_x*(-i*\hbar*\tau*\nabla_{xM} - Q_B*V_{xM})$
$\qquad\qquad\qquad + q_y*(-i*\hbar*\tau*\nabla_{yM} - Q_B*V_{yM}) + q_z*(-i*\hbar*\tau*\nabla_{zM} - Q_B*V_{zM})$

"$_4\nabla^{sn}$" requires wave function "$_1T$" so that the above component equations apply to something that is both a particle and a wave.

Anti-Matter Visualized as the Space-Negative.
Anti-matter was modeled with "$\alpha_{S/M} = \pm i*\pi$". Here, anti-matter is suggested to be the space-negative. The two models for anti-matter are independent, such that matter could be modeled with the space-negative.

SPECIAL ALGEBRA FOR SPECIAL RELATIVITY

4.2 Development of the Dirac Equation

<u>Algebraic Matrix Equation</u> for mechanical components.

$1^2 = \exp(-q*\alpha_M)*\exp(q*\alpha_M)$
$1^2 = (\cosh\alpha_M - q*\sinh\alpha_M)*(\cosh\alpha_M + q*\sinh\alpha_M)$
$1^2 = \cosh^2\alpha_M - q^2*\sinh^2\alpha_M$
$1^2 - \cosh^2\alpha_M = -q^2*\sinh^2\alpha_M$
$-(1^2 - \cosh^2\alpha_M) = q^2*\sinh^2\alpha_M$

"$-(1^2 - \cosh^2\alpha_M) = q^2*\sinh^2\alpha_M$" becomes four equations.

Ψ_+: $(1 + \cosh\alpha_M)*(-1 + \cosh\alpha_M) = (q*\sinh\alpha_M)*(q*\sinh\alpha_M)$
Φ_+: $(1 - \cosh\alpha_M)*(-1 - \cosh\alpha_M) = (-q*\sinh\alpha_M)*(-q*\sinh\alpha_M)$
Φ_-: $(1 - \cosh\alpha_M)*(-1 - \cosh\alpha_M) = (q*\sinh\alpha_M)*(q*\sinh\alpha_M)$
Ψ_-: $(1 + \cosh\alpha_M)*(-1 + \cosh\alpha_M) = (-q*\sinh\alpha_M)*(-q*\sinh\alpha_M)$

"Φ_+" is redundant to "Ψ_+", and "Φ_-" to "Ψ_-". Address both "Ψ_+" and "Ψ_-" with a "$\pm$" sign. The "Ψ" equation is split by introducing enabler functions "PP_M" and "QQ_M".

$((1 + \cosh\alpha_M)*PP_M)*((-1 + \cosh\alpha_M)*-QQ_M) =$
$\qquad ((\pm q*\sinh\alpha_M)*-QQ_M)*((\pm q*\sinh\alpha_M)*PP_M)$

$((-1 + \cosh\alpha_M)*-QQ_M) = ((\pm q*\sinh\alpha_M)*PP_M)$
$((1 + \cosh\alpha_M)*PP_M) = ((\pm q*\sinh\alpha_M)*-QQ_M)$

$(1 + \cosh\alpha_M)*PP_M + (\pm q*\sinh\alpha_M)*QQ_M = 0$
$(\pm q*\sinh\alpha_M)*PP_M + (-1 + \cosh\alpha_M)*QQ_M = 0$

$$\begin{array}{cc} (1 + \cosh\alpha_M) & (\pm q*\sinh\alpha_M) \\ (\pm q*\sinh\alpha_M) & (-1 + \cosh\alpha_M) \end{array} * \begin{array}{c} PP_M \\ QQ_M \end{array} = \begin{array}{c} 0 \\ 0 \end{array}$$

"$\pm$" sign in front of "$q*\sinh\alpha_M$" ("$\pm q*\sinh\alpha_M = q*\sinh(\pm\alpha_M)$" and "$\cosh\alpha_M = \cosh(\pm\alpha_M)$") relates to motion being right or left per "$v_M/c = \pm\tanh\alpha_M = \tanh(\pm\alpha_M)$" because the "$-$" of "$\pm$" refers to space-negative.

"0"'s on the right make it a "singular" algebraic matrix equation.

WAVES

<u>Algebraic Solutions to the Matrix Equation.</u> In general, an algebraic 2x2 singular matrix equation has two independent solution pairs: pair "$_1$" and pair "$_2$". Because of angle identities, there are three ways of writing the two pairs.

$\cosh(\alpha_M/2) = \cosh(\alpha_M - \alpha_M/2) = \cosh\alpha_M \cdot \cosh(\alpha_M/2) - \sinh\alpha_M \cdot \sinh(\alpha_M/2)$

$\sinh(\alpha_M/2) = \sinh(\alpha_M - \alpha_M/2) = \sinh\alpha_M \cdot \cosh(\alpha_M/2) - \cosh\alpha_M \cdot \sinh(\alpha_M/2)$

$(1 + \cosh\alpha_M)/\sinh\alpha_M = \sinh\alpha_M/(-1 + \cosh\alpha_M) = \cosh(\alpha_M/2)/\sinh(\alpha_M/2)$

$((PP_M, QQ_M)_1)_{\text{not-half-anglesA}} = (\pm(-q \cdot \sinh\alpha_M),\ 1 + \cosh\alpha_M)$
$((PP_M, QQ_M)_2)_{\text{not-half-anglesA}} = (\sinh\alpha_M,\ \pm(-q) \cdot (1 + \cosh\alpha_M))$

$((PP_M, QQ_M)_1)_{\text{not-half-anglesB}} = (\pm(-q) \cdot (-1 + \cosh\alpha_M),\ \sinh\alpha_M)$
$((PP_M, QQ_M)_2)_{\text{not-half-anglesB}} = (-1 + \cosh\alpha_M,\ \pm(-q \cdot \sinh\alpha_M))$

$(PP_M, QQ_M)_1 = (\pm(-q \cdot \sinh(\alpha_M/2)),\ \cosh(\alpha_M/2))$
$(PP_M, QQ_M)_2 = (\sinh(\alpha_M/2),\ \pm(-q \cdot \cosh(\alpha_M/2)))$

"$\pm(-q)$" is preferred to be a factor on "$\sinh\alpha_M$". But, as is obvious in the three forms above, this preference cannot be satisfied.

"$\sinh(\alpha_M/2)$" and "$\cosh(\alpha_M/2)$" together form mechanical momentum per "$\sinh\alpha_M = 2 \cdot \sinh(\alpha_M/2) \cdot \cosh(\alpha_M/2)$" and mechanical energy per "$\cosh\alpha_M = \cosh^2(\alpha_M/2) + \sinh^2(\alpha_M/2)$".

A general solution for a singular algebraic matrix equation is a linear combination formed by multiplying an arbitrary constant, "f" or "g", by each of the two solutions.

$PP_{M1} = \pm(-q) \cdot PP_{M2}$ and $QQ_{M1} = \pm(-q) \cdot QQ_{M2}$

$(PP_M, QQ_M) = f \cdot (PP_M, QQ_M)_1 + g \cdot (PP_M, QQ_M)_2$
$= f \cdot (PP_M, QQ_M)_1 \pm g \cdot (-q) \cdot (PP_M, QQ_M)_1$
$= (f \pm (-q) \cdot g) \cdot (PP_M, QQ_M)_1$

The half angle is useful because a half in an exponent represents a square root operation. It implies a need for a square operation later in the analysis. The square operation, in the form of a complex number multiplied by its conjugate, is required for calculating a measurable particle property from a quantum mechanics wave function solution.

SPECIAL ALGEBRA FOR SPECIAL RELATIVITY

Substitute Differential Operators into the Matrix Equation. First, multiply by rest mass "m_B*c".

$$\begin{pmatrix} (m_B*c + m_B*c*\cosh\alpha_M) & (\pm q*m_B*c*\sinh\alpha_M) \\ (\pm q*m_B*c*\sinh\alpha_M) & (-m_B*c + m_B*c*\cosh\alpha_M) \end{pmatrix} * \begin{pmatrix} PP_M \\ QQ_M \end{pmatrix} = \begin{pmatrix} 0 \\ 0 \end{pmatrix}$$

$$\begin{pmatrix} (m_B*c + i*\tau*\hbar*\nabla_{tM} - Q_B*V_{tM}) & \pm(-i*\tau*\hbar*(q_x*\nabla_{xM}+q_y*\nabla_{yM}+q_z*\nabla_{zM})-Q_B*(q_x*V_{xM}+q_y*V_{yM}+q_z*V_{zM})) \\ \pm(-i*\tau*\hbar*(q_x*\nabla_{xM}+q_y*\nabla_{yM}+q_z*\nabla_{zM})-Q_B*(q_x*V_{xM}+q_y*V_{yM}+q_z*V_{zM})) & (-m_B*c + i*\tau*\hbar*\nabla_{tM} - Q*V_{tM}) \end{pmatrix} * \begin{pmatrix} PP_M \\ QQ_M \end{pmatrix} = \begin{pmatrix} 0 \\ 0 \end{pmatrix}$$

"$_1T = \exp(i*\tau*(\pm k_{xM}*x_M - \omega_M*t_M))$" had a handedness specified by "$\tau = +1$". Opposite handedness is specified by "$\tau = -1$". For completeness, "τ" has remained in the analysis until now, when it will be identified as irrelevant. A change in handedness (in this development of the Dirac Equation) is a change from matter to anti-matter (or from anti-matter to matter). To make that change: Change "τ" to "$-\tau$". Multiply all four components of the above 2x2 matrix by "-1". Swap the sign of the electric charge so that "Q_B" is replaced by its negative "$-Q_B$" and swap the sign of the rest mass so that "m_B" is replaced by "$-m_B$". Look at what remains. See it is the same as what was started with. The choice of "τ" was irrelevant and therefore "$\tau = +1$" is used.

$$\begin{pmatrix} (m_B*c + i*\hbar*\nabla_{tM} - Q_B*V_{tM}) & \pm(-i*\hbar*(q_x*\nabla_{xM}+q_y*\nabla_{yM}+q_z*\nabla_{zM})-Q_B*(q_x*V_{xM}+q_y*V_{yM}+q_z*V_{zM})) \\ \pm(-i*\hbar*(q_x*\nabla_{xM}+q_y*\nabla_{yM}+q_z*\nabla_{zM})-Q_B*(q_x*V_{xM}+q_y*V_{yM}+q_z*V_{zM})) & (-m_B*c + i*\hbar*\nabla_{tM} - Q_B*V_{tM}) \end{pmatrix} * \begin{pmatrix} PP_M \\ QQ_M \end{pmatrix} = \begin{pmatrix} 0 \\ 0 \end{pmatrix}$$

The transformation of the mathematical model from particle to wave occurred when mechanical energy and mechanical momentum were replaced by total energy (minus electrical) and total momentum (minus electrical), respectively, by use of differential operators. Now that the equation pertains to waves and not to particles, the either-or plus or minus separately sign "$\pm$" is replaced by the both plus and minus but also neither plus nor minus separately sign "$\pm_N$".

WAVES

$$(m_B*c + i*\hbar*\nabla_{tM} - Q_B*V_{tM})$$
$$\pm_N(-i*\hbar*(q_x*\nabla_{xM}+q_y*\nabla_{yM}+q_z*\nabla_{zM})-Q_B*(q_x*V_{xM}+q_y*V_{yM}+q_z*V_{zM})) \quad PP_M \quad 0$$
$$*\quad =$$
$$\pm_N(-i*\hbar*(q_x*\nabla_{xM}+q_y*\nabla_{yM}+q_z*\nabla_{zM})-Q_B*(q_x*V_{xM}+q_y*V_{yM}+q_z*V_{zM}))$$
$$(-m_B*c + i*\hbar*\nabla_{tM} - Q_B*V_{tM}) \quad QQ_M \quad 0$$

Substitute-in Matrix Isomorphs. "q_x", "q_y" and "q_z" have 2x2 matrix isomorph equivalents which are very similar to Pauli Spin Matrices (with the difference being "R_2" is negative). "$R_1 = G_1/i$", "$R_2 = G_2/i$", "$R_3 = G_3/i$", and "$1 = i/i$".

$$G_1 \Rightarrow \begin{matrix} 0 & i \\ i & 0 \end{matrix} \quad G_2 \Rightarrow \begin{matrix} 0 & -1 \\ 1 & 0 \end{matrix} \quad G_3 \Rightarrow \begin{matrix} i & 0 \\ 0 & -i \end{matrix}$$

$$R_1 \Rightarrow \begin{matrix} 0 & 1 \\ 1 & 0 \end{matrix} \quad R_2 \Rightarrow \begin{matrix} 0 & k \\ -k & 0 \end{matrix} \quad R_3 \Rightarrow \begin{matrix} 1 & 0 \\ 0 & -1 \end{matrix} \quad 1 \Rightarrow \begin{matrix} 1 & 0 \\ 0 & 1 \end{matrix}$$

("k" is the same as "i" but is different symbolically for tracking.)

The traditional substitution is "$q_x \Rightarrow R_1$", "$q_y \Rightarrow R_2$", and "$q_z \Rightarrow R_3$". Enabler functions are replaced by column vectors. Dirac Equation:

$$m_B*c* \begin{matrix} 1 & 0 & 0 & 0 \\ 0 & 1 & 0 & 0 \\ 0 & 0 & -1 & 0 \\ 0 & 0 & 0 & -1 \end{matrix} + (i*\hbar*\partial/\partial ct_M - Q_B*V_{tM})* \begin{matrix} 1 & 0 & 0 & 0 \\ 0 & 1 & 0 & 0 \\ 0 & 0 & 1 & 0 \\ 0 & 0 & 0 & 1 \end{matrix} \quad \pm_N(-i*\hbar*\partial/\partial x_M - Q_B*V_{xM})* \begin{matrix} 0 & 0 & 0 & 1 \\ 0 & 0 & 1 & 0 \\ 0 & 1 & 0 & 0 \\ 1 & 0 & 0 & 0 \end{matrix}$$

$$\pm_N(-i*\hbar*\partial/\partial y_M - Q_B*V_{yM})* \begin{matrix} 0 & 0 & 0 & k \\ 0 & 0 & -k & 0 \\ 0 & k & 0 & 0 \\ -k & 0 & 0 & 0 \end{matrix} \quad \pm_N(-i*\hbar*\partial/\partial z_M - Q_B*V_{zM})* \begin{matrix} 0 & 0 & 1 & 0 \\ 0 & 0 & 0 & -1 \\ 1 & 0 & 0 & 0 \\ 0 & -1 & 0 & 0 \end{matrix} * \begin{matrix} \Psi_{1M} \\ \Psi_{2M} \\ \Psi_{3M} \\ \Psi_{4M} \end{matrix} = \begin{matrix} 0 \\ 0 \\ 0 \\ 0 \end{matrix}$$

The "Dirac Spinor" is the column-vector with four components "Ψ_{1M}", "Ψ_{2M}", "Ψ_{3M}", "Ψ_{4M}". These four component symbols are short-hand for more expanded symbols "$\Psi_{z\pm 1M}$", "$\Psi_{z\pm 2M}$", "$\Psi_{z\pm 3M}$", "$\Psi_{z\pm 4M}$". "z" identifies "q_z" has non-zero major diagonal elements. "$\pm$" identifies "$\pm_N$" is included.

The Dirac Equation is a differential 4x4 singular matrix equation with eight solutions: Four for matter and four for anti-matter.

SPECIAL ALGEBRA FOR SPECIAL RELATIVITY

4.3 Solutions to the Dirac Equation

In general, finding solutions to the Dirac Equation is difficult. But finding solutions is easy for the specific case of motion in the positive "x" direction with no external voltage applied. For that case, the Dirac Equation is reduced to two simple pairs of equations of identical form.

$$\Psi_{1M} + (i\hbar/m_Bc)*\partial\Psi_{1M}/\partial ct_M - \pm_N(i\hbar/m_Bc)*\partial\Psi_{4M}/\partial x_M = 0$$
$$-\Psi_{4M} + (i\hbar/m_Bc)*\partial\Psi_{4M}/\partial ct_M - \pm_N(i\hbar/m_Bc)*\partial\Psi_{1M}/\partial x_M = 0$$

$$\Psi_{2M} + (i\hbar/m_Bc)*\partial\Psi_{2M}/\partial ct_M - \pm_N(i\hbar/m_Bc)*\partial\Psi_{3M}/\partial x_M = 0$$
$$-\Psi_{3M} + (i\hbar/m_Bc)*\partial\Psi_{3M}/\partial ct_M - \pm_N(i\hbar/m_Bc)*\partial\Psi_{2M}/\partial x_M = 0$$

The first pair has four Dirac Spinor solutions analogous, one for one, with the four electromagnetic spiral waves.

First Dirac Spinor Solution (for "$\Psi_{z+1M\text{-first}}$" and "$\Psi_{z+4M\text{-first}}$")

$\Psi_{1M\text{-first}} = \Psi_{amp}*\cosh(\alpha_M/2)*\exp(i*(\pm_N k_{xM\text{-am}}*x_M + \omega_{M\text{-am}}*t_M))$
$\Psi_{4M\text{-first}} = \Psi_{amp}*\sinh(\alpha_M/2)*\exp(i*(\pm_N k_{xM\text{-am}}*x_M + \omega_{M\text{-am}}*t_M))$
$\Psi_{4M\text{-first}} = \tanh(\alpha_M/2)*\Psi_{1M\text{-first}}$

Second Dirac Spinor Solution (for "$\Psi_{z+1M\text{-second}}$" and "$\Psi_{z+4M\text{-second}}$")

$\Psi_{1M\text{-second}} = -\Psi_{amp}*\sinh(\alpha_M/2)*\exp(i*(\pm_N k_{xM\text{-m}}*x_M - \omega_{M\text{-m}}*t_M))$
$\Psi_{4M\text{-second}} = \Psi_{amp}*\cosh(\alpha_M/2)*\exp(i*(\pm_N k_{xM\text{-m}}*x_M - \omega_{M\text{-m}}*t_M))$
$\Psi_{4M\text{-second}} = -\coth(\alpha_M/2)*\Psi_{1M\text{-second}}$

Third Dirac Spinor Solution (for "$\Psi_{z+1M\text{-third}}$" and "$\Psi_{z+4M\text{-third}}$")

$\Psi_{1M\text{-third}} = \Psi_{amp}*\sinh(\alpha_M/2)*\exp(-i*(\pm_N k_{xM\text{-am}}*x_M + \omega_{M\text{-am}}*t_M))$
$\Psi_{4M\text{-third}} = \Psi_{amp}*\cosh(\alpha_M/2)*\exp(-i*(\pm_N k_{xM\text{-am}}*x_M + \omega_{M\text{-am}}*t_M))$
$\Psi_{4M\text{-third}} = \coth(\alpha_M/2)*\Psi_{1M\text{-third}}$

Fourth Dirac Spinor Solution (for "$\Psi_{z+1M\text{-fourth}}$" and "$\Psi_{z+4M\text{-fourth}}$")

$\Psi_{1M\text{-fourth}} = \Psi_{amp}*\cosh(\alpha_M/2)*\exp(-i*(\pm_N k_{xM\text{-m}}*x_M - \omega_{M\text{-m}}*t_M))$
$\Psi_{4M\text{-fourth}} = -\Psi_{amp}*\sinh(\alpha_M/2)*\exp(-i*(\pm_N k_{xM\text{-m}}*x_M - \omega_{M\text{-m}}*t_M))$
$\Psi_{4M\text{-fourth}} = -\tanh(\alpha_M/2)*\Psi_{1M\text{-fourth}}$

WAVES

(For engineering, "$\pm_N$" will be "+". For theory development retain "$\pm_N$", and do not place it as in the below alternative, as explained in the next chapter.)

Alternative First Dirac Spinor Solution   (do not use)

$\Psi_{1M\text{-first}} = \Psi_{amp} * \cosh(\alpha_M/2) * \exp(i*(k_{xM\text{-am}}*x_M + \omega_{M\text{-am}}*t_M))$

$\Psi_{4M\text{-first}} = \pm_N \Psi_{amp} * \sinh(\alpha_M/2) * \exp(i*(k_{xM\text{-am}}*x_M + \omega_{M\text{-am}}*t_M))$

$\Psi_{4M\text{-first}} = \pm_N \tanh(\alpha_M/2) * \Psi_{1M\text{-first}}$

Proof the First Solution is Correct.

$\cosh\alpha_M = \hbar*\omega_M/(m_B*c^2)$; $\sinh\alpha_M = \hbar*k_{xM}/(m_B*c)$ (for $_4V = 0$)

Substitute the first solution into the pair of differential equations. "Ψ_{amp}" and "$\exp(i*(\pm_N k_{xM}*x_M + \omega_M*t_M))$" divide out.

$0 = \Psi_{1M\text{-first}} + (i\hbar/mc)*\partial\Psi_{1M\text{-first}}/\partial ct_M - \pm_N(i\hbar/mc)*\partial\Psi_{4M\text{-first}}/\partial x_M$
$= \cosh(\alpha_M/2) + (i\hbar/mc^2)*(i*\omega_M)*\cosh(\alpha_M/2) - \pm_N(i\hbar/mc)*(i*\pm_N k_{xM})*\sinh(\alpha_M/2)$
$= \cosh(\alpha_M/2) - \cosh\alpha_M*\cosh(\alpha_M/2) + \sinh\alpha_M*\sinh(\alpha_M/2) = 0$

$0 = -\Psi_{4M\text{-first}} + (i\hbar/mc)*\partial\Psi_{4M\text{-first}}/\partial ct_M - \pm_N(i\hbar/mc)*\partial\Psi_{1M\text{-first}}/\partial x_M$
$= -\sinh(\alpha_M/2) + (i\hbar/mc^2)*(i*\omega_M)*\sinh(\alpha_M/2) - \pm_N(i\hbar/mc)*(i*\pm_N k_{xM})*\cosh(\alpha_M/2)$
$= -\sinh(\alpha_M/2) - \cosh\alpha_M*\sinh(\alpha_M/2) + \sinh\alpha_M*\cosh(\alpha_M/2) = 0$

Single Speed in this Simple Example.
Regardless of there being only one speed represented by "α_M", the concept of an interference group applies.

Matter and Anti-matter.
The plus "+" or the minus "-" sign (for "$\pm_N$" "+") in front of "ω_M" determines if the solution is anti-matter or is matter, respectively. "$k_{xM\text{-am}}$" and "$\omega_{M\text{-am}}$" are for anti-matter. "$k_{xM\text{-m}}$" and "$\omega_{M\text{-m}}$" are for matter. The "$_{am}$" subscript is only a reminder that anti-matter requires a space-negative in a Lorentz Transformation. The actual numerical value of "k_{xM}" and "ω_M" is not dependent on it being matter or anti-matter.

$k_{xM\text{-am}} = k_{xM\text{-m}} = k_{xM}$; $\omega_{M\text{-am}} = \omega_{M\text{-m}} = \omega_M$

SPECIAL ALGEBRA FOR SPECIAL RELATIVITY

Matter. Second and fourth solutions pertain to matter (for "$\pm_N$" "+") with "$m_B > 0$" and "$Q_B < 0$". "$k_{xM-m}*x_M - \omega_{M-m}*t_M = {-_4}k^{*j}\bullet_4 r$" describes wave crests and nodes that move in the positive "x_M" direction at phase speed "v_{pM}". Phase speed "v_{pM}/c", for the simple case of no voltage, is the reciprocal of the electron particle's group speed "$v_M/c = \tanh\alpha_M$".

$$v_{pM-m} = \omega_M/k_{xM} = \omega_{M-m}/k_{xM-m} = c*\cosh\alpha_M/\sinh\alpha_M$$
$$v_{pM-m}/c = c/v_{M-m} = \coth\alpha_M \quad \text{(matter)}$$

Anti-Matter. First and third solutions (for "$\pm_N$" "+") have "$m_B < 0$" and "$Q_B > 0$" for an anti-matter electron (positron). "$k_{xM-am}*x_M + \omega_{M-am}*t_M = {_4}k^{*jsn}\bullet_4 r$" describes wave crests and nodes that move in the negative "x_M" direction at speed "v_{pM}".

$$v_{pM-am}/c = (\omega_M/c)/(-k_{xM}) = (\omega_{M-am}/c)/(-k_{xM-am}) = \cosh\alpha_M/-\sinh\alpha_M$$
$$= c/v_{M-am} = -\coth\alpha_M \quad \text{(anti-matter)}$$

In the headlight/tail-light visualization both electrons and positrons have headlights pointing toward more positive "x" (right). And, electric current for both electrons and positrons is the same.

Spin. The "+"/"−" sign in front of "*i*" determines if angular momentum spin is right-hand or left-hand. In the car visualization a right-side steering wheel represents right-hand spin. A positron with a left steering wheel is the anti-matter counterpart to an electron with a right steering wheel, and that pairs first with fourth and second with third.

Glove Visualization for Both Anti-matter and Spin. Second/fourth solutions (matter) are visualized as a left/right glove pair with fingers pointing to positive "x". Pull the gloves inside out, and now the gloves represent the first/third solutions (anti-matter) pointing to negative "x". Second became first and fourth became third. Rotate these to point to positive "x" and place them inside the original gloves, first inside fourth and third inside second. Excess energy separates them with second/fourth moving to more positive "x" and third/first moving to more negative "x", for equal and opposite linear momentum and angular momentum.

WAVES

4.4 Particle Properties

Because Dirac Spinor waves cannot be measured directly, the solution is post-processed to replace wave nature with particle nature, using the method proposed by Max Born for Schrödinger's Equation solutions.

Multiply the complex-conjugate of the Dirac Spinor solution wave function, "$_4\Psi^{*i}$", by the Dirac Equation, and perform the opposite operation. Add the two and, alternatively, subtract the two. Subtraction drops the mathematically real terms. Addition drops imaginary terms.

Subtraction Equation Resulting in Electric Current Density.

The sum equals zero because of the zeros on the right side of the Dirac Equation. ("M" subscript is dropped to not clutter the equation.)

$0 \quad = \Psi_1^{*i}(mc^*\Psi_1 + i\hbar^*\partial\Psi_1/\partial ct - QV_t^*\Psi_1$
$- \pm_N i\hbar^*\partial\Psi_4/\partial x - \pm_N QV_x^*\Psi_4 - \pm_N k^* i\hbar^*\partial\Psi_4/\partial y - \pm_N k^* QV_y^*\Psi_4 - \pm_N i\hbar^*\partial\Psi_3/\partial z - \pm_N QV_z^*\Psi_3)$
$+ \Psi_2^{*i}(mc^*\Psi_2 + i\hbar^*\partial\Psi_2/\partial ct - QV_t^*\Psi_2$
$- \pm_N i\hbar^*\partial\Psi_3/\partial x - \pm_N QV_x^*\Psi_3 - \pm_N(-k)^* i\hbar^*\partial\Psi_3/\partial y - \pm_N(-k)^* QV_y^*\Psi_3 - \pm_N(-1)^* i\hbar^*\partial\Psi_4/\partial z$
$\qquad - \pm_N(-1)^* QV_z^*\Psi_4)$
$+ \Psi_3^{*i}(-mc^*\Psi_3 + i\hbar^*\partial\Psi_3/\partial ct - QV_t^*\Psi_3$
$- \pm_N i\hbar^*\partial\Psi_2/\partial x - \pm_N QV_x^*\Psi_2 - \pm_N k^* i\hbar^*\partial\Psi_2/\partial y - \pm_N k^* QV_y^*\Psi_2 - \pm_N i\hbar^*\partial\Psi_1/\partial z - \pm_N QV_z^*\Psi_1)$
$+ \Psi_4^{*i}(-mc^*\Psi_4 + i\hbar^*\partial\Psi_4/\partial ct - QV_t^*\Psi_4$
$- \pm_N i\hbar^*\partial\Psi_1/\partial x - \pm_N(-k)^* i\hbar^*\partial\Psi_1/\partial y - \pm_N(-k)^* QV_y^*\Psi_1 - \pm_N(-1)^* i\hbar^*\partial\Psi_2/\partial z$
$\qquad - \pm_N(-1)^* QV_z^*\Psi_2)$

$- (mc^*\Psi_1^{*i} - i\hbar^*\partial\Psi_1^{*i}/\partial ct - QV_t^*\Psi_1^{*i} + \pm_N i\hbar^*\partial\Psi_4^{*i}/\partial x - \pm_N QV_x^*\Psi_4^{*i}$
$+ \pm_N(-k)^* i\hbar^*\partial\Psi_4^{*i}/\partial y - \pm_N(-k)^* QV_y^*\Psi_4^{*i} + \pm_N i\hbar^*\partial\Psi_3^{*i}/\partial z - \pm_N QV_z^*\Psi_3^{*i})^*\Psi_1$
$- (mc^*\Psi_2^{*i} - i\hbar^*\partial\Psi_2^{*i}/\partial ct - QV_t^*\Psi_2^{*i} + \pm_N i\hbar^*\partial\Psi_3^{*i}/\partial x - \pm_N QV_x^*\Psi_3^{*i}$
$+ \pm_N(+k)^* i\hbar^*\partial\Psi_3^{*i}/\partial y - \pm_N(k)^* QV_y^*\Psi_3^{*i} + \pm_N(-1)^* i\hbar^*\partial\Psi_4^{*i}/\partial z - \pm_N(-1)^* QV_z^*\Psi_4^{*i})^*\Psi_2$
$- (mc^*\Psi_3^{*i} - i\hbar^*\partial\Psi_3^{*i}/\partial ct - QV_t^*\Psi_3^{*i} + \pm_N i\hbar^*\partial\Psi_2^{*i}/\partial x - \pm_N QV_x^*\Psi_2^{*i}$
$+ \pm_N(-k)^* i\hbar^*\partial\Psi_2^{*i}/\partial y - \pm_N(-k)^* QV_y^*\Psi_2^{*i} + \pm_N i\hbar^*\partial\Psi_1^{*i}/\partial z - \pm_N QV_z^*\Psi_1^{*i})^*\Psi_3$
$- (mc^*\Psi_4^{*i} - i\hbar^*\partial\Psi_4^{*i}/\partial ct - QV_t^*\Psi_4^{*i} + \pm_N i\hbar^*\partial\Psi_1^{*i}/\partial x - \pm_N QV_x^*\Psi_1^{*i}$
$+ \pm_N(k)^* i\hbar^*\partial\Psi_1^{*i}/\partial y - \pm_N(k)^* QV_y^*\Psi_1^{*i} + \pm_N(-1)^* i\hbar^*\partial\Psi_2^{*i}/\partial z - \pm_N(-1)^* QV_z^*\Psi_2^{*i})^*\Psi_4$

"mc" and "QV" terms subtract away. Only "$i\hbar$" terms remain.

$0 = \Psi_1^{*i}(i\hbar^*\partial\Psi_1/\partial ct - \pm_N i\hbar^*\partial\Psi_4/\partial x - \pm_N k^* i\hbar^*\partial\Psi_4/\partial y - \pm_N i\hbar^*\partial\Psi_3/\partial z)$
$+ \Psi_2^{*i}(i\hbar^*\partial\Psi_2/\partial ct - \pm_N i\hbar^*\partial\Psi_3/\partial x - \pm_N(-k)^* i\hbar^*\partial\Psi_3/\partial y - \pm_N(-1)^* i\hbar^*\partial\Psi_4/\partial z)$
$+ \Psi_3^{*i}(i\hbar^*\partial\Psi_3/\partial ct - \pm_N i\hbar^*\partial\Psi_2/\partial x - \pm_N k^* i\hbar^*\partial\Psi_2/\partial y - \pm_N i\hbar^*\partial\Psi_1/\partial z)$
$+ \Psi_4^{*i}(i\hbar^*\partial\Psi_4/\partial ct - \pm_N i\hbar^*\partial\Psi_1/\partial x - \pm_N(-k)^* i\hbar^*\partial\Psi_1/\partial y - \pm_N(-1)^* i\hbar^*\partial\Psi_2/\partial z)$

SPECIAL ALGEBRA FOR SPECIAL RELATIVITY

$- (-i\hbar * \partial \Psi_1^{*i}/\partial ct + \pm_N i\hbar * \partial \Psi_4^{*i}/\partial x + \pm_N(-k) * i\hbar * \partial \Psi_4^{*i}/\partial y + \pm_N i\hbar * \partial \Psi_3^{*i}/\partial z) * \Psi_1$
$- (-i\hbar * \partial \Psi_2^{*i}/\partial ct + \pm_N i\hbar * \partial \Psi_3^{*i}/\partial x + \pm_N k * i\hbar * \partial \Psi_3^{*i}/\partial y \pm_N -i\hbar * \partial \Psi_4^{*i}/\partial z) * \Psi_2$
$- (-i\hbar * \partial \Psi_3^{*i}/\partial ct + \pm_N i\hbar * \partial \Psi_2^{*i}/\partial x + \pm_N(-k) * i\hbar * \partial \Psi_2^{*i}/\partial y + \pm_N i\hbar * \partial \Psi_1^{*i}/\partial z) * \Psi_3$
$- (-i\hbar * \partial \Psi_4^{*i}/\partial ct + \pm_N(i\hbar * \partial \Psi_1^{*i}/\partial x + \pm_N k * i\hbar * \partial \Psi_1^{*i}/\partial y \pm_N -i\hbar * \partial \Psi_2^{*i}/\partial z) * \Psi_4$

"$i\hbar$" terms are combined using the chain rule. For example:

$$\Psi_4^{*i} * (\partial \Psi_2/\partial z) + (\partial \Psi_4^{*i}/\partial z) * \Psi_2 = \partial(\Psi_4^{*i} * \Psi_2)/\partial z$$

$$\Psi_4^{*i} * (-(-1) * i\hbar * \partial \Psi_2/\partial z) + (-(-1) * i\hbar * \partial \Psi_4^{*i}/\partial z) * \Psi_2 = -(-1) * i\hbar * \partial(\Psi_4^{*i} * \Psi_2)/\partial z$$

$0 = i\hbar * \partial(\Psi_{1M}^{*i} * \Psi_{1M} + \Psi_{2M}^{*i} * \Psi_{2M} + \Psi_{3M}^{*i} * \Psi_{3M} + \Psi_{4M}^{*i} * \Psi_{4M})/\partial ct_M$
$\quad - \pm_N i\hbar * \partial(\Psi_{1M}^{*i} * \Psi_{4M} + \Psi_{2M}^{*i} * \Psi_{3M} + \Psi_{3M}^{*i} * \Psi_{2M} + \Psi_{4M}^{*i} * \Psi_{1M})/\partial x_M$
$\quad - \pm_N k * i\hbar * \partial(\Psi_{1M}^{*i} * \Psi_{4M} - \Psi_{2M}^{*i} * \Psi_{3M} + \Psi_{3M}^{*i} * \Psi_{2M} - \Psi_{4M}^{*i} * \Psi_{1M})/\partial y_M$
$\quad - \pm_N i\hbar * \partial(\Psi_{1M}^{*i} * \Psi_{3M} - \Psi_{2M}^{*i} * \Psi_{4M} + \Psi_{3M}^{*i} * \Psi_{1M} - \Psi_{4M}^{*i} * \Psi_{2M})/\partial z_M$

Substitute "+" for "$\pm_N$" because the subtraction equation is a particle equation and is not a wave equation.

$0 = \nabla_{tM} * J_{tM} + \nabla_{xM} * J_{xM} + \nabla_{yM} * J_{yM} + \nabla_{zM} * J_{zM}$
$\quad = \partial J_{tM}/\partial ct_M + \partial J_{xM}/\partial x_M + \partial J_{yM}/\partial y_M + \partial J_{zM}/\partial z_M$

$J_{tM} = Q_B * (\Psi_{1M}^{*i} * \Psi_{1M} + \Psi_{2M}^{*i} * \Psi_{2M} + \Psi_{3M}^{*i} * \Psi_{3M} + \Psi_{4M}^{*i} * \Psi_{4M})$
$J_{xM} = -Q_B * (\Psi_{1M}^{*i} * \Psi_{4M} + \Psi_{2M}^{*i} * \Psi_{3M} + \Psi_{3M}^{*i} * \Psi_{2M} + \Psi_{4M}^{*i} * \Psi_{1M})$
$J_{yM} = -Q_B * k * (\Psi_{1M}^{*i} * \Psi_{4M} - \Psi_{2M}^{*i} * \Psi_{3M} + \Psi_{3M}^{*i} * \Psi_{2M} - \Psi_{4M}^{*i} * \Psi_{1M})$
$J_{zM} = -Q_B * (\Psi_{1M}^{*i} * \Psi_{3M} - \Psi_{2M}^{*i} * \Psi_{4M} + \Psi_{3M}^{*i} * \Psi_{1M} - \Psi_{4M}^{*i} * \Psi_{2M})$

"Ψ" components are "$\sqrt{\text{particle count per volume}}$" and "$Q_B$" is "electric charge per particle" to make measurement units on "J" "electric charge per volume".

First Dirac Spinor Solution example:

$J_{tM} = Q_B * (\Psi_{1M}^{*i} * \Psi_{1M} + \Psi_{2M}^{*i} * \Psi_{2M} + \Psi_{3M}^{*i} * \Psi_{3M} + \Psi_{4M}^{*i} * \Psi_{4M})$
$\quad = Q_B * \Psi_{amp}^2 * (\cosh^2(\alpha_M/2) + \sinh^2(\alpha_M/2))$
$\quad = Q_B * \Psi_{amp}^2 * \cosh\alpha_M$

WAVES

$$J_{xM} = -Q_B*(\Psi_{1M}^{*i}*\Psi_{4M} + \Psi_{2M}^{*i}*\Psi_{3M} + \Psi_{3M}^{*i}*\Psi_{2M} + \Psi_{4M}^{*i}*\Psi_{1M})$$
$$= -Q_B*\Psi_{amp}^2*(2*\cosh(\alpha_M/2)*\sinh(\alpha_M/2))$$
$$= -Q_B*\Psi_{amp}^2*\sinh\alpha_M$$

$$J_{yM} = 0 \qquad J_{zM} = 0$$

$$_4J = Q_B*\Psi_{amp}^2*(1_M*\cosh\alpha_M - q_{xM}*\sinh\alpha_M)$$
$$= Q_B*\Psi_{amp}^2*1_M*\exp(-q_x*\alpha_M)$$

The first solution above was for a positron that moves to negative "x" for "$\alpha_M > 0$", per the "$-q_x*\alpha_M$". "$Q_B > 0$" so net current is to negative "x".

Addition Equation has the last four lines of the subtraction equation started by a "+" rather than a "-". It pertains to the field outside the classical radius of the electron.

$$0 = \Psi_1^{*i}*(mc*\Psi_1 - QV_t*\Psi_1 - \pm_N QV_x*\Psi_4 - \pm_N k*QV_y*\Psi_4 - \pm_N QV_z*\Psi_3)$$
$$+ \Psi_2^{*i}*(mc*\Psi_2 - QV_t*\Psi_2 - \pm_N QV_x*\Psi_3 - \pm_N(-k)*QV_y*\Psi_3 - \pm_N(-1)*QV_z*\Psi_4)$$
$$+ \Psi_3^{*i}*(-mc*\Psi_3 - QV_t*\Psi_3 - \pm_N QV_x*\Psi_2 - \pm_N k*QV_y*\Psi_2 - \pm_N QV_z*\Psi_1)$$
$$+ \Psi_4^{*i}*(-mc*\Psi_4 - QV_t*\Psi_4 - \pm_N QV_x*\Psi_1 - \pm_N(-k)*QV_y*\Psi_1 - \pm_N(-1)*QV_z*\Psi_2)$$

$$+ (mc*\Psi_1^{*i} - QV_t*\Psi_1^{*i} - \pm_N QV_x*\Psi_4^{*i} - \pm_N(-k)*QV_y*\Psi_4^{*i} - \pm_N QV_z*\Psi_3^{*i})*\Psi_1$$
$$+ (mc*\Psi_2^{*i} - QV_t*\Psi_2^{*i} - \pm_N QV_x*\Psi_3^{*i} - \pm_N k*QV_y*\Psi_3^{*i} - \pm_N(-1)*QV_z*\Psi_4^{*i})*\Psi_2$$
$$+ (mc*\Psi_3^{*i} - QV_t*\Psi_3^{*i} - \pm_N QV_x*\Psi_2^{*i} - \pm_N(-k)*QV_y*\Psi_2^{*i} - \pm_N QV_z*\Psi_1^{*i})*\Psi_3$$
$$+ (mc*\Psi_4^{*i} - QV_t*\Psi_4^{*i} - \pm_N QV_x*\Psi_1^{*i} - \pm_N k*QV_y*\Psi_1^{*i} - \pm_N(-1)*QV_z*\Psi_2^{*i})*\Psi_4$$

$$+ \Psi_1^{*i}*(i\hbar*\partial\Psi_1/\partial ct - \pm_N i\hbar*\partial\Psi_4/\partial x - \pm_N k*i\hbar*\partial\Psi_4/\partial y - \pm_N i\hbar*\partial\Psi_3/\partial z)$$
$$+ \Psi_2^{*i}*(i\hbar*\partial\Psi_2/\partial ct - \pm_N i\hbar*\partial\Psi_3/\partial x - \pm_N(-k)*i\hbar*\partial\Psi_3/\partial y - \pm_N(-1)*i\hbar*\partial\Psi_4/\partial z)$$
$$+ \Psi_3^{*i}*(i\hbar*\partial\Psi_3/\partial ct - \pm_N i\hbar*\partial\Psi_2/\partial x - \pm_N k*i\hbar*\partial\Psi_2/\partial y - \pm_N i\hbar*\partial\Psi_1/\partial z)$$
$$+ \Psi_4^{*i}*(i\hbar*\partial\Psi_4/\partial ct - \pm_N i\hbar*\partial\Psi_1/\partial x - \pm_N(-k)*i\hbar*\partial\Psi_1/\partial y - \pm_N(-1)*i\hbar*\partial\Psi_2/\partial z)$$

$$+ (-i\hbar*\partial\Psi_1^{*i}/\partial ct + \pm_N i\hbar*\partial\Psi_4^{*i}/\partial x + \pm_N(-k)*i\hbar*\partial\Psi_4^{*i}/\partial y + \pm_N i\hbar*\partial\Psi_3^{*i}/\partial z)*\Psi_1$$
$$+ (-i\hbar*\partial\Psi_2^{*i}/\partial ct + \pm_N i\hbar*\partial\Psi_3^{*i}/\partial x + \pm_N k*i\hbar*\partial\Psi_3^{*i}/\partial y - \pm_N i\hbar*\partial\Psi_4^{*i}/\partial z)*\Psi_2$$
$$+ (-i\hbar*\partial\Psi_3^{*i}/\partial ct + \pm_N i\hbar*\partial\Psi_2^{*i}/\partial x + \pm_N(-k)*i\hbar*\partial\Psi_2^{*i}/\partial y + \pm_N i\hbar*\partial\Psi_1^{*i}/\partial z)*\Psi_3$$
$$+ (-i\hbar*\partial\Psi_4^{*i}/\partial ct + \pm_N i\hbar*\partial\Psi_1^{*i}/\partial x + \pm_N k*i\hbar*\partial\Psi_1^{*i}/\partial y - \pm_N i\hbar*\partial\Psi_2^{*i}/\partial z)*\Psi_4$$

SPECIAL ALGEBRA FOR SPECIAL RELATIVITY

$= 2*(m*c)*(\Psi_1^{*i}*\Psi_1 + \Psi_2^{*i}*\Psi_2 - \Psi_3^{*i}*\Psi_3 - \Psi_4^{*i}*\Psi_4)$
$+ 2*(-Q*V_t)*(\Psi_1^{*i}*\Psi_1 + \Psi_2^{*i}*\Psi_2 + \Psi_3^{*i}*\Psi_3 + \Psi_4^{*i}*\Psi_4)$
$+ 2*(-QV_x)*(\Psi_1^{*i}*\Psi_4 + \Psi_2^{*i}*\Psi_3 + \Psi_3^{*i}*\Psi_2 + \Psi_4^{*i}*\Psi_1)$
$+ 2*k*(-QV_y)*(+\Psi_1^{*i}*\Psi_4 - \Psi_2^{*i}*\Psi_3 + \Psi_3^{*i}*\Psi_2 - \Psi_4^{*i}*\Psi_1)$
$+ 2*(-QV_z)*(\Psi_1^{*i}*\Psi_3 - \Psi_2^{*i}*\Psi_4 + \Psi_3^{*i}*\Psi_1 - \Psi_4^{*i}*\Psi_2)$
$+ \Psi_1^{*i}*(i\hbar*\partial\Psi_1/\partial ct - \pm_N i\hbar*\partial\Psi_4/\partial x - \pm_N k*i\hbar*\partial\Psi_4/\partial y - \pm_N i\hbar*\partial\Psi_3/\partial z)$
$+ \Psi_2^{*i}*(i\hbar*\partial\Psi_2/\partial ct - \pm_N i\hbar*\partial\Psi_3/\partial x - \pm_N(-k)*i\hbar*\partial\Psi_3/\partial y - \pm_N(-1)*i\hbar*\partial\Psi_4/\partial z)$
$+ \Psi_3^{*i}*(i\hbar*\partial\Psi_3/\partial ct - \pm_N i\hbar*\partial\Psi_2/\partial x - \pm_N k*i\hbar*\partial\Psi_2/\partial y - \pm_N i\hbar*\partial\Psi_1/\partial z)$
$+ \Psi_4^{*i}*(i\hbar*\partial\Psi_4/\partial ct - \pm_N i\hbar*\partial\Psi_1/\partial x - \pm_N(-k)*i\hbar*\partial\Psi_1/\partial y - \pm_N(-1)*i\hbar*\partial\Psi_2/\partial z)$
$+ (-i\hbar*\partial\Psi_1^{*i}/\partial ct + \pm_N i\hbar*\partial\Psi_4^{*i}/\partial x + \pm_N(-k)*i\hbar*\partial\Psi_4^{*i}/\partial y + \pm_N i\hbar*\partial\Psi_3^{*i}/\partial z)*\Psi_1$
$+ (-i\hbar*\partial\Psi_2^{*i}/\partial ct + \pm_N i\hbar*\partial\Psi_3^{*i}/\partial x + \pm_N k*i\hbar*\partial\Psi_3^{*i}/\partial y - \pm_N i\hbar*\partial\Psi_4^{*i}/\partial z)*\Psi_2$
$+ (-i\hbar*\partial\Psi_3^{*i}/\partial ct + \pm_N i\hbar*\partial\Psi_2^{*i}/\partial x + \pm_N(-k)*i\hbar*\partial\Psi_2^{*i}/\partial y + \pm_N i\hbar*\partial\Psi_1^{*i}/\partial z)*\Psi_3$
$+ (-i\hbar*\partial\Psi_4^{*i}/\partial ct + \pm_N i\hbar*\partial\Psi_1^{*i}/\partial x + \pm_N k*i\hbar*\partial\Psi_1^{*i}/\partial y - \pm_N i\hbar*\partial\Psi_2^{*i}/\partial z)*\Psi_4$

Specific to the Fourth Dirac Spinor Solution, the addition equation results in the following equation.

$0 = 2*(m*c)*(\Psi_{1\text{M-fourth}}^{*i}*\Psi_{1\text{M-fourth}} - \Psi_{4\text{M-fourth}}^{*i}*\Psi_{4\text{M-fourth}})$
$+ i\hbar*\Psi_{1\text{M-fourth}}^{*i}*\partial\Psi_{1\text{M-fourth}}/\partial ct - \pm_N i\hbar*\Psi_{1\text{M-fourth}}^{*i}*\partial\Psi_{4\text{M-fourth}}/\partial x$
$+ i\hbar*\Psi_{4\text{M-fourth}}^{*i}*\partial\Psi_{4\text{M-fourth}}/\partial ct - \pm_N i\hbar*\Psi_{4\text{M-fourth}}^{*i}*\partial\Psi_{1\text{M-fourth}}/\partial x$
$+ -i\hbar*\partial\Psi_{1\text{M-fourth}}^{*i}/\partial ct*\Psi_{1\text{M-fourth}} + \pm_N i\hbar*\partial\Psi_{4\text{M-fourth}}^{*i}/\partial x*\Psi_{1\text{M-fourth}}$
$+ -i\hbar*\partial\Psi_{4\text{M-fourth}}^{*i}/\partial ct*\Psi_{4\text{M-fourth}} + \pm_N i\hbar*\partial\Psi_{1\text{M-fourth}}^{*i}/\partial x*\Psi_{4\text{M-fourth}}$

$= 2*(m*c)*\Psi_{\text{amplitude}}^2*(\cosh^2(\alpha_M/2) - \sinh^2(\alpha_M/2))$
$- \hbar*(\omega_M/c)*\Psi_{\text{amplitude}}^2*(\cosh^2(\alpha_M/2) + \cosh^2(\alpha_M/2))$
$+ \pm_N \hbar*(k_{xM})*\Psi_{\text{amplitude}}^2*(\pm_N\cosh(\alpha_M/2)*\sinh(\alpha_M/2) + \pm_N\sinh(\alpha_M/2)*\cosh(\alpha_M/2))$
$+ \pm_N \hbar*(k_{xM})*\Psi_{\text{amplitude}}^2*(\pm_N\cosh(\alpha_M/2)*\sinh(\alpha_M/2) + \pm_N\sinh(\alpha_M/2)*\cosh(\alpha_M/2))$
$- \hbar*(\omega_M/c)*\Psi_{\text{amplitude}}^2*(\sinh^2(\alpha_M/2) + \sinh^2(\alpha_M/2))$

$= 2*(m*c)*\Psi_{\text{amplitude}}^2$
$+ \hbar*(\omega_M/c)*2*\Psi_{\text{amplitude}}^2*\cosh(\alpha_M) - \hbar*(k_{xM})*2*\Psi_{\text{amplitude}}^2*\sinh(\alpha_M)$

$= 2*(m*c)*\Psi_{\text{amplitude}}^2 + 2*(m*c)*\Psi_{\text{amplitude}}^2*(\cosh^2(\alpha_M) - \sinh^2(\alpha_M))$
$= 2*(m*c)*\Psi_{\text{amplitude}}^2*(1 - (\cosh^2(\alpha_M) - \sinh^2(\alpha_M)))$
$= 0$

The addition equation pertains to energy and momentum which are phenomena outside the classic radius of the electron. In contrast, the subtraction equation pertains to electric charge inside the classical radius.

WAVES

4.5 Two Alternative Arrangements

<u>The "x" Arrangement</u> has "$q_y => R_1$", "$q_z => R_2$", and "$q_x => R_3$".

$$m*c* \begin{pmatrix} 1 & 0 & 0 & 0 \\ 0 & 1 & 0 & 0 \\ 0 & 0 & -1 & 0 \\ 0 & 0 & 0 & -1 \end{pmatrix} + (i*\hbar*\partial/\partial ct_M - q*V_{tM})* \begin{pmatrix} 1 & 0 & 0 & 0 \\ 0 & 1 & 0 & 0 \\ 0 & 0 & 1 & 0 \\ 0 & 0 & 0 & 1 \end{pmatrix} \pm_N (-i*\hbar*\partial/\partial y_M - q*V_{yM})* \begin{pmatrix} 0 & 0 & 0 & 1 \\ 0 & 0 & 1 & 0 \\ 0 & 1 & 0 & 0 \\ 1 & 0 & 0 & 0 \end{pmatrix}$$

$$\pm_N (-i*\hbar*\partial/\partial z_M - q*V_{zM})* \begin{pmatrix} 0 & 0 & 0 & k \\ 0 & 0 & -k & 0 \\ 0 & k & 0 & 0 \\ -k & 0 & 0 & 0 \end{pmatrix} \pm_N (-i*\hbar*\partial/\partial x_M - q*V_{xM})* \begin{pmatrix} 0 & 0 & 1 & 0 \\ 0 & 0 & 0 & -1 \\ 1 & 0 & 0 & 0 \\ 0 & -1 & 0 & 0 \end{pmatrix} * \begin{pmatrix} \Psi_{x\pm1M} \\ \Psi_{x\pm2M} \\ \Psi_{x\pm3M} \\ \Psi_{x\pm4M} \end{pmatrix} = \begin{pmatrix} 0 \\ 0 \\ 0 \\ 0 \end{pmatrix}$$

For the case of "x" direction motion and no external voltage:

$\Psi_{x\pm1M} + (i\hbar/mc)*\partial\Psi_{x\pm1M}/\partial ct_M - \pm_N(i\hbar/mc)*\partial\Psi_{x\pm3M}/\partial x_M = 0$
$-\Psi_{x\pm3M} + (i\hbar/mc)*\partial\Psi_{x\pm3M}/\partial ct_M - \pm_N(i\hbar/mc)*\partial\Psi_{x\pm1M}/\partial x_M = 0$

$\Psi_{x\pm2M} + (i\hbar/mc)*\partial\Psi_{x\pm2M}/\partial ct_M + \pm_N(i\hbar/mc)*\partial\Psi_{x\pm4M}/\partial x_M = 0$
$-\Psi_{x\pm4M} + (i\hbar/mc)*\partial\Psi_{x\pm4M}/\partial ct_M + \pm_N(i\hbar/mc)*\partial\Psi_{x\pm2M}/\partial x_M = 0$

First Dirac Spinor Solution
$\Psi_{x\pm1M\text{-first}} = \Psi_{amp}*\cosh(\alpha_M/2)*\exp(i*(\pm_N k_{xM}*x_M + \omega_M*t_M))$
$\Psi_{x\pm3M\text{-first}} = \Psi_{amp}*\sinh(\alpha_M/2)*\exp(i*(\pm_N k_{xM}*x_M + \omega_M*t_M))$
$\Psi_{x\pm3M\text{-first}} = \Psi_{x\pm1M\text{-first}}*\tanh(\alpha_M/2)$

Second Dirac Spinor Solution
$\Psi_{x\pm1M\text{-second}} = \Psi_{amp}*\sinh(\alpha_M/2)*\exp(i*(\pm_N k_{xM}*x_M - \omega_M*t_M))$
$\Psi_{x\pm3M\text{-second}} = -\Psi_{amp}*\cosh(\alpha_M/2)*\exp(i*(\pm_N k_{xM}*x_M - \omega_M*t_M))$
$\Psi_{x\pm3M\text{-second}} = -\Psi_{x\pm1M\text{-second}}*\coth(\alpha_M/2)$

Third Dirac Spinor Solution
$\Psi_{x\pm1M\text{-third}} = \Psi_{amp}*\sinh(\alpha_M/2)*\exp(-i*(\pm_N k_{xM}*x_M + \omega_M*t_M))$
$\Psi_{x\pm3M\text{-third}} = \Psi_{amp}*\cosh(\alpha_M/2)*\exp(-i*(\pm_N k_{xM}*x_M + \omega_M*t_M))$
$\Psi_{x\pm3M\text{-third}} = \Psi_{x\pm1M\text{-third}}*\coth(\alpha_M/2)$

SPECIAL ALGEBRA FOR SPECIAL RELATIVITY

Fourth Dirac Spinor Solution

$\Psi_{x\pm1M\text{-fourth}} = \Psi_{amp}*\cosh(\alpha_M/2)*\exp(-i*(\pm_N k_{xM}*x_M - \omega_M*t_M))$

$\Psi_{x\pm3M\text{-fourth}} = -\Psi_{amp}*\sinh(\alpha_M/2)*\exp(-i*(\pm_N k_{xM}*x_M - \omega_M*t_M))$

$\Psi_{x\pm3M\text{-fourth}} = -\Psi_{x\pm1M\text{-fourth}}*\tanh(\alpha_M/2)$

<u>The "y" Arrangement</u> has "$q_z => R_1$", "$q_x => R_2$", and "$q_y => R_3$".

$$m*c* \begin{matrix} 1 & 0 & 0 & 0 \\ 0 & 1 & 0 & 0 \\ 0 & 0 & -1 & 0 \\ 0 & 0 & 0 & -1 \end{matrix} + (i*\hbar*\partial/\partial ct_M - q*V_{tM})* \begin{matrix} 1 & 0 & 0 & 0 \\ 0 & 1 & 0 & 0 \\ 0 & 0 & 1 & 0 \\ 0 & 0 & 0 & 1 \end{matrix} \pm_N (-i*\hbar*\partial/\partial z_M - q*V_{zM})* \begin{matrix} 0 & 0 & 0 & 1 \\ 0 & 0 & 1 & 0 \\ 0 & 1 & 0 & 0 \\ 1 & 0 & 0 & 0 \end{matrix}$$

$$\pm_N (-i*\hbar*\partial/\partial x_M - q*V_{xM})* \begin{matrix} 0 & 0 & 0 & k \\ 0 & 0 & -k & 0 \\ 0 & k & 0 & 0 \\ -k & 0 & 0 & 0 \end{matrix} \pm_N (-i*\hbar*\partial/\partial y_M - q*V_{yM})* \begin{matrix} 0 & 0 & 1 & 0 \\ 0 & 0 & 0 & -1 \\ 1 & 0 & 0 & 0 \\ 0 & -1 & 0 & 0 \end{matrix} * \begin{matrix} \Psi_{y\pm1M} \\ \Psi_{y\pm2M} \\ \Psi_{y\pm3M} \\ \Psi_{y\pm4M} \end{matrix} = \begin{matrix} 0 \\ 0 \\ 0 \\ 0 \end{matrix}$$

For the case of "x" direction motion and no external voltage:

$\Psi_{y\pm1M} + (i\hbar/mc)*\partial\Psi_{y\pm1M}/\partial ct_M - \pm_N k*(i\hbar/mc)*\partial\Psi_{y\pm4M}/\partial x_M = 0$

$-\Psi_{y\pm4M} + (i\hbar/mc)*\partial\Psi_{y\pm4M}/\partial ct_M \pm_N k*(i\hbar/mc)*\partial\Psi_{y\pm1M}/\partial x_M = 0$

$\Psi_{y\pm2M} + (i\hbar/mc)*\partial\Psi_{y\pm2M}/\partial ct_M \pm_N k*(i\hbar/mc)*\partial\Psi_{y\pm3M}/\partial x_M = 0$

$-\Psi_{y\pm3M} + (i\hbar/mc)*\partial\Psi_{y\pm3M}/\partial ct_M - \pm_N k*(i\hbar/mc)*\partial\Psi_{y\pm2M}/\partial x_M = 0$

First Dirac Spinor Solution

$\Psi_{y\pm1M\text{-first}} = \Psi_{amp}*\cosh(\alpha_M/2)*\exp(i*(\pm_N k_{xM}*x_M + \omega_M*t_M))$

$\Psi_{y\pm4M\text{-first}} = -k*\Psi_{amp}*\sinh(\alpha_M/2)*\exp(i*(\pm_N k_{xM}*x_M + \omega_M*t_M))$

$\Psi_{y\pm4M\text{-first}} = -k*\Psi_{y\pm1M\text{-first}}*\tanh(\alpha_M/2)$

Second Dirac Spinor Solution

$\Psi_{y\pm1M\text{-second}} = \Psi_{amp}*\sinh(\alpha_M/2)*\exp(i*(\pm_N k_{xM}*x_M - \omega_M*t_M))$

$\Psi_{y\pm4M\text{-second}} = -(-k)*\Psi_{amp}*\cosh(\alpha_M/2)*\exp(i*(\pm_N k_{xM}*x_M - \omega_M*t_M))$

$\Psi_{y\pm4M\text{-second}} = k*\Psi_{y\pm1M\text{-second}}*\coth(\alpha_M/2)$

Third Dirac Spinor Solution

$\Psi_{y\pm 1M\text{-third}} = \Psi_{amp}*\sinh(\alpha_M/2)*\exp(-i*(\pm_N k_{xM}*x_M + \omega_M*t_M))$

$\Psi_{y\pm 4M\text{-third}} = -k*\Psi_{amp}*\cosh(\alpha_M/2)*\exp(-i*(\pm_N k_{xM}*x_M + \omega_M*t_M))$

$\Psi_{y\pm 4M\text{-third}} = -k*\Psi_{y\pm 1M\text{-third}}*\coth(\alpha_M/2)$

Fourth Dirac Spinor Solution

$\Psi_{y\pm 1M\text{-fourth}} = \Psi_{amp}*\cosh(\alpha_M/2)*\exp(-i*(\pm_N k_{xM}*x_M - \omega_M*t_M))$

$\Psi_{y\pm 4M\text{-fourth}} = -(-k)*\Psi_{amp}*\sinh(\alpha_M/2)*\exp(-i*(\pm_N k_{xM}*x_M - \omega_M*t_M))$

$\Psi_{y\pm 4M\text{-fourth}} = k*\Psi_{y\pm 1M\text{-fourth}}*\tanh(\alpha_M/2)$

"z", "x", and "y" arrangements appear redundant because they result in the same measurable charge density space-time invariant "$_2J$".

4.6 Lorentz Transformation of a Dirac Spinor

The Lorentz Transformation for a Dirac Spinor uses half angle "$\alpha_M/2$". Proof of validity of the Lorentz Transformation is that the Dirac Spinor in "S" can be obtained by either the Lorentz Transformation of the Dirac Spinor directly, or by solving the Dirac Equation with its component variables specified in "S" rather than in "M". "$_4J$", too, must be properly affected by the Lorentz Transformation.

The fourth solution is the example because it is for matter.

"$\Psi_{1M\text{-fourth}}$" and "$\Psi_{4M\text{-fourth}}$" are associated with compound-label-numbers "e_{1M}" and "e_{4M}", respectively, with "$e_{1M} = q_x*e_{4M}$".

$$e_1 = q_x*e_4 = \begin{matrix} 1 & 0 & 0 & 0 \\ 0 & 0 & 0 & 1 \\ 0 & 0 & 1 & 0 \\ 0 & 1 & 0 & 0 \end{matrix} = \begin{matrix} 1 \\ 0 \\ 0 \\ 0 \end{matrix} * \begin{matrix} 0 \\ 0 \\ 0 \\ 1 \end{matrix}$$

In the next chapter is an explanation for anti-commutative operations "$e_{1M} = -e_{4M}*q_x$" and "$e_{4M} = -e_{1M}*q_x$".

Condensing components into a single expression for "$_4\Psi\text{-fourth}$".

$_4\Psi\text{-fourth} = e_{1M}*\Psi_{1M} + e_{4M}*\Psi_{4M}$
$\qquad = e_{1M}*\Psi_{1M\text{-fourth}} + e_{4M}*\Psi_{4M\text{-fourth}}$

SPECIAL ALGEBRA FOR SPECIAL RELATIVITY

$$= (e_{1M}*\Psi_{amp}*\cosh(\alpha_M/2)*\exp(-i*(\pm_N k_{xM-m}*x_M - \omega_{M-m}*t_M)))$$
$$+ (e_{4M}*\Psi_{amp}*-\sinh(\alpha_M/2)*\exp(-i*(\pm_N k_{xM-m}*x_M - \omega_{M-m}*t_M)))$$

$$= (e_{1M}*\Psi_{amp}*\cosh(\alpha_M/2)*\exp(-i*(\pm_N k_{xM-m}*x_M - \omega_{M-m}*t_M)))$$
$$+ (-e_{1M}*q_x*\Psi_{amp}*-\sinh(\alpha_M/2)*\exp(-i*(\pm_N k_{xM-m}*x_M - \omega_{M-m}*t_M)))$$

$$= e_{1M}*\Psi_{amp}*\exp(q_x*\alpha_M/2)*\exp(-i*(\pm_N k_{xM-m}*x_M - \omega_{M-m}*t_M))$$

"$e_{1M}*\Psi_{amp}*\exp(q_x*\alpha_M/2)*\exp(-i*(\pm_N k_{xM}*x_M - \omega_M*t_M))$" is convenient because "e_{1M}" changes to "e_{1S}", "α_M" changes to "$\alpha_S = \alpha_M + \alpha_{S/M}$", and "$\pm_N k_{xM-m}*x_M - \omega_{M-m}*t_M$" to "$\pm_N k_{xS-m}*x_S - \omega_{S-m}*t_S$". We begin with the general form of the Lorentz Transformation.

$$_4\Psi_{\text{-fourth}} = e_{1M}*\Psi_{amp}*\exp(q_x*\alpha_M/2)*\exp(-i*(\pm_N k_{xM-m}*x_M - \omega_{M-m}*t_M))$$

$$= e_{1M}*\Psi_{amp}*\exp(q_x*\alpha_M/2)*\exp(-i*(\pm_N k_{xM-m}*x_M - \omega_{M-m}*t_M))$$
$$*\exp(q_x*\alpha_{S/M}/2)/\exp(q_x*\alpha_{S/M}/2)$$

$$= (e_{1M}/\exp(q_x*\alpha_{S/M}/2))$$
$$*\Psi_{amp}*\exp(q_x*\alpha_M/2)*\exp(q_x*\alpha_{S/M}/2)*\exp(-i*(\pm_N k_{xM-m}*x_M - \omega_{M-m}*t_M))$$

$$= e_{1S}*\Psi_{amp}*\exp(q_x*(\alpha_M + \alpha_{S/M})/2)*\exp(-i*(\pm_N k_{xM-m}*x_M - \omega_{M-m}*t_M))$$
$$= e_{1S}*\Psi_{amp}*\exp(q_x*\alpha_S/2)*\exp(-i*(\pm_N k_{xM-m}*x_M - \omega_{M-m}*t_M))$$
$$= e_{1S}*\Psi_{amp}*\exp(q_x*\alpha_S/2)*\exp(-i*(\pm_N k_{xS-m}*x_S - \omega_{S-m}*t_S))$$

$$= e_{1S}*\Psi_{1S\text{-fourth}} + e_{4S}*\Psi_{4S\text{-fourth}}$$

$\cosh(-\alpha_M/2)*\cosh(-\alpha_{S/M}/2) + \sinh(-\alpha_M/2)*\sinh(-\alpha_{S/M}/2) = \cosh(-\alpha_M/2 - \alpha_{S/M}/2)$
$\cosh(-\alpha_M/2)*\sinh(-\alpha_{S/M}/2) + \sinh(-\alpha_M/2)*\cosh(-\alpha_{S/M}/2) = \sinh(-\alpha_M/2 - \alpha_{S/M}/2)$

$e_1*\Psi_{1S\text{-fourth}} + e_4*\Psi_{4S\text{-fourth}}$
$$= (e_1*\Psi_{1M\text{-fourth}} + e_4*\Psi_{4M\text{-fourth}})*\exp(q_x*\alpha_{S/M}/2)$$
$$= (e_1*\Psi_{1M\text{-fourth}} + e_4*\Psi_{4M\text{-fourth}})*(\cosh(\alpha_{S/M}/2) + q_x*\sinh(\alpha_{S/M}/2))$$

$$= e_1*\Psi_{1M\text{-fourth}}*\cosh(\alpha_{S/M}/2) + e_4*q_x*\Psi_{4M\text{-fourth}}*\sinh(\alpha_{S/M}/2)$$
$$+ e_1*q_x*\Psi_{1M\text{-fourth}}*\sinh(\alpha_{S/M}/2) + e_4*\Psi_{4M\text{-fourth}}*\cosh(\alpha_{S/M}/2)$$

$$= e_1*(\Psi_{1M\text{-fourth}}*\cosh(\alpha_{S/M}/2) - \Psi_{4M\text{-fourth}}*\sinh(\alpha_{S/M}/2))$$
$$+ e_4*(-\Psi_{1M\text{-fourth}}*\sinh(\alpha_{S/M}/2) + \Psi_{4M\text{-fourth}}*\cosh(\alpha_{S/M}/2))$$

WAVES

$\Psi_{1S\text{-fourth}} = \Psi_{1M\text{-fourth}}*\cosh(\alpha_{S/M}/2) - \Psi_{4M\text{-fourth}}*\sinh(\alpha_{S/M}/2)$

$= \Psi_{amp}*\cosh(-\alpha_M/2)*\cosh(\alpha_{S/M}/2)*\exp(-i*(\pm_N k_{xM\text{-}m}*x_M - \omega_{M\text{-}m}*t_M))$
$- \Psi_{amp}*\sinh(-\alpha_M/2)*\sinh(\alpha_{S/M}/2)*\exp(-i*(\pm_N k_{xM\text{-}m}*x_M - \omega_{M\text{-}m}*t_M))$

$= \Psi_{amp}*\cosh(-\alpha_M/2)*\cosh(-\alpha_{S/M}/2)*\exp(-i*(\pm_N k_{xM\text{-}m}*x_M - \omega_{M\text{-}m}*t_M))$
$+ \Psi_{amp}*\sinh(-\alpha_M/2)*\sinh(-\alpha_{S/M}/2)*\exp(-i*(\pm_N k_{xM\text{-}m}*x_M - \omega_{M\text{-}m}*t_M))$

$= \Psi_{amp}*\cosh(-\alpha_M/2 - \alpha_{S/M}/2)*\exp(-i*(\pm_N k_{xM\text{-}m}*x_M - \omega_{M\text{-}m}*t_M))$
$= \Psi_{amp}*\cosh(-\alpha_S/2)*\exp(-i*(\pm_N k_{xM\text{-}m}*x_M - \omega_{M\text{-}m}*t_M))$
$= \Psi_{amp}*\cosh(-\alpha_S/2)*\exp(-i*(\pm_N k_{xS\text{-}m}*x_S - \omega_{S\text{-}m}*t_S))$
$= \Psi_{amp}*\cosh(\alpha_S/2)*\exp(-i*(\pm_N k_{xS\text{-}m}*x_S - \omega_{S\text{-}m}*t_S))$

$\Psi_{4S\text{-fourth}} = \Psi_{1M\text{-fourth}}*\sinh(-\alpha_{S/M}/2) + \Psi_{4M\text{-fourth}}*\cosh(-\alpha_{S/M}/2)$

$= \Psi_{amp}*\cosh(-\alpha_M/2)*\sinh(-\alpha_{S/M}/2)*\exp(-i*(\pm_N k_{xM\text{-}m}*x_M - \omega_{M\text{-}m}*t_M))$
$+ \Psi_{amp}*\sinh(-\alpha_M/2)*\cosh(-\alpha_{S/M}/2)*\exp(-i*(\pm_N k_{xM\text{-}m}*x_M - \omega_{M\text{-}m}*t_M))$

$= \Psi_{amp}*\sinh(-\alpha_M/2 - \alpha_{S/M}/2)*\exp(-i*(\pm_N k_{xM\text{-}m}*x_M - \omega_{M\text{-}m}*t_M))$
$= \Psi_{amp}*\sinh(-\alpha_S/2)*\exp(-i*(\pm_N k_{xM\text{-}m}*x_M - \omega_{M\text{-}m}*t_M))$
$= \Psi_{amp}*\sinh(-\alpha_S/2)*\exp(-i*(\pm_N k_{xS\text{-}m}*x_S - \omega_{S\text{-}m}*t_S))$
$= -\Psi_{amp}*\sinh(\alpha_S/2)*\exp(-i*(\pm_N k_{xS\text{-}m}*x_S - \omega_{S\text{-}m}*t_S))$

$e_{1M} = e_1*\exp(-q_x*\varsigma/2) = e_1*\cosh(\varsigma/2) - e_1*q_x*\sinh(\varsigma/2)$
$\qquad\qquad\qquad\qquad\quad = e_1*\cosh(\varsigma/2) + e_4*\sinh(\varsigma/2)$

$e_{4M} = e_4*\exp(-q_x*\varsigma/2) = -e_4*q_x*\sinh(\varsigma/2) + e_4*\cosh(\varsigma/2)$
$\qquad\qquad\qquad\qquad\quad = e_1*\sinh(\varsigma/2) + e_4*\cosh(\varsigma/2)$

$e_{1S} = e_{1M}/\exp(q_x*\alpha_{S/M}/2) = e_{1M}*\cosh(\alpha_{S/M}/2) + e_{4M}*\sinh(\alpha_{S/M}/2)$
$e_{4S} = e_{4M}/\exp(q_x*\alpha_{S/M}/2) = e_{1M}*\sinh(\alpha_{S/M}/2) + e_{4M}*\cosh(\alpha_{S/M}/2)$

$e_{1S} = -e_{4S}*q_x$
$e_{4S} = -e_{1S}*q_x$

Complex-conjugate is needed for electric current density.

SPECIAL ALGEBRA FOR SPECIAL RELATIVITY

$_4\Psi_{\text{-fourth}}{}^{*i} = (\exp(-i*(\pm_N k_{xM-m}*x_M - \omega_{M-m}*t_M)))^{*i}*(\exp(-q_x*\alpha_M/2))^{*i}*\Psi_{\text{amp}}{}^{*i}*e_{1M}{}^{*i}$

$\Psi_{1S\text{-fourth}}{}^{*i} = \Psi_{1M\text{-fourth}}{}^{*i}*\cosh(-\alpha_{S/M}/2) + \Psi_{4M\text{-fourth}}{}^{*i}*\sinh(-\alpha_{S/M}/2)$
$\Psi_{4S\text{-fourth}}{}^{*i} = \Psi_{1M\text{-fourth}}{}^{*i}*\sinh(-\alpha_{S/M}/2) + \Psi_{4M\text{-fourth}}{}^{*i}*\cosh(-\alpha_{S/M}/2)$

$J_{tS} = Q_B*(\Psi_{1S}{}^{*i}*\Psi_{1S} + \Psi_{2S}{}^{*i}*\Psi_{2S} + \Psi_{3S}{}^{*i}*\Psi_{3S} + \Psi_{4S}{}^{*i}*\Psi_{4S})$
$\phantom{J_{tS}} = Q_B*(\Psi_{1S\text{-fourth}}{}^{*i}*\Psi_{1S\text{-fourth}} + \Psi_{4S\text{-fourth}}{}^{*i}*\Psi_{4S\text{-fourth}})$

$\phantom{J_{tS}} = Q_B*((\Psi_{1M\text{-fourth}}{}^{*i}*\cosh(-\alpha_{S/M}/2) + \Psi_{4M\text{-fourth}}{}^{*i}*\sinh(-\alpha_{S/M}/2))$
$ *(\Psi_{1M\text{-fourth}}*\cosh(-\alpha_{S/M}/2) + \Psi_{4M\text{-fourth}}*\sinh(-\alpha_{S/M}/2))$
$ + (\Psi_{1M\text{-fourth}}{}^{*i}*\sinh(-\alpha_{S/M}/2) + \Psi_{4M\text{-fourth}}{}^{*i}*\cosh(-\alpha_{S/M}/2))$
$ *(\Psi_{1M\text{-fourth}}*\sinh(-\alpha_{S/M}/2) + \Psi_{4M\text{-fourth}}*\cosh(-\alpha_{S/M}/2)))$

$\phantom{J_{tS}} = Q_B*(\Psi_{1M\text{-fourth}}{}^{*i}*\Psi_{1M\text{-fourth}}*\cosh^2(\alpha_{S/M}/2)$
$ + \Psi_{1M\text{-fourth}}{}^{*i}*\Psi_{4M\text{-fourth}}*\cosh(\alpha_{S/M}/2)*\sinh(-\alpha_{S/M}/2)$
$ + \Psi_{4M\text{-fourth}}{}^{*i}*\Psi_{4M\text{-fourth}}*\sinh^2(\alpha_{S/M}/2)$
$ + \Psi_{4M\text{-fourth}}{}^{*i}*\Psi_{1M\text{-fourth}}*\cosh(\alpha_{S/M}/2)*\sinh(-\alpha_{S/M}/2)$
$ + \Psi_{1M\text{-fourth}}{}^{*i}*\Psi_{1M\text{-fourth}}*\sinh^2(\alpha_{S/M}/2)$
$ + \Psi_{1M\text{-fourth}}{}^{*i}*\Psi_{4M\text{-fourth}}*\cosh(\alpha_{S/M}/2)*\sinh(-\alpha_{S/M}/2)$
$ + \Psi_{4M\text{-fourth}}{}^{*i}*\Psi_{4M\text{-fourth}}*\cosh^2(\alpha_{S/M}/2)$
$ + \Psi_{4M\text{-fourth}}{}^{*i}*\Psi_{1M\text{-fourth}}*\cosh(\alpha_{S/M}/2)*\sinh(-\alpha_{S/M}/2)$

$\phantom{J_{tS}} = Q_B*(\Psi_{1M\text{-fourth}}{}^{*i}*\Psi_{1M\text{-fourth}}*(\cosh^2(\alpha_{S/M}/2) + \sinh^2(\alpha_{S/M}/2))$
$ + \Psi_{1M\text{-fourth}}{}^{*i}*\Psi_{4M\text{-fourth}}*2*\cosh(\alpha_{S/M}/2)*\sinh(-\alpha_{S/M}/2)$
$ + \Psi_{4M\text{-fourth}}{}^{*i}*\Psi_{4M\text{-fourth}}*(\sinh^2(\alpha_{S/M}/2) + \cosh^2(\alpha_{S/M}/2))$
$ + \Psi_{4M\text{-fourth}}{}^{*i}*\Psi_{1M\text{-fourth}}*2*\cosh(\alpha_{S/M}/2)*\sinh(-\alpha_{S/M}/2)$

$\phantom{J_{tS}} = Q_B*(\Psi_{1M\text{-fourth}}{}^{*i}*\Psi_{1M\text{-fourth}} + \Psi_{4M\text{-fourth}}{}^{*i}*\Psi_{4M\text{-fourth}})*\cosh\alpha_{S/M}$
$ + Q_B*(\Psi_{1M\text{-fourth}}{}^{*i}*\Psi_{4M\text{-fourth}} + \Psi_{4M\text{-fourth}}{}^{*i}*\Psi_{1M\text{-fourth}})*\sinh(-\alpha_{S/M})$

$= Q_B*\Psi_{\text{amp}}^2*(\cosh\alpha_M/2*\cosh\alpha_M/2 + \sinh(-\alpha_M/2)*\sinh(-\alpha_M/2))*\cosh\alpha_{S/M}$
$+ Q_B*\Psi_{\text{amp}}^2*(\cosh\alpha_M/2*\sinh(-\alpha_M/2) + \sinh(-\alpha_M/2)*\cosh\alpha_M/2)*\sinh(-\alpha_{S/M})$

$ = Q_B*\Psi_{\text{amp}}^2*\cosh\alpha_M*\cosh\alpha_{S/M} + Q_B*\Psi_{\text{amp}}^2*\sinh(-\alpha_M)*\sinh(-\alpha_{S/M})$
$ = Q_B*\Psi_{\text{amp}}^2*\cosh(-(\alpha_M + \alpha_{S/M}))$
$ = Q_B*\Psi_{\text{amp}}^2*\cosh(-\alpha_S)$

$J_{xS} = -Q_B*(\Psi_{1S}{}^{*i}*\Psi_{4S} + \Psi_{2S}{}^{*i}*\Psi_{3S} + \Psi_{3S}{}^{*i}*\Psi_{2S} + \Psi_{4S}{}^{*i}*\Psi_{1S})$

WAVES

$$= -Q_B*(\Psi_{1S\text{-fourth}}{}^{*i}*\Psi_{4S\text{-fourth}} + \Psi_{4S\text{-fourth}}{}^{*i}*\Psi_{1S\text{-fourth}})$$

$$\begin{aligned}= -Q_B*(&(\Psi_{1M\text{-fourth}}{}^{*i}*\cosh(\alpha_{S/M}/2) + \Psi_{4M\text{-fourth}}{}^{*i}*\sinh(-\alpha_{S/M}/2))\\ &*(\Psi_{1M\text{-fourth}}*\sinh(-\alpha_{S/M}/2) + \Psi_{4M\text{-fourth}}*\cosh(\alpha_{S/M}/2))\\ &+ (\Psi_{1M\text{-fourth}}{}^{*i}*\sinh(-\alpha_{S/M}/2) + \Psi_{4M\text{-fourth}}{}^{*i}*\cosh(\alpha_{S/M}/2))\\ &*(\Psi_{1M\text{-fourth}}*\cosh(\alpha_{S/M}/2) + \Psi_{4M\text{-fourth}}*\sinh(-\alpha_{S/M}/2)))\end{aligned}$$

$$\begin{aligned}= -Q_B*(&\Psi_{1M\text{-fourth}}{}^{*i}*\Psi_{4M\text{-fourth}}*\cosh^2(\alpha_{S/M}/2)\\ &+ \Psi_{1M\text{-fourth}}{}^{*i}*\Psi_{1M\text{-fourth}}*\cosh(\alpha_{S/M}/2)*\sinh(-\alpha_{S/M}/2)\\ &+ \Psi_{4M\text{-fourth}}{}^{*i}*\Psi_{1M\text{-fourth}}*\sinh^2(\alpha_{S/M}/2)\\ &+ \Psi_{4M\text{-fourth}}{}^{*i}*\Psi_{4M\text{-fourth}}*\cosh(\alpha_{S/M}/2)*\sinh(-\alpha_{S/M}/2)\\ &+ \Psi_{1M\text{-fourth}}{}^{*i}*\Psi_{4M\text{-fourth}}*\sinh^2(\alpha_{S/M}/2)\\ &+ \Psi_{1M\text{-fourth}}{}^{*i}*\Psi_{1M\text{-fourth}}*\cosh(\alpha_{S/M}/2)*\sinh(-\alpha_{S/M}/2)\\ &+ \Psi_{4M\text{-fourth}}{}^{*i}*\Psi_{1M\text{-fourth}}*\cosh^2(\alpha_{S/M}/2)\\ &+ \Psi_{4M\text{-fourth}}{}^{*i}*\Psi_{4M\text{-fourth}}*\cosh(\alpha_{S/M}/2)*\sinh(-\alpha_{S/M}/2))\end{aligned}$$

$$\begin{aligned}= -Q_B*(&\Psi_{1M\text{-fourth}}{}^{*i}*\Psi_{4M\text{-fourth}}*(\cosh^2(\alpha_{S/M}/2) + \sinh^2(\alpha_{S/M}/2))\\ &+ \Psi_{1M\text{-fourth}}{}^{*i}*\Psi_{1M\text{-fourth}}*2*\cosh(\alpha_{S/M}/2)*\sinh(-\alpha_{S/M}/2)\\ &+ \Psi_{4M\text{-fourth}}{}^{*i}*\Psi_{1M\text{-fourth}}*(\sinh^2(\alpha_{S/M}/2) + \cosh^2(\alpha_{S/M}/2))\\ &+ \Psi_{4M\text{-fourth}}{}^{*i}*\Psi_{4M\text{-fourth}}*2*\cosh(\alpha_{S/M}/2)*\sinh(-\alpha_{S/M}/2))\end{aligned}$$

$$\begin{aligned}= &-Q_B*(\Psi_{1M\text{-fourth}}{}^{*i}*\Psi_{4M\text{-fourth}} + \Psi_{4M\text{-fourth}}{}^{*i}*\Psi_{1M\text{-fourth}})*\cosh\alpha_{S/M}\\ &- Q_B*(\Psi_{1M\text{-fourth}}{}^{*i}*\Psi_{1M\text{-fourth}} + \Psi_{4M\text{-fourth}}{}^{*i}*\Psi_{4M\text{-fourth}})*\sinh(-\alpha_{S/M})\end{aligned}$$

$$\begin{aligned}= &-Q_B*(\Psi_{amp}{}^2(\cosh\alpha_M/2*\sinh(-\alpha_M/2) + \sinh(-\alpha_M/2)*\cosh\alpha_M/2)*\cosh\alpha_{S/M}\\ &+ Q_B*(\Psi_{amp}{}^2(\cosh\alpha_M/2*\cosh\alpha_M/2 + \sinh(-\alpha_M/2)*\sinh(-\alpha_M/2))*\sinh(-\alpha_{S/M})\end{aligned}$$

$$= -Q_B*\Psi_{amp}{}^2*\sinh(-\alpha_M)*\cosh\alpha_{S/M} + Q_B*\Psi_{amp}{}^2*\cosh\alpha_M*\sinh(-\alpha_{S/M})$$
$$= -Q_B*\Psi_{amp}{}^2*\sinh(-(\alpha_M + \alpha_{S/M}))$$
$$= -Q_B*\Psi_{amp}{}^2*\sinh(-\alpha_S)$$

$$\begin{aligned}{}_4J &= Q_B*\Psi_{amp}{}^2*(1_S*\cosh\alpha_S - q_{xS}*\sinh(-\alpha_S))\\ &= Q_B*\Psi_{amp}{}^2*1_S*\exp(q_x*\alpha_S) \quad (\pm_N \text{ becomes + for a particle})\end{aligned}$$

In "$_4J = Q_B*\Psi_{amp}{}^2*1_S*\exp(q_x*\alpha_S)$", "$\Psi_{amp}{}^2 > 0$" and "$Q_B < 0$". "$\exp(q_x*\alpha_S)$" represents motion to positive "x" (right) for positive "α_S" and "Q_B" makes current-density of that motion negative. Similarly, "$Q_B < 0$" makes the charge density time term negative.

The conclusion is that the Lorentz Transformation was correct.

SPECIAL ALGEBRA FOR SPECIAL RELATIVITY

<u>Anti-Matter</u>. For anti-matter there is an "$_{\text{-am}}$" subscript.

$$\begin{pmatrix} \omega_{\text{S-am}}/c \\ k_{x\text{S-am}} \end{pmatrix} = \begin{pmatrix} \cosh\alpha_{\text{S/M}} & -\sinh\alpha_{\text{S/M}} \\ -\sinh\alpha_{\text{S/M}} & \cosh\alpha_{\text{S/M}} \end{pmatrix} * \begin{pmatrix} \omega_{\text{M-am}}/c \\ k_{x\text{M-am}} \end{pmatrix}$$

$$\begin{pmatrix} \omega_{\text{S-m}}/c \\ k_{x\text{S-m}} \end{pmatrix} = \begin{pmatrix} \cosh\alpha_{\text{S/M}} & \sinh\alpha_{\text{S/M}} \\ \sinh\alpha_{\text{S/M}} & \cosh\alpha_{\text{S/M}} \end{pmatrix} * \begin{pmatrix} \omega_{\text{M-m}}/c \\ k_{x\text{M-m}} \end{pmatrix}$$

The above process can be repeated for the First Dirac Spinor Solution (that is, for anti-matter), to the result "$\alpha_S = \alpha_M - \alpha_{S/M}$".

4.7 Exercises

<u>Text Comprehension Exercises</u>.

1) Start with "$_4p$" and "Q_B*_4V", and end with

$$1*m_B*c*\cosh\alpha_M = i*\hbar*\tau*\nabla_{tM} - Q_B*V_{tM}$$

$$q*m_B*c*\sinh\alpha_M = q_x*(-i*\hbar*\tau*\nabla_{xM} - Q_B*V_{xM})$$
$$+ q_y*(-i*\hbar*\tau*\nabla_{yM} - Q_B*V_{yM}) + q_z*(-i*\hbar*\tau*\nabla_{zM} - Q_B*V_{zM})$$

2) Start with "$1^2 = \exp(-q*\alpha_M)*\exp(q*\alpha_M)$" and end with

$$\begin{pmatrix} (m_B*c + i*\hbar*\nabla_{tM} - Q_B*V_{tM}) & \pm_N(-i*\hbar*(q_x*\nabla_{xM}+q_y*\nabla_{yM}+q_z*\nabla_{zM})-Q_B*(q_x*V_{xM}+q_y*V_{yM}+q_z*V_{zM})) \\ \pm_N(-i*\hbar*(q_x*\nabla_{xM}+q_y*\nabla_{yM}+q_z*\nabla_{zM})-Q_B*(q_x*V_{xM}+q_y*V_{yM}+q_z*V_{zM})) & (-m_B*c + i*\hbar*\nabla_{tM} - Q_B*V_{tM}) \end{pmatrix} * \begin{pmatrix} P_M \\ Q_M \end{pmatrix} = \begin{pmatrix} 0 \\ 0 \end{pmatrix}$$

3) Write the Dirac Equation as four first order differential equations.

4) Prove the Fourth Dirac Spinor Solution to the Dirac Equation.

WAVES

5) Make the analogy that a matter electron with left-hand spin is a car with its steering wheel on the left, as in France. Two very high energy photons collide over the channel between Dover and Dunkerque to create a matter car and an anti-matter car. Explain the analogy between cars and electrons.

6) Find electric charge current density space-time invariant "$_4J$" for the Second Dirac Spinor Solution.

$\Psi_{1M\text{-second}} = -\Psi_{amp}*\sinh(\alpha_M/2)*\exp(i*(\pm_N k_{xM-m}*x_M - \omega_{M-m}*t_M))$
$\Psi_{4M\text{-second}} = \Psi_{amp}*\cosh(\alpha_M/2)*\exp(i*(\pm_N k_{xM-m}*x_M - \omega_{M-m}*t_M))$

7) Find the four solutions for the second pair of equations for "x" direction motion for "x"-arrangement and "y"-arrangement Dirac Equations. Prove correct and calculate current density.

8) Write the general form of the Lorentz Transformation for the Second Dirac Spinor Solution. Use "$e_{1M} = -e_{4M}*q_x$" and "$e_{4M} = -e_{1M}*q_x$".

$\Psi_{1M\text{-second}} = -\Psi_{amp}*\sinh(\alpha_M/2)*\exp(i*(\pm_N k_{xM-m}*x_M - \omega_{M-m}*t_M))$
$\Psi_{4M\text{-second}} = \Psi_{amp}*\cosh(\alpha_M/2)*\exp(i*(\pm_N k_{xM-m}*x_M - \omega_{M-m}*t_M))$

Select Exercise Solutions.

1) $_4p = \exp(-\kappa*\varsigma/2)*(E_M/c + q_x*p_{xM} + q_y*p_{yM} + q_z*p_{zM})*\exp(-\kappa*\varsigma/2)$
$= \exp(-\kappa*\varsigma/2)*(m_B*c*\cosh\alpha_M + q*m_B*c*\sinh\alpha_M)*\exp(-\kappa*\varsigma/2)$
$= 1_M*m_B*c*\cosh\alpha_M + q_M*m_B*c*\sinh\alpha_M$

$\hbar*_4k = {}_4p + Q_B*_4V = i*\tau*\hbar*_4\nabla^{sn} = i*\tau*\hbar*(_1\nabla) - i*\tau*\hbar*(_3\nabla)$
$\hbar*_4k^{sn} = {}_4p^{sn} + Q_B*_4V^{sn} = i*\tau*\hbar*(_4\nabla^{sn})^{sn} = i*\tau*\hbar*_4\nabla$
$= i*\tau*\hbar*(_1\nabla) + i*\tau*\hbar*(_3\nabla)$

$1_M*m*c*\cosh\alpha_M = i*\tau*\hbar*(_1\nabla) - Q_B*_1V$

$1*m_B*c*\cosh\alpha_M = i*\hbar*\tau*\nabla_{tM} - Q_B*V_{tM}$

$q_M*m*c*\sinh\alpha_M = -i*\tau*\hbar*(_3\nabla) - Q_B*_3V$
$q*m_B*c*\sinh\alpha_M = q_x*(-i*\hbar*\tau*\nabla_{xM} - Q_B*V_{xM})$
$+ q_y*(-i*\hbar*\tau*\nabla_{yM} - Q_B*V_{yM}) + q_z*(-i*\hbar*\tau*\nabla_{zM} - Q_B*V_{zM})$

SPECIAL ALGEBRA FOR SPECIAL RELATIVITY

2) $1^2 = \exp(-q*\alpha_M)*\exp(q*\alpha_M)$
   $1^2 = (\cosh\alpha_M - q*\sinh\alpha_M)*(\cosh\alpha_M + q*\sinh\alpha_M)$
   $1^2 = \cosh^2\alpha_M - q^2*\sinh^2\alpha_M$
   $1^2 - \cosh^2\alpha_M = -q^2*\sinh^2\alpha_M$
   $-(1^2 - \cosh^2\alpha_M) = q^2*\sinh^2\alpha_M$

   $(1 + \cosh\alpha_M)*(-1 + \cosh\alpha_M) = (\pm q*\sinh\alpha_M)*(\pm q*\sinh\alpha_M)$

   $((1 + \cosh\alpha_M)*P_M)*((-1 + \cosh\alpha_M)*-Q_M) =$
   $\quad\quad\quad\quad ((\pm q*\sinh\alpha_M)*-Q_M)*((\pm q*\sinh\alpha_M)*P_M)$

   $(1 + \cosh\alpha_M)*P_M = (\pm q*\sinh\alpha_M)*(-Q_M) + 0$
   $(-1 + \cosh\alpha_M)*(-Q_M) = (\pm q*\sinh\alpha_M)*P_M - 0$

$$\begin{pmatrix} (1 + \cosh\alpha_M) & (\pm q*\sinh\alpha_M) \\ (\pm q*\sinh\alpha_M) & (-1 + \cosh\alpha_M) \end{pmatrix} * \begin{pmatrix} P_M \\ Q_M \end{pmatrix} = \begin{pmatrix} 0 \\ 0 \end{pmatrix}$$

$$\begin{pmatrix} (m_B*c + m_B*c*\cosh\alpha_M) & (\pm q*m_B*c*\sinh\alpha_M) \\ (\pm q*m_B*c*\sinh\alpha_M) & (-m_B*c + m_B*c*\cosh\alpha_M) \end{pmatrix} * \begin{pmatrix} P_M \\ Q_M \end{pmatrix} = \begin{pmatrix} 0 \\ 0 \end{pmatrix}$$

$$\begin{pmatrix} (m_B*c + i*\hbar*\nabla_{tM} - Q_B*V_{tM}) & \pm_N(-i*\hbar*(q_x*\nabla_{xM}+q_y*\nabla_{yM}+q_z*\nabla_{zM})-Q_B*(q_x*V_{xM}+q_y*V_{yM}+q_z*V_{zM})) \\ \pm_N(-i*\hbar*(q_x*\nabla_{xM}+q_y*\nabla_{yM}+q_z*\nabla_{zM})-Q_B*(q_x*V_{xM}+q_y*V_{yM}+q_z*V_{zM})) & (-m_B*c + i*\hbar*\nabla_{tM} - Q_B*V_{tM}) \end{pmatrix} * \begin{pmatrix} P_M \\ Q_M \end{pmatrix} = \begin{pmatrix} 0 \\ 0 \end{pmatrix}$$

3) Solution:

$m_B*c*\Psi_{1M} + i*\hbar*\partial\Psi_{1M}/\partial c t_M - Q_B*V_{tM}*\Psi_{1M} \pm_N -i*\hbar*\partial\Psi_{4M}/\partial x_M \pm_N -Q_B*V_{xM}*\Psi_{4M}$
$\pm_N -i*\hbar*k*\partial\Psi_{4M}/\partial y_M \pm_N -Q_B*V_{yM}*k*\Psi_{4M} \pm_N -i*\hbar*\Psi_{3M}\partial/\partial z_M \pm_N -Q_B*V_{zM}*\Psi_{3M} = 0$

$m_B*c*\Psi_{2M} + i*\hbar*\partial\Psi_{2M}/\partial c t_M - Q_B*V_{tM}*\Psi_{2M} \pm_N -i*\hbar*\partial\Psi_{3M}/\partial x_M \pm_N -Q_B*V_{xM}*\Psi_{3M}$
$\pm_N -i*\hbar*-k*\partial\Psi_{3M}/\partial y_M \pm_N -Q_B*V_{yM}*-k*\Psi_{3M} \pm_N -i*\hbar*-\Psi_{4M}\partial/\partial z_M \pm_N -Q_B*V_{zM}*-\Psi_{4M} = 0$

WAVES

$m_B*c*\text{-}\Psi_{3M} + i*\hbar*\partial\Psi_{3M}/\partial ct_M - Q_B*V_{tM}*\Psi_{3M} \pm_N -i*\hbar*\partial\Psi_{2M}/\partial x_M \pm_N -Q_B*V_{xM}*\Psi_{2M}$
$\pm_N -i*\hbar*k*\partial\Psi_{2M}/\partial y_M \pm_N -Q_B*V_{yM}*k*\Psi_{2M} \pm_N -i*\hbar*\Psi_{1M}\partial/\partial z_M \pm_N -Q_B*V_{zM}*\Psi_{1M} = 0$

$m_B*c*\text{-}\Psi_{4M} + i*\hbar*\partial\Psi_{4M}/\partial ct_M - Q_B*V_{tM}*\Psi_{4M} \pm_N -i*\hbar*\partial\Psi_{1M}/\partial x_M \pm_N -Q_B*V_{xM}*\Psi_{1M}$
$\pm_N -i*\hbar*\text{-}k*\partial\Psi_{1M}/\partial y_M \pm_N -Q_B*V_{yM}*\text{-}k*\Psi_{1M} \pm_N -i*\hbar*\text{-}\Psi_{2M}\partial/\partial z_M \pm_N -Q_B*V_{zM}*\text{-}\Psi_{2M} = 0$

4) $\Psi_{1M\text{-fourth}} + (i\hbar/m_Bc)*\partial\Psi_{1M\text{-fourth}}/\partial ct_M - \pm_N(i\hbar/m_Bc)*\partial\Psi_{4M\text{-fourth}}/\partial x_M = 0$

$\cosh(\alpha_M/2)$
$+ (i\hbar/m_Bc^2)*(--i*\omega_M)*\cosh(\alpha_M/2) - \pm_N(i\hbar/m_Bc)*(-i*\pm_Nk_{xM})*\text{-}\sinh(\alpha_M/2) = 0$

$\cosh(\alpha_M/2) - \cosh\alpha_M*\cosh(\alpha_M/2) + \sinh\alpha_M*\sinh(\alpha_M/2) = 0$
$0 = 0$

$\text{-}\Psi_{4M\text{-fourth}} + (i\hbar/m_Bc)*\partial\Psi_{4M\text{-fourth}}/\partial ct_M - \pm_N(i\hbar/m_Bc)*\partial\Psi_{1M\text{-fourth}}/\partial x_M = 0$

$--\sinh(\alpha_M/2)$
$+ (i\hbar/m_Bc^2)*(--i*\omega_M)*\text{-}\sinh(\alpha_M/2) - \pm_N(i\hbar/m_Bc)*(-i*\pm_Nk_{xM})*\cosh(\alpha_M/2) = 0$

$\sinh(\alpha_M/2) + \cosh\alpha_M*\sinh(\alpha_M/2) - \sinh\alpha_M*\cosh(\alpha_M/2) = 0$
$0 = 0$

5) Car Visualization for Both Anti-matter and Spin: In the pair production event, a left-hand spin matter electron is modeled as a car made of matter with steering wheel on the left. The car faces Dunkerque, France. In the same physical space at the instant of production a reverse-parity second car is also produced. This second car made of anti-matter also has steering wheel on the left and faces Dunkerque, France. Any energy in excess of the rest energy of the two cars is equally applied as kinetic energy that pushes the matter car toward Dunkerque and the anti-matter car toward Dover. The steering wheel on the left of both cars means both cars have a left-hand spin, for zero total angular momentum (per the analogy of a nut and bolt that unscrew and separate, retaining their spin), just as there needs to be zero total linear momentum. The anti-matter car has reverse-parity. To un-reverse the reverse-parity of the anti-matter car, flatten it front to back and go further to stretch it full length, so that the front points toward Dover, England, and see the steering wheel is on the right, as cars are in England. The right steering wheel means right-handed spin. Reverse parity has reverse spin.

SPECIAL ALGEBRA FOR SPECIAL RELATIVITY

6) $\Psi_{1M\text{-second}}{}^{*i} = -\Psi_{amp}*\sinh(\alpha_M/2)*\exp(-i*(\pm_N k_{xM\text{-}m}*x_M - \omega_{M\text{-}m}*t_M))$
   $\Psi_{4M\text{-second}}{}^{*i} = \Psi_{amp}*\cosh(\alpha_M/2)*\exp(-i*(\pm_N k_{xM\text{-}m}*x_M - \omega_{M\text{-}m}*t_M))$

   $J_{tM} = Q_B*(\Psi_{1M}{}^{*i}*\Psi_{1M} + \Psi_{2M}{}^{*i}*\Psi_{2M} + \Psi_{3M}{}^{*i}*\Psi_{3M} + \Psi_{4M}{}^{*i}*\Psi_{4M})$
   $= Q_B*\Psi_{amp}{}^2*(\sinh^2(\alpha_M/2) + \cosh^2(\alpha_M/2))$
   $= Q_B*\Psi_{amp}{}^2*\cosh\alpha_M$

   $J_{xM} = -Q_B*(\Psi_{1M}{}^{*i}*\Psi_{4M} + \Psi_{2M}{}^{*i}*\Psi_{3M} + \Psi_{3M}{}^{*i}*\Psi_{2M} + \Psi_{4M}{}^{*i}*\Psi_{1M})$
   $= -Q_B*\Psi_{amp}{}^2*(-2*\cosh(\alpha_M/2)*\sinh(\alpha_M/2))$
   $= Q_B*\Psi_{amp}{}^2*\sinh\alpha_M$

   $J_{yM} = 0 \qquad J_{zM} = 0$

   $_4J = Q_B*\Psi_{amp}{}^2*(1_M*\cosh\alpha_M + q_{xM}*\sinh\alpha_M)$
   $= Q_B*\Psi_{amp}{}^2*1_M*\exp(q_x*\alpha_M)$

7) Solution not given. See text for similar solutions.

8) $_4\Psi_{\text{-second}} = e_{1M}*\Psi_{1M} + e_{4M}*\Psi_{4M}$
   $= e_{1M}*\Psi_{1M\text{-second}} + e_{4M}*\Psi_{4M\text{-second}}$

   $= (e_{1M}*-\Psi_{amp}*\sinh(\alpha_M/2)*\exp(i*(\pm_N k_{xM\text{-}m}*x_M - \omega_{M\text{-}m}*t_M)))$
   $+ (e_{4M}*\Psi_{amp}*\cosh(\alpha_M/2)*\exp(i*(\pm_N k_{xM\text{-}m}*x_M - \omega_{M\text{-}m}*t_M)))$

   $= (-e_{4M}*q_x*-\Psi_{amp}*\sinh(\alpha_M/2)*\exp(i*(\pm_N k_{xM\text{-}m}*x_M - \omega_{M\text{-}m}*t_M)))$
   $+ (e_{4M}*\Psi_{amp}*\cosh(\alpha_M/2)*\exp(i*(\pm_N k_{xM\text{-}m}*x_M - \omega_{M\text{-}m}*t_M)))$

   $= e_{4M}*\Psi_{amp}*\exp(q_x*\alpha_M/2)*\exp(i*(\pm_N k_{xM\text{-}m}*x_M - \omega_{M\text{-}m}*t_M))$

   $= e_{4M}*\Psi_{amp}*\exp(q_x*\alpha_M/2)*\exp(i*(\pm_N k_{xM\text{-}m}*x_M - \omega_{M\text{-}m}*t_M))$
   $*\exp(q_x*\alpha_{S/M}/2)/\exp(q_x*\alpha_{S/M}/2)$

   $= e_{4S}*\Psi_{amp}*\exp(q_x*(\alpha_M + \alpha_{S/M})/2)*\exp(i*(\pm_N k_{xM\text{-}m}*x_M - \omega_{M\text{-}m}*t_M))$
   $= e_{4S}*\Psi_{amp}*\exp(q_x*\alpha_S/2)*\exp(i*(\pm_N k_{xM\text{-}m}*x_M - \omega_{M\text{-}m}*t_M))$
   $= e_{4S}*\Psi_{amp}*\exp(q_x*\alpha_S/2)*\exp(i*(\pm_N k_{xS\text{-}m}*x_S - \omega_{S\text{-}m}*t_S))$

   $= e_{1S}*\Psi_{1S\text{-second}} + e_{4S}*\Psi_{4S\text{-second}}$

WAVES

Further Thought

1) Perform the Lorentz Transformation of the Fourth Dirac Spinor Solution with "$\alpha_{S/M} = (i - j_x)*\pi/2$" (for the transformation of time-like to space-like), and again with "$\alpha_{S/M} = i*\pi/2$" (for sub-light-speed motion of a Dirac Spinor wave crests and nodes) with a factor "$-i$" applied to the hyperbolic-radius, and then again with "$\alpha_{S/M} = \pm i*\pi$" (for anti-matter) with a factor "-1" applied to the hyperbolic-radius. Explain what the results represent physically and speculate if anything in nature fits that description.

2) The equation that relates the current-density invariant to the Dirac Spinor invariant has components of the Dirac Spinor invariant multiplied by each other. The implication is that Dirac Spinor space is a square root of geometric space. If this implication is valid, then other four-component invariants of our geometric-vector space should also be able to be expressed as a form of a square of Dirac Spinor space. Try to quantify this thought by creating a more general theory that relates Dirac Spinor space to geometric space.

3) Geometry of Dirac Spinor Space. The mathematical existence of the four "Ψ" components of the Dirac Equation suggest there is a translation from numbers to geometry that creates a physically real "Ψ" space. We expect some sort of physical reality to "Ψ" space because the intention of theory-development-algebra is to mathematically model physics in all its intricate detail, and physics is real, a reality that we traditionally suppose is geometrically real. In this book we ignore the possible physical reality of "Ψ" space per the excuse of the Process from Descartes, in which we do not revert to the geometry of step 3 until prepared to take a measurement. This excuse supposes there is no geometry other than what we perceive because numbers are fundamental, not geometry. Forget the Process from Descartes and make a guess at what the geometry of "Ψ" Dirac Spinor Space is. How might the guess be verified by experiment? If there is a geometry applicable to Dirac Spinor "Ψ" space, then a translation to geometry is needed inside Dirac Spinor "Ψ" space. Think about this further after reading the next chapter in which electromagnetic field theory is combined into the Dirac Equation through the requirement that precision resolution is restricted to a finite value in geometry, any geometry, including Dirac spinor space.

Special Algebra for Special Relativity

Chapter 5 – Proposed Theory

Real numbers are redefined by altering the axioms in Axiomatic Set Theory that pertain to infinity. The altered axioms remove holes in logic created by Cantor's actual infinities aleph null and aleph one by removing those two actual infinities. Aleph null is replaced with the largest number yet counted-to and aleph one is replaced with the reciprocal of zero.

Two proposed new axioms pertain to division by zero. The *axiom of reciprocal of zero* states that no division by zero can result in a non-zero finite number. This axiom replaces "do not divide by zero" and "division by zero is indeterminate and undefined" with "1/0 + 7 = 1/0", "(1/0)*7 = 1/0", and "2^(1/0) = 3^(1/0) = either 1/0 or 0". The indefinite magnitude of "1/0" per that algebra mirrors the indefinite magnitude of Cantor's aleph one. The second axiom that pertains to division by zero states "1/0" is the quantity of a set.

The two proposed axioms that pertain to division by zero along with the *axiom of choice* permit the definition of idealized real numbers. Idealized real numbers have a spacing of integer zero one to the next. The set of idealized real numbers on a finite segment of the number-line has quantity "1/0".

The next modification to Axiomatic Set Theory is the removal of the *axiom of choice*. Without the *axiom of choice*, the only reference to infinity is the *axiom of infinity* by which nested null sets are counted in a derivation of the unboundedness of natural numbers used in counting. This creates a time dependency because the largest number yet counted-to is a transient finite number, and it substitutes for Cantor's aleph null infinity.

Cantor's Continuum Hypothesis is a conjecture presently used as an axiom in Axiomatic Set Theory, but with the proposed changes to axioms, a Modified Continuum Hypothesis is a theorem that states there are no positive numbers as sets between the largest number yet counted-to and the reciprocal of zero. In other words, there are no actual infinities as quantities of sets.

The Modified Continuum Hypothesis creates practical real numbers. A practical real number has place-value digits after the decimal point that extend only so far as the largest number yet counted-to. Beyond that, the place-value digits are unknown and unknowable and create finite magnitude imprecision. Similarly, zeros to the left end at the largest number yet counted-to and beyond that count unknown and unknowable place-value digits create large-scale imprecision. Large-scale imprecision applied to the Dirac Equation derives Maxwell's Equations. That application creates a new model of physics, and it formally transitions Axiomatic Set Theory from pure mathematics, which is an experimental forum, into applied mathematics, which is real.

SPECIAL ALGEBRA FOR SPECIAL RELATIVITY

5.1 Cantor's Theory of Infinite Sets

Cantor's Contribution. Cantor proposed a positive countable infinity that was the quantity of natural numbers and a larger positive uncountable infinity that was the quantity of real numbers. The two infinities became actual infinities per his Continuum Hypothesis: No set can have a quantity of members between the two infinities.

Cantor proposed his theory of infinite sets late in the 1800's, a few decades before 1905, when Einstein presented his theory of Special Relativity. Special Relativity, like every other mathematical model of physics, does not directly use Cantor's two infinities. The disconnect between Cantor's infinities and Special Relativity prompted the author's search for a proper replacement in applied mathematics for the positive actual infinity of pure mathematics.

Cantor's infinity theory was critically important for several reasons:

- Cantor created the foundation for evolving more theory on how natural numbers and real numbers are constructed.

- Cantor's proofs showed the importance of defining numbers using place-value digits, as opposed to ratios for rational numbers, or nebulously for irrational numbers.

- Cantor's eventual discovery of base two for exponents suggested a switch from base ten to base two was needed.

- Cantor placed positive actual infinity into a formal theory against which alternative theories of infinity could be contrasted.

- Cantor formalized the two thousand years of thought about infinity that preceded him.

- Cantor's proposed infinities generated plenty of literature so that several perspectives on the theory could be found by reading, rather than by inventing.

- Cantor's work evolved into Axiomatic Set Theory with his Continuum Hypothesis effectively being another axiom. Into that structure, an alternative axiomatic definition for properties of infinity could be introduced.

CHAPTER 5 - PROPOSED THEORY OF NON-FINITE NUMBERS

- Cantor's infinities were omitted from Group Theory, which had evolved around the same time, to make evident there was no means of performing binary operations (other than what he had specified) on his two infinities.

- Cantor's two infinities were absent from mathematical models of physics and that meant they probably weren't real and highlighted that an improvement in the theory of infinity was needed and was possible.

- Cantor's proofs, for example the diagonal proof and the string of digits proof, were for finite numbers only, and required a person to make a conceptually questionable leap to infinity, and that seemed acceptable due to an inability to count to infinity, but felt wrong, regardless.

- Cantor's work and the Axiomatic Set Theory that followed his work had holes in logic, and that meant improvement was not only possible, but was needed if infinity was to be accepted into applied mathematics.

By the early 1870's the Middle Ages were gone, and the modern world had come with all its glory, with trains and steamboats and science and amazing new ideas. In this vibrant and awakened world, Cantor and his associate Dedekind took it on themselves to make workable the great unknowable unknown in mathematics, the infinity. Feel not only their excitement but also their burden of responsibility. Feel grit they needed while they endured attacks on their reputations. All the scattered bits of speculation on infinity, from two millenniums and more of great thinkers who preceded them, had to be made useable so that future math enthusiasts could build from it. More than that, they wanted to successfully solve the mystery of what infinity really was, not only because modern times demanded it but also, I think, to satisfy personal curiosity.

Only compliments go to Cantor. No criticisms. In this book read about a proposed modification to Cantor's work, but while doing so, note that nothing said in this book detracts from Cantor's valuable insights. Cantor and over a hundred years of mathematicians who followed him put formality to infinity by building a strong foundation of concepts.

Each math enthusiast knows, just like the author of this book knows, that no person's work is the final say, and that's because each subsequent math enthusiast looks for some new way to contribute, and in doing so might make an aspect of previous work obsolete. No negativity is ever intended because each math enthusiast naturally and politely appreciates all help given, and specifically with respect to the writing of this book, Cantor's help was exceptional.

SPECIAL ALGEBRA FOR SPECIAL RELATIVITY

Cantor's Theory of Infinite Sets – Countable Sets. Positive actual infinity "N_0" (aleph null, the countable infinity) is the quantity of members in "N", the natural numbers. The set below has counting numbers "1, 2, 3, ...", terminated at aleph null. Natural numbers are what Cantor called "cardinal numbers", which he defined as the quantity of a set.

$N = \{1, 2, 3, ..., N_0\}$

Alternatively, "1, 2, 3, ..." can be "ordinal numbers" spoken as first, second, third, ..." in which case the set does not end at cardinal number "N_0".

Cantor counted integers to prove the set of integers "Z" has the same quantity "N_0" as does the set of natural numbers "N": "0" was first, "1" was second, then "-1", was third, "2" was fourth, "-2" was fifth, etc. The person reading the proof must extrapolate from a finite count using ordinal numbers to an actual infinity quantity that is a cardinal number. The word "countable" means members of the set can be identified one-to-one with the set of natural numbers, at least for the finite number portion of the set as given in the proof. Examples of countable sets are integers, prime numbers, rational numbers (per Cantor's diagonal proof), squares of natural numbers, even numbers, and products of two integers. The set of real numbers is not "countable".

Dedekind Cut. The Dedekind Cut was the precedent to Cantor's theory of infinite sets. The Dedekind Cut is best described by Dedekind himself in a quote from *Essays on the Theory of Numbers* by Richard Dedekind, 1963 (originally 1901) by Dover Publications, Inc.

Page 15: "From now on, therefore, to every definite cut there corresponds a definite rational or irrational number, and we regard two numbers as *different* or *unequal* always and only when they correspond to essentially different cuts."

If the cut is at a rational number, then the high side number is the rational number and the low side number is infinitely close, but just less. Alternatively, if the cut is at an irrational number, then both the high side and low side numbers are infinitely close to each other, and the irrational number is between them. Per the definition of the Dedekind Cut, we ignore the removed material of a saw cut, and we pretend the cut is a scissors cut (which does not remove material) by making the two numbers "essentially" the same number.

A true scissors cut would be an idealized cut because high and low numbers differ by integer zero, with the example of one as both high and low numbers. The problem was that if a cut of integer zero separates real numbers, then a

CHAPTER 5 - PROPOSED THEORY OF NON-FINITE NUMBERS

quantity of reciprocal zero real numbers as individual points form a finite number-line segment, and that quantity violated the "do not divide by zero" rule.

In contrast, a non-idealized cut would be a saw cut that removes a finite interval from the number-line. A succession of finite intervals places real numbers a finite distance from each other, and there is a finite quantity of real numbers as finite-length points that form a number-line segment.

The Dedekind Cut is something between the idealized scissors cut and non-idealized saw cut because Dedekind made the cut positive actual infinitesimal in width, so that sequential real numbers had a positive actual infinitesimal interval from one to the next and there were a positive actual infinity quantity of real numbers spanning a finite interval of the number-line. The Dedekind Cut is a saw cut that is "essentially" a scissors cut, and it is the word "essentially" that creates the concept for positive actual infinity, which is what he was trying to define with his cut.

To visualize the Dedekind cut, pretend the positive actual infinity is "3" and use base two. A Dedekind Cut at half (".100" in base two) has ".100" on the high side and ".100 - $2^{\wedge}$-infinity = .100 - .001 = .011" on the low side. Notice that we assume zeros for all the place-value digits after the count of infinity, which was "3" in this example. To satisfy the word "essentially", the difference "$2^{\wedge}$-infinity" must be smaller than a rational number, and that is satisfied if "infinity" is a positive actual infinity that is a quantity of a set. For the zeros to exist as place-value digits far to the right, "infinity + 1" must exist as a quantity of a set.

In that visualization, we assumed infinity is positive such that it is greater than zero, and such that infinity is not the ideal extreme of "1/0" for which the interval would be "$2^{\wedge}$-(1/0)", which appears equal to "0".

As another example, a Dedekind Cut at irrational number square-root-of-two "$\sqrt{2}$ = 1.414213..." has high/low number pairs at the cut (1.4, 1.5), (1.41, 1.42), (1.414, 1.415), (1.4142, 1.4143), etc., for the value of infinity increasing through "1", "2", "3", "4", etc., respectively. When infinity reaches an actual infinity, then the high and low numbers and all numbers between them, including the irrational number square-root-of-two, are essentially the same number, because the span between the low and high numbers is the positive infinitesimal "($10^{\wedge}$-infinity) > 0".

Define "Lmax" as the largest number yet counted-to. If sequential rational numbers "1/Lmax" and "1/(Lmax - 1)" differ by "1/(Lmax*(Lmax - 1))", then a difference "$2^{\wedge}$(-Lmax)" of sequential real numbers is smaller, and more so if "$2^{\wedge}$(-Lmax)" becomes "$2^{\wedge}$-infinity". This inequality "1/(Lmax*(Lmax – 1)) >> $2^{\wedge}$(-Lmax) might have been the justification for the word "essentially".

SPECIAL ALGEBRA FOR SPECIAL RELATIVITY

Uncountable Sets. "Cardinality" of a set is the quantity of members in the set. Cardinality is given operator notation of two "|" lines. A textbook for Cantor's theory of infinite sets is *Mathematical Proofs, A Transition to Advanced Mathematics, Second Edition* by Chartrand, Polimeni, and Zhang, Pearson Addison Wesley, 2008, beginning on Page 221. From the top of Page 236:

"Indeed, if A is any denumerable set, then $|A| = N_0$. The set **R** of real numbers is also referred to as the continuum and its cardinality is denoted by c. Hence $|\mathbf{R}| = c$ and from what we have seen, $N_0 < c$. It was the German mathematician Georg Cantor who helped to put the theory of sets on a firm foundation."

Cantor later changed nomenclature from a German script "c" to the Hebrew aleph one "N_1". Also, "denumerable" is another word meaning "countable". The quote summarizes Cantor's theory of real numbers by stating, in other words, that there is a quantity "N_1" of real numbers over an infinite ("N_0") interval of the number-line with "$N_1 > N_0$".

Cantor proved "$N_1 > N_0$" using quantity "N_1" real numbers spanning zero (inclusive) to one (exclusive) and using quantity "N_0" of place-value digits after the decimal point. He created several different real numbers by writing random place-value digits after the decimal point.

0.5928609813...	0.5232...
0.7290165316...	0.8991...
0.7831994831...	
0.8482401809...	

Cantor took the first place-value digit from the first number, second from the next, and so on, to form a new number, for example "0.5232...". And he varied the process to create other new numbers, for example "0.8991...".

Because new numbers could be formed by this process, he concluded "$N_1 > N_0$". The person must extrapolate from finite to infinite.

As a refinement to "$N_1 > N_0$", Cantor proposed "$N_1 = 2\wedge N_0$". To visualize "$N_1 = 2\wedge N_0$", substitute "3" for "N_0". The set of three members "{.1, .01, .001}" implies eight sets per "$8 = 2^3$":

{}, {.1}, {.01}, {.001}, {.1, .01}, {.1, .001}, {.01, .001}, {.1, .01, .001}

There are "$2 = 2\wedge 1$" numbers "0.0" and "0.1" if we consider only the first place-value digit after the decimal point. "$4 = 2\wedge 2$" numbers "0.00", "0.01", "0.10" and "0.11" for the first two. "$8 = 2\wedge 3$" as given above for the first three. This visualization requires base two, and that's why base two is in "$N_1 = 2\wedge N_0$". It doesn't work for base ten.

CHAPTER 5 - PROPOSED THEORY OF NON-FINITE NUMBERS

"$N_1 = 2^{\wedge}N_0$" as a quantity of real numbers applies to the span from "0" to "1" and also the span from negative infinity "$-N_0$" to positive infinity "N_0" so that the quantity "N_1" applies to any uncountable set, in analogy to "N_0" applying to any countable set.

Properties of the two Actual Infinities. From Cantor's proofs, we identify the following properties for "N_0" and "N_1". Finite numbers satisfy these properties.

- "$N_0 < N_1$" per Cantor's proof, and "$N_1 = 2^{\wedge}N_0$"

- "N_0" and "N_1" have no contribution after the decimal point, in analogy to natural numbers used in counting, and so are quantities of sets.

- "N_0" and "N_1" are positive, also in analogy to natural numbers used in counting, and as required by "$N_0 < N_1$" and "$N_1 = 2^{\wedge}N_0$"

Perhaps this other property also applies:

- "N_1" is an even number because "$N_1 = 2^{\wedge}N_0$", by furthering the analogy of Cantor's infinities with finite numbers

Attempt to place a "1" at the "N_0"th position before the decimal point by writing an ever-increasing string of zeros left of a decimal point to form "$N_1 = 2^{\wedge}N_0$" in base two. Per this visualization, "$N_1 = 2^{\wedge}N_0$" cannot be negative and therefore cannot equal "1/0". Per an analogous visualization, "$1/N_1 = 2^{\wedge}-N_0$" cannot equal integer zero.

Cantor's Continuum Hypothesis. Included in Cantor's theory of infinite sets is the conjecture that there is no set "S" for which

$N_0 < |S| < N_1$

Inequality "$N_0 < |S| < N_1$" states the quantity of members in a set may equal "N_0", as applies to countable sets, and may equal "N_1", as applies to uncountable sets, but the quantity of members in a set cannot be "$|S|$" between "N_0" and "N_1".

There cannot be a count up from "N_0" toward "N_1" because any number in a count is the maximum number in the set of numbers from one to that number. There is no "$N_0 + 1$", "$2*N_0$", "$N_1/2$", or "$N_1 - 1$".

SPECIAL ALGEBRA FOR SPECIAL RELATIVITY

The Continuum Hypothesis (that "$N_0 < |S| < N_1$" is impossible) cannot apply if "N_0" and/or "N_1" are finite. Therefore, it is the Continuum Hypothesis that creates mathematically the property of positive actual infinity for "N_0" and for "N_1". And it is the Continuum Hypothesis that justifies multi-set-applicability of "N_0" and of "N_1".

Notice in the Dedekind Cut it was supposed that zeros began at quantity "infinity + 1" place-value digits after the decimal point. Because of the Continuum Hypothesis, "infinity + 1" as "$N_0 + 1$" does not exist, and that means the proposed visualization for the Dedekind Cut does not apply in Cantor's theory of infinity. It means "N_0" should not be visualized as a finite number.

The Continuum Hypothesis is only a conjecture because it has not been proven. Attempts have been made. In the referenced textbook (middle of page 236) is a description of the attempts:

"However, in 1931 the Austrian mathematician Kurt Gödel proved that it was impossible to disprove the Continuum Hypothesis from the axioms on which the theory of sets is based. In 1963 the American mathematician Paul Cohen took it one step further by showing that it was also impossible to *prove* the Continuum Hypothesis from these axioms. Thus the Continuum Hypothesis is independent of the axioms of set theory."

Axiomatic Set Theory. Cantor's theory of infinite sets became further formalized between 1904 and 1908 and was renamed "Axiomatic Set Theory". A very readable summary of the effect the Continuum Hypothesis has had on Axiomatic Set Theory is given in the article "Dispute over Infinity Divides Mathematicians – To determine the nature of infinity, mathematicians face a choice between two new logical axioms. What they decide could help shape the future of mathematical truth" by Natalie Wolchover, Quanta Magazine, December 3, 2013.

Another reference is *THE PHILOSOPHY OF SET THEORY, An Historical Introduction to Cantor's Paradise*, by Mary Tiles, from Dover Publications, Inc, 2004 (Originally 1989 from Basil Blackwell Ltd.). The "axioms of set theory" are listed in Mary Tiles' book, Pages 121-123.

Axiom of extensionality   *Axiom of foundation*
Null set axiom   *Subset axiom and replacement axiom*
Pair set axiom   *Power set axiom*
Sum set axiom   *Power Axiom of choice*
Axiom of infinity

CHAPTER 5 - PROPOSED THEORY OF NON-FINITE NUMBERS

Axiom of Infinity. Per page 125, the *axiom of infinity* identifies the unboundedness of natural numbers because one can always be added. The thing that is added is a null set, so that nothing is added to nothing, repeatedly, as nested null sets. To justify why null sets are counted rather than objects, think of numbers as distinctly separate from physical reality, so that one, two, three, and so on exist independently as if they are physical nothings. Regardless of counting somethings or nothings, the *axiom of infinity* pertains only to the largest finite quantity yet counted-to, and not to Cantor's countable infinity and not to the actual infinity Dedekind imagined.

Axiom of Choice. Imagine two irrational numbers are multiplied to form a product "$\log_2 3 * \sqrt{2}$". During the process of counting place-value digits right of the decimal point, we specify the integer number of each place-value digit while calculating the result of that multiplication operation. Per that method, and through utilization of the Axiom of Infinity, we have specified the product "$\log_2 3 * \sqrt{2}$" only up to the largest finite natural number yet counted-to as a series of integers to the right of the decimal point. But per a proof of irrationality, it is known there are more than a finite quantity of digits to an irrational number. By what axiomatic permission do we specify the remaining place-value digits of "$\log_2 3 * \sqrt{2}$"?

The *axiom of choice* lets us specify the remaining place-value digits. Why call it the axiom of "choice"? Bertrand Russell explains why in this quote, "To choose one sock from each of infinitely many pairs of socks requires the Axiom of Choice, but for shoes the Axiom is not needed." To understand this quote, think in base 2, which is 0's and 1's, for the place-value digits of numbers "$\log_2 3$", "$\sqrt{2}$", and their product "$\log_2 3 * \sqrt{2}$". In base 2 for an irrational number product "$\log_2 3 * \sqrt{2}$", each 0 or 1 must be selected. Visualize the selection occurring after the decimal point one place-value digit at a time. A duration of time, however, is finite, and there are infinitely many place-value digits that need that 0 or 1 selected to complete the number "$\log_2 3 * \sqrt{2}$". The *axiom of choice* lets us choose 0 or 1 for all the place-value digits counting has not gotten to, to the completion of irrational number "$\log_2 3 * \sqrt{2}$", whatever "completion" means.

For "$\log_2 3$" and "$\sqrt{2}$" we can suppose all the digits exist, with the analogy being that shoes are already left 0 or right 1 and can be assigned in a bulk operation without a sequential count operation selection process. Similarly, the shoe analogy applies to a rational number's digits beyond what can be counted, because we know the rule for assigning 0 or 1.

Gödel proved *axiom of choice* cannot be disproven by other axioms, 1939. In 1963 Cohen proved the *axiom of choice* is independent of the other axioms.

SPECIAL ALGEBRA FOR SPECIAL RELATIVITY

It is the selection of properties for infinity that define an algebra. For Group Theory, it was assumed all numbers were finite and therefore there was no actual infinity. Group Theory had no holes in logic because it only pertained to finite, and that meant Group Theory was acceptable in applied mathematics. Conversely, Axiomatic Set Theory address actual infinity with the *axiom of choice* (which is a follow-on to the *axiom of infinity*) and with Cantor's Continuum Hypothesis, to create an algebra that has holes in logic and cannot be used in applied mathematics.

Axiomatic Set Theory, Holes in Logic. A hole in logic is a paradox or an inconsistency in a deductive conclusion. These are annoying, but sometimes holes are so bad they disqualify the logic. In Axiomatic Set Theory, the holes in logic disqualify the logic as is evident by its lack of use in models of physics. Examples of holes in logic are below.

- "$\sqrt{2}$" is maybe not included in Cantor's set of real numbers. There are "N_1" real numbers from "0" to "$\sqrt{2}$" and also from "0" to "1", therefore "$\sqrt{2} = N_1/N_1$". "N_1" is even per "$N_1 = 2^{\wedge}N_0$", therefore "N_1" cannot be divided by 2 until it becomes odd, and so "$\sqrt{2}$" does not equal "N_1/N_1". But, also, "$N_1/2$" does not exist, and that means the "possible to be odd" requirement might not apply.

- "$(\sqrt{2})^2$" should equal integer "2". But if "$\sqrt{2}$" has place-value digits only to the positive countable infinity, then likewise "$(\sqrt{2})^2$" only has place-value digits as zeros to the positive countable infinity, and that differs from "2" which has zeros extending beyond a positive quantity. In other words, "$(\sqrt{2})^2$" is specific to a span of positive interval "$1/N_1$" and "2" is specific to a point of integer zero interval.

- "$\log_2 3$" is maybe not included in Cantor's set of real numbers. There are "N_1" real numbers from "0" to "$\log_2 3$" and also from "0" to "1", therefore "$\log_2 3 = N_1/N_1$", from which $2^{\wedge}N_1 = 3^{\wedge}N_1$. There is no algebra by which "$3^{\wedge}N_1$" can be evaluated. Therefore, no, we cannot prove "$\log_2 3$" is included in Cantor's set of real numbers.

- Cantor assigned some algebra rules for "N_0" and "N_1", but otherwise, there is not enough algebra for inverse operations and a variety of not-assigned operations. Cantor gave us: "$N_0 = N_0/2$", "$N_1 = N_1/2$", "$N_0 = N_0 + 1$", "$N_1 = N_1 + 1$", "$N_1 = N_0 * N_1$", "$N_1 = 2^{\wedge}N_0$". There is no basis for these operations other than Cantor's assignment of them, and therefore we cannot generalize.

CHAPTER 5 - PROPOSED THEORY OF NON-FINITE NUMBERS

For example, it seems reasonable to want a result for "$\log_2 N_0$" and for "$3 \wedge N_0$", but there isn't a result.

- If calculation "$N_1 = 2 \wedge N_0$" is accepted, then, it follows "$N_0 = \ln(N_1)/\ln(2)$" should be, too. That algebraic manipulation leads to "$N_0 * \ln(2) + \gamma = 1 + 1/2 + 1/3 + \ldots + 1/(N_1 - 1)$", but that equation requires a violation of the Continuum Hypothesis.

- Cantor later proposed a cascade of infinities: "$N_1 = 2 \wedge N_0$", "$N_2 = 2 \wedge N_1$", "$N_3 = 2 \wedge N_2$", "$N_4 = 2 \wedge N_3$", ... Is there a "$N_{N1} = 2 \wedge N_{N1-1}$", and "$N_{N1subN1}$", etc.? If the infinities follow natural numbers, then the countable infinity should be in the cascade, maybe.

- Conceptually, if "N_0" is a static number-like thing, maybe a region of numbers, then counting should be able to reach it, eventually.

Blurry fuzziness or fluffiness created by subtly confounding logic, coupled with a lack of alternatives, has, perhaps, kept alive the notion of positive actual infinity as the quantity of a set.

5.2 Idealized Real Numbers

<u>Idealized Compared to Practical</u>. An "idealized real number" has all its place-value digits. The word "all" will be defined by a quantity "1/0". An "all" quantity only exists in our imagination and because it cannot exist physically, is somewhat ambiguous and can be argued about.

In contrast, a "practical real number" has the quantity of place-value digits that are known or knowable limited to however many have been counted-to, both after the decimal point and the long string of zeros before the decimal point. The word "practical" refers to our natural, physical, real world, and as such arguments are settled through applications in mathematical models of physics that match measurements.

<u>Six Dots</u>. A real number written in decimal format can be rewritten as a power series. The advantage the power series format provides is an explicit order-of-magnitude factor on each of the place-value digits. As an example, consider "$\sqrt{2} = 1.4142\ldots$" written as "$\sqrt{2} = 1*10^0 + 4*10^{-1} + 1*10^{-2} + 4*10^{-3} + 2*10^{-4} + \ldots$". Per Group Theory, the three dots extend only as far as has been counted-to.

SPECIAL ALGEBRA FOR SPECIAL RELATIVITY

Alternatively, per Cantor's theory, the three dots extend to the positive countable infinity, "N_0". As a third interpretation, per Axiomatic Set Theory, the three dots extend to all the digits which we justify because of the *axiom of choice*. Axiomatic Set Theory is intertwined wth Cantor's theory, and Cantor's theory so stick to Group Theory, so that three dots extend only to a finite count. A finite count is created by the operation of adding one to the previous number and so is time dependent.

To address Axiomatic Set Theory, consider for the moment the operation of creating integer "1" from "$\sqrt{2} = 1.4142...$". This operation has "1.4142..." truncated (that is, cut) at the decimal point, and everything to the right of the decimal point is discarded. This truncation and discard operation is a bulk operation for which we did not perform the operation of counting. Because a bulk operation did not require counting, we should use a different symbol, and that symbol is proposed to be six dots, "......". The example is an integer which has all zeros after the decimal point, not as far as has been counted, but larger, "all": "$1 = 1.0000......$".

If "Lmax" is the largest number yet counted-to, and if "1/0" is the symbol for "all", then: "$Lmax = 1 + 1 + 1 + ...$" and "$1/0 = 1 + 1 + 1 + ...$". A power series for "1" is now more correctly written "$1 = 1*10^0 + 0*10^{-1} + 0*10^{-2} + ... + 0*10^{-Lmax} + ... + 0*10^{-(1/0)}$". If the last term factor "$10^{-(1/0)}$" equals integer zero, then "1" is a point on the number-line, and "1" is not an interval.

Division by Zero is formally prohibited in the definition of rational numbers in Group Theory. The prohibition against division by zero is generalized to "do not divide by zero", and this generalization is an emotional reaction due to the use of "1/0" as a trick. An example of "1/0" used as a trick is: "$x = y$, $x*x = x*y$, $x*x - y*y = x*y - y*y$, $(x - y)*(x + y) = (x - y)*y$, $x + y = y$". Because of tricks like that, math with "1/0" has the name "pseudo-mathematics" (false math). This name-calling and shunning of "1/0" is perhaps why Dedekind and Cantor used a positive infinity ("N_1") rather than "1/0" as the quantity of points in a line segment. (The "quantity of points in a line" is another way of saying "the quantity of real numbers in a segment of the number-line".)

To get past the paradigm of shunning "1/0", below is a proposed algebra for "1/0" that avoids the errors "1/0" can cause and is not as restrictive as the informal "do not divide by zero". Proposed algebra for "1/0":

- A finite magnitude number can be added to "1/0" without changing "1/0". For example "$1/0 + 7 = 1/0 + (0*7)/0 = (1 + 0*7)/0 = 1/0$". Other examples: "$1/0 = 1/0 + \sqrt{2}$", "$1/0 = 1/0 + 7/5$".

CHAPTER 5 - PROPOSED THEORY OF NON-FINITE NUMBERS

- A finite magnitude non-zero number can be multiplied by "1/0" without changing "1/0". For example "(1/0)*7 = 1/(0*1/7) = 1/(0/7) = 1/0". Other examples: "1/0 = √2*1/0", "1/0 = (7/5)*1/0".

- A finite magnitude non-zero natural number can have "1/0" as the exponent, and equal either "1/0" or "0". For example: "2^(1/0) = 3^(1/0) = 10^(1/0) = (7/5)^(1/0) = either 1/0 or 0".

- Division by "1/0" is multiplication by "0". For example "6/(1/0) = 6*0 = 0".

- "1/0" is the largest magnitude of numbers.

- "1/0" is both/either positive and/or negative.

- Difference "1/0 - 1/0" does not have a result unless it is specified to equal zero.

- Ratio "0/0" does not have a result.

Axiom of Reciprocal of Zero. In Group Theory, formal permission to divide by zero requires "1/0" not be a rational number, and other than that, Axiomatic Set Theory does not mention division by zero. Therefore, to divide by zero, propose a new axiom to be called "*axiom of reciprocal of zero*": "No operation that includes division by zero may possibly result in a non-zero finite number." The proposed axiom introduces "1/0" to a modified Axiomatic Set Theory and does so constructively. The proposed algebra for "1/0" can be expanded and perfected by experience while using it, for example inverse operations such as logarithms may be added. Conceptually, this change feels correct because now there is a mathematical equivalent to the word "vertical".

Idealized Real Numbers. An idealized real number, for example "√2", can now be expressed with "all" its place-value digits as "√2 = 1.4142......" and placed into a power series format as

$$\sqrt{2} = 1*10^0 + 4*10^{-1} + 1*10^{-2} + \ldots a_{Lmax}*10^{-Lmax} + \ldots + a_{1/0}*10^{-1/0}$$

Special Algebra for Special Relativity

In this power series, "Lmax" increases with time as the largest number yet counted-to, and "1/0" is indefinite because of the proposed algebra, such that it can be added to and multiplied-by without changing. Also, the exponent factor "$10^{-1/0}$" is equal to zero, so that the tail end place-value digits "$a_{1/0}$" have integer zero contribution to the value of "$\sqrt{2}$". Alternatively, "$10^{-1/0}$" can equal "1/0", and that, too, might have relevance, but under the excuse idealized real numbers cannot be measured, we will, for now, ignore it. The indefinite magnitude property of "1/0" per the proposed algebra mirrors what Cantor tried to create for "N_1" and that suggests "1/0" should substitute for uncountable infinity "N_1".

Notice in the power series that "1/0" is used as the quantity of a set. This property of "1/0" being the quantity of a set again mirrors "N_1", and it persists even after adding an irrational number to "1/0", for example "1/0 = 1/0 + $\sqrt{2}$". This property of being a quantity of a set does not necessarily follow from the proposed axiom, and so a second axiom is proposed: *axiom that reciprocal of zero is a quantity of a set*: "The division reciprocal of zero is a quantity of a set."

With this proposed formality as a modification to Axiomatic Set Theory, integers are idealized real numbers, as are the other rational numbers that have a repeating pattern for "all" their place-value digits. And irrational numbers with seemingly random place-value digits for "all" their place-value digits are idealized real numbers. To prove "$\sqrt{2}$" is an idealized real number, the ratio "0/0" is not permitted and is replaced by "$0 = \sqrt{2}*0$" and by "$1/0 = \sqrt{2}*(1/0)$". Because "1/0 = 1/0 + 1", "1/0" is both even and odd, and so satisfies the two observations of both "1/0"s being even and one of the "1/0"s able to be odd. And because "$2\wedge 1/0 = 3\wedge 1/0$", "$\log_2 3$" is also proven included in the set of idealized real numbers.

<u>Proving Irrelevance of "N_1"</u>. Per the proposed axiom that "1/0" is the quantity of a set, the number "1/0" has no contribution after the decimal point, and that means "$1/0 = \sqrt{2}*(1/0)$" states that "1/0" is the smallest number multiplied by irrational number "$\sqrt{2}$" to a product that has no contribution after the decimal point. If that statement is true, then product "$\sqrt{2}*N_1$" has a contribution after the decimal point because "N_1" is positive and so is not so big it is both positive and negative as is "1/0", and therefore product "$\sqrt{2}*N_1$" cannot equal "N_1" because "N_1" is a quantity of a set, and so has no contribution after the decimal point. That is the proof that "N_1" has no purpose and should be discarded.

CHAPTER 5 - PROPOSED THEORY OF NON-FINITE NUMBERS

5.3 Practical Real Numbers

<u>Removal of *Axiom of Choice*</u>. The *axiom of choice* was created to provide a means to assume place-value digits had definite values for the case of those place-value digits being beyond a quantity yet counted-to. And although Bertrand Russell's shoe and sock analogy states the *axiom of choice* only pertains to products of irrational numbers, for example "$\log_2 3 * \sqrt{2}$", without the *axiom of choice*, there is no means of justifying knowing those place-value digit values for any real number, with examples being irrational number "$\sqrt{2}$", rational number "1/3". And integer "4" has zeros only to the largest number yet counted-to, if "4" is placed on the number-line as a real number, to become a number of the continuum.

If the *axiom of choice* is deleted, then, per the more general interpretation of the *axiom of choice*, no numbers of the continuum have place-value digits with actual integer values beyond the largest number yet counted-to, which is the largest number provided by the *axiom of infinity*. Not only does this apply to the place-value digits after the decimal point, but also before the decimal point. It follows that there is no purpose of aleph null, "$\aleph_0$".

<u>Modified Continuum Hypothesis</u> states there is no set with a quantity between the largest number yet counted-to, "Lmax", a number permitted per the *axiom of infinity*, and the reciprocal of zero, "1/0", with "1/0" given properties per the two proposed axioms that pertain to reciprocal of zero. "Lmax" has replaced "$\aleph_0$" and "1/0" has replaced "$\aleph_1$". The substitutions widen the Cantor's Continuum Hypothesis to be so broad that "$\aleph_0$" and "$\aleph_1$" are no longer permitted to exist. Also, the proposed Modified Continuum Hypothesis follows from the *axiom of infinity* and the two proposed axioms for reciprocal of zero, and therefore does not stand alone as conjecture to form its own axiom but rather is a derived theorem, so long as the *axiom of choice* has been removed.

<u>Small-Scale Imprecision</u>. A real number, for example "$\sqrt{2}$", is a practical real number if its known or knowable place-value digits after and before the decimal point end at finite count "Lmax", which is the largest number yet counted-to. For a unit square, a unit side is not integer "1" but rather is integer "1" plus finite imprecision, and the diagonal is not idealized "$\sqrt{2}$" but rather is idealized "$\sqrt{2}$" plus finite imprecision. That finite imprecision must be quantified.

Special Algebra for Special Relativity

Small-scale imprecision " "00" = 0.0000...bbbb......" (double zero) has the name "local-zero" with the word "local" referring to a specific value of "Lmax". The three dots shows the zeros extend only to a count of "Lmax", the largest number yet counted-to. The zeros were placed individually one at a time and are known or knowable. The six dots show the "b"s extend to a quantity "1/0". The "b"s can only be operated-on in a bulk operation and are place-holder place-value digits that are unknown and unknowable. For magnitude in base two, small-scale imprecision absolute value is less than "2^{-Lmax}", " $|"00"| < 2^{-Lmax}$ ". Small-scale imprecision is in contrast to an integer zero (or to an idealized real number zero) because the integer (or idealized real number) form of zero has no "b"s and has no finite magnitude.

In base two, the symbol "b" is both/either a "0" or a "1". The selection of "0" or "1" occurs when an observation is made. The observation is a reference to Schrödinger's Cat. The cat is inside a box and if a radioactive nucleus decays, the emitted particle fractures a vial of poison, and the cat dies. Alive is state "0" and dead is state "1". A radioactive decay is an event that is unique to an observing particle, and that observing particle is not us until we open the box to look. Using that analogy, symbol "b" is for "box" or is for "both".

Each "b" converts to either "0" or else "1" from left to right with time.

If we extrapolate to the beginning of time at the Big Bang, which is perhaps when "Lmax = 0", and we consider the reality of measurements by which the granularity of time is too small to be measured such that "Lmax" is increasing very quickly, then "Lmax" is now very, very large. Per this assumption that Lmax is very large and is increasing very quickly, it appears that it isn't possible to know "Lmax" as a specific number. And, as another property of "Lmax", due to time passing at different rates based on general relativity's gravity, "Lmax" is unique in each particle's observation of each other particle. As another speculative observation, each fermion particle interacts with each other fermion particle by emitting boson particles at the speed-of-light, and when the expansion of the universe has become so fast and vast that emitted boson particles never reach another fermion particle, then time has ended and "Lmax" will have reached "1/0". A cosmological point to notice is that numbers and math in general are being coupled with physics so that applied mathematics is the only real mathematics.

"b" symbols cause small-scale imprecision to be positive. "d" can be used as an alternative, for which "d = b - b". The "b" boxes are still opened one at a time from left to right, but now there are two boxes for each place-value digit, and the result can be negative or positive. Math alone does not presently suggest which is correct, "b"s or "d"s, and perhaps future application will settle which. (The application to electromagnetism suggests "b"s and not "d"s.)

CHAPTER 5 - PROPOSED THEORY OF NON-FINITE NUMBERS

Large-Scale Imprecision. An idealized real number, for example "2", has only zeros extending left of the "2", so that zeros are "all" the place-value digits going left, to a quantity "1/0". In contrast, a practical real number "2" has the zeros extending left only to a finite count "Lmax". Starting at count "Lmax + 1" the place-value digits are unknown and unknowable, and so are not zero.

The total of these unknown and unknowable left-of-decimal place-value digits is called "large-scale imprecision" or "local-infinity", "$\Omega\Omega$" (double omega), " $|\text{"}\Omega\Omega\text{"}| > 2^{\text{Lmax}}$ ".

In contrast to small-scale imprecision, it is difficult to imagine what the total sum of the unknown and unknowable place-value digits is, if they are "b"s or "d"s. A possibility is that large-scale imprecision is the division reciprocal of small-scale imprecision, and that will be the visualization in this textbook, so that, at least, we can create a probability distribution for large-scale imprecision. Later, as applications in physics become available, the applications might better identify the probability distribution of large-scale imprecision and determine if it is positive only or is positive/negative. The application to electromagnetic fields suggests positive only.

Large-scale imprecision cannot be the quantity of a set and so is different from the countable and uncountable actual infinities. And large-scale imprecision is not an actual infinity because it can be evaluate statistically using finite numbers larger than "Lmax".

Truncated Numbers. A truncated number is the rational number portion of a practical real number. The concept is that a measurement is of a truncated number and added to the truncated number is small-scale imprecision or else large-scale imprecision, to form a practical real number.

Use base two. "1/(2^Lmax)" is the smallest positive truncated number, "2^Lmax" is largest, and the quantity of truncated numbers is "((2^Lmax)*(2^Lmax) + 1)*2 - 1". For "Lmax = 0", truncated numbers are "1, 0, and -1". For "Lmax = 1" truncated numbers are "10.0, 01.1, 01.0, 00.1, 00.0, -00.1, -01.0, -01.1, -10.0". The eight ("2^Lmax = 8") truncated numbers between zero (inclusive) and one (exclusive) for "Lmax = 3": ".111, .110, .101, .100, .011, .010, .001, .000".

A rational number with a denominator not a power of two (for example "1/3") has a non-zero repeating pattern and is not in a set of truncated numbers because the repeat pattern would end at a count of "Lmax" place-value digits.

SPECIAL ALGEBRA FOR SPECIAL RELATIVITY

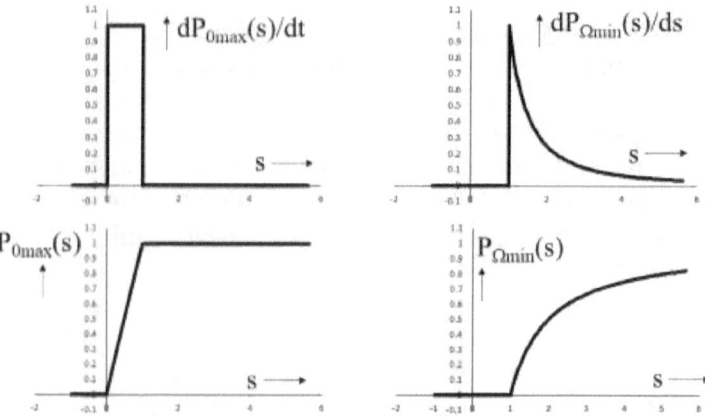

Figure 35. Probability Distribution for "Lmax = 0" for positive-only local-zero ""00" = 0.bbbbb......" left and its division reciprocal ""ΩΩ" = 1/"00"" right. Higher values of "Lmax" make the square thinner and taller and make the reciprocal further to the right.

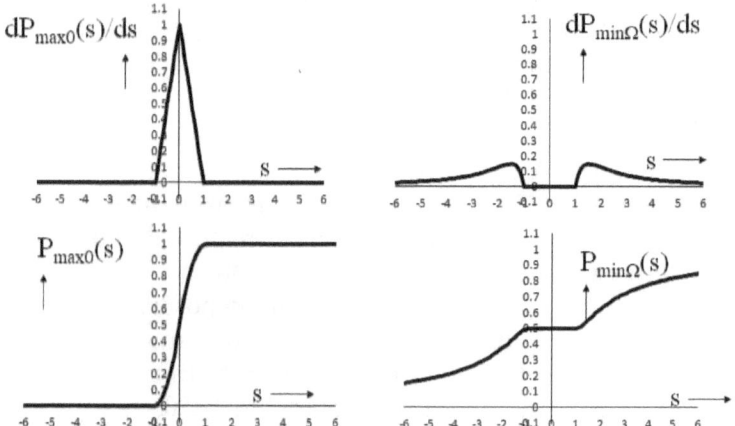

Figure 36. Probability Distribution for "Lmax = 0" for positive-negative local-zero ""00" = 0.ddddd......" left and its division reciprocal ""ΩΩ" = 1/"00"" right. Higher values of "Lmax" make the spike taller and make the two arms further apart.

CHAPTER 5 - PROPOSED THEORY OF NON-FINITE NUMBERS

Either Large-Scale or Small-Scale. A truncated number has both large-scale and small-scale imprecision added to it, but each is added separately. If both were added together, then large-scale imprecision dominates the number.

Finished-Calculation and Final-Result. Truncated numbers with finite precision pertain to the real, natural, physical world of continuum quantities in which the problem to be solved is set up. The problem is then translated into all number algebra that includes small-scale and large-scale imprecision and analysis is performed. The finished calculation has a portion that has extreme magnitude beyond "$2^{\pm Lmax}$" that is the large-scale and/or small-scale imprecision. That extreme magnitude portion is dropped to create the final-result, so that the final-result is a truncated number that can be measured.

- Finished-Calculation is at the end of the all-number second step in the Process from Descartes. It includes contributions beyond "$2^{\pm Lmax}$".

- Final-Result. Impose finite precision to find the physical world "final-result" in the third step in the Process from Descartes by discarding contributions beyond "$2^{\pm Lmax}$".

5.4 Application to Special Relativity

Time-space hyperbolic angle "α_M" (that relates to speed per $v_M = c*\tanh\alpha_M$") is a unitless number inside an exponential function. It is proposed to be a truncated number to which large-scale and small-scale imprecision is added using a Lorentz Transformation.

The addition of small-scale imprecision is trivial because small-scale imprecision is very, very small and in measurements is indistinguishable from zero. In contrast, the addition of large-scale imprecision dominates the calculation of component values because large-scale imprecision is very, very large, infinitely so, compared to a measurable speed "v_M"'s time-space hyperbolic angle "α_M".

Adding Imprecision. Reference frame "M" (for moving) is inside a moving bus. Imagine a baseball moving at positive speed "v_M" from the back of the bus toward the front, as if it is rolling along the floor. A baseball is our visualization of an electron particle.

SPECIAL ALGEBRA FOR SPECIAL RELATIVITY

If an electron is accelerated to the speed-of-light by adding infinity to "α_M", then the electron has infinite relativistic energy and momentum. To counter that impossible phenomenon from happening, when large-scale imprecision is added to "α_M" the relativistic mass remains finite by including a second term in "$\alpha_{S/M}$" in the Lorentz Transformation "$\alpha_S = \alpha_M + \alpha_{S/M}$". The second term has a "q_x" factor so that it only affects mass, or, in general, the hyperbolic radius.

$\alpha_{S/M} = (1 - q_x)*\Xi$ ("Ξ" (ksi, xi), for either large- or small-scale imprecision)
$\phantom{\alpha_{S/M}} = (1 - q_x)*$"$\Omega\Omega$"   Large-Scale Case
or else $= (1 - q_x)*$"00"   Small-Scale Case

Large-scale and small-scale imprecision are Lorentz Transformed separately, such that the imprecision is either large-scale or else is small-scale, and not both added together. If the two imprecisions were added together, then large-scale dominates and there is no sub-light-speed motion.

<u>General Form</u>. "$_2r$" is time-space location invariant. "t_B" is time for a clock mounted on the electron. "1_M" is the time direction compound label-number.

$_2r = 1_M*(c*t_B)*\exp(q_x*\alpha_M)$
$ = 1_M*(c*t_B)*\exp(q_x*\alpha_M)*1$
$ = 1_M*(c*t_B)*\exp(q_x*\alpha_M)*\exp(q_x*\alpha_{S/M})/\exp(q_x*\alpha_{S/M})$
$ = (1_M/\exp(q_x*\alpha_{S/M}))*(c*t_B)*\exp(q_x*\alpha_M)*\exp(q_x*\alpha_{S/M})$
$ = 1_S*(c*t_B)*\exp(q_x*\alpha_M)*\exp(q_x*\alpha_{S/M})$
$ = 1_S*(c*t_B)*\exp(q_x*\alpha_M)*\exp(q_x*(1 - q_x)*\Xi)$
$ = 1_S*(c*t_B)*\exp(q_x*\alpha_M)*\exp(q_x*\Xi)*\exp(-q_x^2*\Xi)$
$ = 1_S*(c*t_B)*\exp(q_x*\alpha_M)*\exp(q_x*\Xi)*\exp(-\Xi)$
$ = 1_S*(c*t_B)*\exp(q_x*\alpha_M)*(\cosh\Xi + q_x*\sinh\Xi)*\exp(-\Xi)$
$ = 1_S*(c*t_B)*\exp(q_x*\alpha_M)*((\exp\Xi + \exp(-\Xi)) + q_x*(\exp\Xi - \exp(-\Xi)))*(\exp(-\Xi))/2$
$ = 1_S*(c*t_B)*\exp(q_x*\alpha_M)*((1 + \exp^2(-\Xi)) + q_x*(1 - \exp^2(-\Xi)))/2$
$ = 1_S*(c*t_B)*\exp(q_x*\alpha_M)*((1 + q_x)/2 + (1 - q_x)*\exp^2(-\Xi))/2$

Large-scale case, for which $\Xi = $ "$\Omega\Omega$":
$_2r = 1_S*(c*t_B)*\exp(q_x*\alpha_M)*((1 + q_x)/2 + (1 - q_x)*\exp^2(-$"$\Omega\Omega$"$)/2)$
$ \approx 1_S*(c*t_B)*\exp(q_x*\alpha_M)*((1 + q_x)/2 + (1 - q_x)*0)/2$
$ = 1_S*(c*t_B)*\exp(q_x*\alpha_M)*(1 + q_x)/2$
$ = 1_S*(c*t_B)*(\cosh\alpha_M + q_x*\sinh\alpha_M)*(1 + q_x)/2$
$ = 1_S*(c*t_B)*((\cosh\alpha_M + q_x^2*\sinh\alpha_M) + q_x*(\cosh\alpha_M + \sinh\alpha_M))/2$
$ = 1_S*(c*t_B)*((\cosh\alpha_M + \sinh\alpha_M) + q_x*(\cosh\alpha_M + \sinh\alpha_M))/2$
$ = 1_S*(c*t_B)*(\exp\alpha_M)*(1 + q_x)/2$

CHAPTER 5 - PROPOSED THEORY OF NON-FINITE NUMBERS

Components – Large-Scale Case
$c*t_S \approx (c*t_B)*(exp\alpha_M)/2$
$x_S \approx (c*t_B)*(exp\alpha_M)/2$

Components for Large-Scale Case. Large-scale case components "$c*t_S$" and "x_S" above are the "final-result". The final-result was created from the finished-calculation at the step with equals sign "=" replaced by "$\approx$" symbol. At that step the very small nearly zero term "$(1 - q_x)*exp^2(-\text{"}\Omega\Omega\text{"})/2$" was dropped because it was below "$2^{-L_{max}}$" magnitude, so that "$c*t_S$" and "x_S" final-results are truncated numbers.

Large-scale case time component "$c*t_S = (c*t_B)*(exp\alpha_M)/2$" equaled space location component "$x_S = (c*t_B)*(exp\alpha_M)/2$". Time and space are equal if the motion is at the speed-of-light, because "$x_S/t_S = c$" follows from "$c*t_S = x_S$".

The equations above were given for time-space location "$_2r$". By replacing hyperbolic radius "$c*t_B$" with "m_B*c" the equations pertain to energy-momentum invariant "$_2p$". After that substitution, large-scale case energy component "$E_S/c = (m_B*c)*(exp\alpha_M)/2$" equals momentum component "$p_S = (m_B*c)*(exp\alpha_M)/2$", and that equivalence also applies to motion at the speed-of-light.

A photon particle travels macroscopic distances at the speed-of-light with energy equal to momentum. It appears that the large-scale case is for bus "M" moving at the speed-of-light relative to roadside observer "S", and for that observer "S", the electron is observed as a photon.

The assumption above was that large-scale imprecision "$\Omega\Omega$" was positive. If large-scale imprecision is negative, then the component values are infinite due to the extremely large term "$exp^2(-\text{"}\Omega\Omega\text{"})$". Infinite component values cannot be measured.

Label Numbers for Large-Case Imprecision. As opposed to components, label numbers cannot be measured. Label numbers are a mathematical bookkeeping tool by which we ensure invariants are invariants and gauge invariance is maintained. Their physical analogy is direction, but any measurement of a direction is of a component, and not of a label number. Confounding the problem is the singular label number factor "$1 + q_x$" or "$1 - q_x$" associated with large-scale imprecision.

Assume a label number has the same modification as components.

SPECIAL ALGEBRA FOR SPECIAL RELATIVITY

Label Numbers – General Case
$$1_S = 1_M/\exp(q_x*\alpha_{S/M})$$
$$= 1_M*\exp(-q_x*(1 - q_x)*\Xi)$$
$$= 1_M*\exp(-q_x*\Xi)*\exp(\Xi)$$
$$= 1_M*(\cosh\Xi - q_x*\sinh\Xi)*\exp(\Xi)$$
$$= 1_M*((\exp\Xi + \exp-\Xi) - q_x*(\exp\Xi - \exp-\Xi))*(\exp\Xi)/2$$
$$= 1_M*((1 + \exp^2\Xi) - q_x*(-1 + \exp^2\Xi))/2$$
$$= 1_M*((1 + q_x)/2 + (1 - q_x)*(\exp^2\Xi)/2)$$

$$1_S^{*j} = 1_M/\exp(q_x*\alpha_{S/M}) = ((1 + q_x^{*j})/2 + (1 - q_x^{*j})*(\exp^2\Xi)/2)*1_M^{*j}$$

$$(1_S^{*j}*1_S) = ((1 + q_x^{*j})/2 + (1 - q_x^{*j})*(\exp^2\Xi)/2)*(1_M^{*j}*1_M)$$
$$*((1 + q_x)/2 + (1 - q_x)*(\exp^2\Xi)/2)$$
$$= ((1 + q_x^{*j})/2 + (1 - q_x^{*j})*(\exp^2\Xi)/2)*((1 + q_x)/2 + (1 - q_x)*(\exp^2\Xi)/2)$$
$$= (1/4)*((1 + q_x^{*j}) + (1 - q_x^{*j})*(\exp^2\Xi))*((1 + q_x) + (1 - q_x)*(\exp^2\Xi))$$
$$= (1/4)*((1 + q_x^{*j})*(1 + q_x) + (1 - q_x^{*j})*(\exp^2\Xi))*(1 + q_x)$$
$$+ (1 + q_x^{*j})*(1 - q_x)*(\exp^2\Xi)) + (1 - q_x^{*j})*(\exp^2\Xi))*(1 - q_x)*(\exp^2\Xi))$$
$$= (1/4)*((1 - q_x)*(1 + q_x) + (1 + q_x)*(\exp^2\Xi)*(1 + q_x)$$
$$+ (1 - q_x)*(1 - q_x)*(\exp^2\Xi) + (1 + q_x)*(\exp^2\Xi)*(1 - q_x)*(\exp^2\Xi))$$
$$= (1/4)*((1 + q_x)*(\exp^2\Xi)*(1 + q_x) + (1 - q_x)*(1 - q_x)*(\exp^2\Xi))$$
$$= (1/4)*(\exp^2\Xi)*((1 + q_x)*(1 + q_x) + (1 - q_x)*(1 - q_x))$$
$$= (1/4)*(\exp^2\Xi)*((1^2 + 2*q_x + q_x^2) + (1^2 - 2*q_x + q_x^2))$$
$$= (1/4)*(\exp^2\Xi)*((1 + 1) + (1 + 1))$$
$$= (1/4)*4*(\exp^2\Xi)$$
$$= \exp^2\Xi$$

Product "$1_S^{*j}*1_S$" is not of unit magnitude because the exotic Lorentz Transformation "$\alpha_{S/M} = (1 - q_x)*\Xi$" includes a "$q_x$" factor which creates a term not affected by the conjugate operation.

For the large-scale case, what gets dropped in the transition from finished-calculation to final-result are the terms that match what was dropped for the components.

$$1_S = 1_M*((1 + q_x)/2 + (1 - q_x)*(\exp^{2"}\Omega\Omega")/2)$$
$$\approx 1_M*(1 + q_x)/2$$

$$q_{xS} = 1_S*q_x = 1_M*(q_x + q_x^2)/2 = 1_M*(1 + q_x)/2 = 1_S$$

CHAPTER 5 - PROPOSED THEORY OF NON-FINITE NUMBERS

$$(1_S^{*j} * 1_S) = ((1 + q_x^{*j})/2) * (1_M^{*j} * 1_M) * ((1 + q_x)/2)$$
$$= (1/4) * (1 + q_x^{*j}) * (1 + q_x)$$
$$= (1/4) * (1 - 1) = 0$$

Product "$1_S^{*j} * 1_S$" is not of unit magnitude because it is zero. Regardless, "1_S" appears to be unit magnitude because it is half of "1" plus half of "q_x", each of which is unit magnitude.

Hyperbolic Radius Check for the General Case:

$$_4r^{*j} *_4r = (1_S*(c*t_B)*\exp(q_x*\alpha_M)*((1 + q_x)/2 + (1 - q_x)*\exp^2(-\Xi))/2)^{*j}$$
$$*(1_S*(c*t_B)*\exp(q_x*\alpha_M)*((1 + q_x)/2 + (1 - q_x)*\exp^2(-\Xi))/2)$$
$$= ((c*t_B)*\exp(q_x^{*j}*\alpha_M)*((1 + q_x^{*j})/2 + (1 - q_x^{*j})*\exp^2(-\Xi))/2)$$
$$*(1_S^{*j}*1_S)*(c*t_B)*\exp(q_x*\alpha_M)*((1 + q_x)/2 + (1 - q_x)*\exp^2(-\Xi))/2)$$
$$= ((c*t_B)*\exp(q_x^{*j}*\alpha_M)*((1 + q_x^{*j})/2 + (1 - q_x^{*j})*\exp^2(-\Xi))/2)$$
$$*(\exp^2\Xi)*(c*t_B)*\exp(q_x*\alpha_M)*((1 + q_x)/2 + (1 - q_x)*\exp^2(-\Xi))/2)$$
$$= (c*t_B)^2*(\exp^2\Xi)*(\exp(q_x^{*j}*\alpha_M)*((1 + q_x^{*j})/2 + (1 - q_x^{*j})*\exp^2(-\Xi))/2)$$
$$*\exp(q_x*\alpha_M)*((1 + q_x)/2 + (1 - q_x)*\exp^2(-\Xi))/2)$$
$$= (c*t_B)^2*(\exp^2\Xi)*((1 + q_x^{*j})/2 + (1 - q_x^{*j})*\exp^2(-\Xi))/2)$$
$$*((1 + q_x)/2 + (1 - q_x)*\exp^2(-\Xi))/2)$$
$$= (c*t_B)^2*(\exp^2\Xi)*(1/2)^2*((1 + q_x^{*j}) + (1 - q_x^{*j})*\exp^2(-\Xi))$$
$$*((1 + q_x) + (1 - q_x)*\exp^2(-\Xi))$$
$$= (c*t_B)^2*(\exp^2\Xi)*(1/2)^2*((1 + q_x^{*j})*(1 - q_x)*\exp^2(-\Xi)$$
$$+ (1 - q_x^{*j})*\exp^2(-\Xi)*(1 + q_x))$$
$$= (c*t_B)^2*(1/2)^2*((1 + q_x^{*j})*(1 - q_x) + (1 - q_x^{*j})*(1 + q_x))$$
$$= (c*t_B)^2*(1/2)^2*((1 - q_x)*(1 - q_x) + (1 + q_x)*(1 + q_x))$$
$$= (c*t_B)^2*(1/2)^2*((1^2 - 2*q_x + q_x^2) + (1^2 + 2*q_x + q_x^2))$$
$$= (c*t_B)^2*(1/2)^2*((1 + 1) + (1 + 1))$$
$$= (c*t_B)^2$$

The hyperbolic radius "$c*t_B$" calculation shows the general case for the exotic Lorentz Transformation was performed correctly. Notice in the analysis that factor "$\exp^2\Xi$" from the label numbers is countered by factor "$\exp^2(-\Xi)$" from the components. That pair of factors is removed for the large-scale case, but then the singular property of the label numbers dominates so that the hyperbolic radius is zero.

SPECIAL ALGEBRA FOR SPECIAL RELATIVITY

$$_4r^{*j}*_4r = (c*t_S*1_S^{*j} + x_S*q_{xS}^{*j})*(1_S*c*t_S + q_{xS}*x_S)$$
$$= (c*t_B)^2*(\exp^2\alpha_M)*(1/2)^2*(1_S^{*j} + q_{xS}^{*j})*(1_S + q_{xS})$$
$$= (c*t_B)^2*(\exp^2\alpha_M)*(1/2)^2*(1/2)^2*((1+q_x)^{*j} + (1+q_x)^{*j})*((1+q_x) + (1+q_x))$$
$$= (c*t_B)^2*(\exp^2\alpha_M)*(1/2)^2*(1/2)*(1+q_x)^{*j}*(1+q_x)$$
$$= (c*t_B)^2*(\exp^2\alpha_M)*(1/2)^2*(1/2)*0$$
$$= 0$$

Small-Scale Precision Case. Because "00" is smaller magnitude than "2^{-Lmax}", terms that have "00" are too small to be retained after the transition from step 2 finished-calculation to step 3 final-result.

Small-scale case
$$= 1_S*(c*t_B)*\exp(q_x*\alpha_M)*((1+q_x)/2 + (1-q_x)*\exp^2(-\text{"00"}))/2$$
$$\approx 1_S*(c*t_B)*\exp(q_x*\alpha_M)*((1+q_x)/2 + (1-q_x)*1)/2$$
$$= 1_S*(c*t_B)*\exp(q_x*\alpha_M)*(2/2)$$
$$= 1_S*(c*t_B)*\exp(q_x*\alpha_M)$$
$$= 1_M*(c*t_B)*\exp(q_x*\alpha_M)$$

Components – Small-Scale Case
$c*t_S \approx (c*t_B)*\cosh\alpha_M = c*t_M$
$x_S \approx (c*t_B)*\sinh\alpha_M = x_M$

Label Numbers – Small-Scale Case
$1_S = 1_M*((1+q_x)/2 + (1-q_x)*(\exp^2\text{"00"})/2)$
$\approx 1_M*((1+q_x)/2 + (1-q_x)/2)$
$= 1_M$

$q_{xS} = 1_S*q_x = q_{xM}$

The small-scale case had no change in time or in space, as applies to the originally measured motion of the sub-light speed particle, the electron.

Factor "1/2" in Large-Scale Case Components. Factor "1/2" in large-scale case components "$c*t_S \approx (c*t_B)*(\exp\alpha_M)/2$" and "$x_S \approx (c*t_B)*(\exp\alpha_M)/2$" is due to a phenomenon called "incident transmission rate" (which is a new concept first described in this book). As an example, a stationary electron "$v_M = 0$" hypothetically emits a photon that gets absorbed three time units later, at "$c*t_S = 3$". Because "$v_M = 0$" and "$\alpha_M = 0$", "$\exp\alpha_M = 1$", which is an inconsequential factor, so that "$t_S = t_B/2$", and that means if "$c*t_S = 3$" then the absorption happened at "$c*t_B = 6$".

CHAPTER 5 - PROPOSED THEORY OF NON-FINITE NUMBERS

The small-scale case for this example is a stationary electron for which "$t_S = t_B$". At "$c*t_B = 6$" the absorption incident occurs in "B", and the electron recoils in response. The electron/photon combined particle is one particle and reacts as one particle, both by the photon being absorbed and the electron recoiling in response, at the same time "$c*t_B = 6$".

The observer in "S", which is you or me, witnessed the photon absorption at "$c*t_S = 3$" and assumed there was a signal at the speed-of-light back to the electron that arrived at "$c*t_S = 6$" telling it to recoil in response. The speed-of-light is the incident transmission rate for this example.

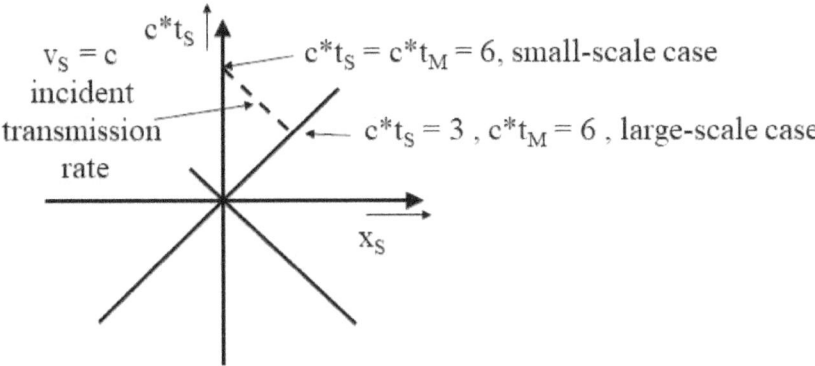

Figure 37. For a stationary electron "$c*t_M = c*t_S$" per the small-scale case, and for its emitted photon "$c*t_M/2 = c*t_S$" per the large-scale case. An incident of photon observation occurs at "$c*t_M = 6$", which is "$c*t_S = 3$". In "S" it appears an incident transmission rate at the speed-of-light occurs to alert the electron at "$c*t_S = 6$" to recoil. But with respect to the electron/photon combined particle, both events are one event that occurred at "$c*t_M = 6$".

The assumption is that the electron/photon combined particle is a particle only when observed as being a particle. This assumption is based on the results of the EPR experiment. The EPR experiment has two aspects. The first aspect is that when the EPR experiment is performed with two particles of the same type, for example two electrons or else two photons, the signal of particle properties is transmitted instantly from one particle to its entangled counterpart particle, not at the speed-of-light "$v = c$" but instantly, "$v = 1/0$", called "spooky action at a distance". An instant speed of "$v = 1/0$" for the incident transmission rate is appropriate for two particles of the same kind because both particles have the same correlation between "t_S" and "t_B". The underlying assumption to this interpretation of the EPR experiment is that entangled particles are the same particle.

SPECIAL ALGEBRA FOR SPECIAL RELATIVITY

The EPR experiment has a second aspect and that is that particle properties are realized at the moment of observation and not at the moment of creation. It means the electron does not recoil by emitting a photon at "$t_B = 0$" and later the photon gets absorbed. Rather, it means the observation event of absorbing the photon makes the photon real, and then the electron recoils with respect to the observer "S" because the photon is now real with respect to the observer "S".

With respect to General Relativity and its expanding universe cosmology, a photon that traverses past an event horizon, be it a black hole or be it light reaching us from a quasar, remains one with the electron that emitted it, but the speed-of-light is insufficient as a means of incident transmission rate for the electron to recoil. This is a clue to the development of more theory.

Conclusion. The conclusion from the Lorentz Transformation that added large-scale and small-scale imprecision to the time-space hyperbolic angle that represented an electron's speed is that the proposed added angle "$\alpha_{S/M} = (1 - q_x)*\Xi$" correctly modelled a photon particle's motion at the speed-of-light and, simultaneously, correctly modelled the electron's motion at a sub-light speed. Both were modelled simultaneously by considering large-scale imprecision as a separate case from small-scale imprecision.

The rest of this chapter builds on that success, first by expanding the application to other invariants including electric charge density, and then by applying the same Lorentz Transformation to matter waves for a derivation of a quantum dynamics model of electromagnetic fields.

Speed Calculation. A direct calculation of speed confirms large-scale imprecision models motion at the speed-of-light and small-scale imprecision models motion at a sub-light speed. "$\alpha_{S/M} = (1 - q_x)*\Xi$"

$v_{S/M}/c = \tanh\alpha_{S/M} = \tanh((1 - q_x)*\Xi)$
$= (\tanh\Xi - q_x*\tanh\Xi)/(1 - q_x*\tanh\Xi*\tanh\Xi)$
$= (1 - q_x)*\tanh\Xi/(1 - q_x*\tanh^2\Xi)$
$= (1 - q_x)*(\sinh\Xi/\cosh\Xi)/(1 - q_x*\sinh^2\Xi/\cosh^2\Xi)$
$= (1 - q_x)*((\exp\Xi - \exp(-\Xi))/(\exp\Xi + \exp(-\Xi)))$
  $/(1 - q_x*(\exp^2\Xi - 2 + \exp^2(-\Xi))/(\exp^2\Xi + 2 + \exp^2(-\Xi)))$

$v_{S/M\text{-LargeCase}}/c = (1 - q_x)*((\exp"\Omega\Omega" - \exp(-"\Omega\Omega"))/(\exp"\Omega\Omega" + \exp(-"\Omega\Omega")))$
  $/(1 - q_x*(\exp^{2}"\Omega\Omega" - 2 + \exp^2(-"\Omega\Omega"))/(\exp^{2}"\Omega\Omega" + 2 + \exp^2(-"\Omega\Omega")))$
$\approx (1 - q_x)*(\exp"\Omega\Omega"/\exp"\Omega\Omega")/(1 - q_x*\exp^{2}"\Omega\Omega"/\exp^{2}"\Omega\Omega")$
$= (1 - q_x)/(1 - q_x)$
$= 1$

CHAPTER 5 - PROPOSED THEORY OF NON-FINITE NUMBERS

$$v_{S/M\text{-SmallCase}}/c = (1 - q_x)*((\exp\text{``}00\text{''} - \exp(-\text{``}00\text{''}))/(\exp\text{``}00\text{''} + \exp(-\text{``}00\text{''})))$$
$$/(1 - q_x*(\exp^2\text{``}00\text{''} - 2 + \exp^2(-\text{``}00\text{''}))/(\exp^2\text{``}00\text{''} + 2 + \exp^2(-\text{``}00\text{''})))$$
$$\approx (1 - q_x)*(1 - 1)/(1 + 1)/(1 - q_x*(1 - 2 + 1)/(1 + 2 + 1))$$
$$= (1 - q_x)*0/(1 - q_x*0)$$
$$= 0/1 = 0$$

There is only one roadside "$_S$" where the observer stands. Also, there is only one bus "$_M$" where a different observer sits. There are two "$_{S/M}$" speeds for the bus: "$_{S/M\text{-LargeCase}}$" and "$_{S/M\text{-SmallCase}}$". Because of the two "$_{S/M}$" speeds, there are two separate observations from the one roadside "$_S$" of the one bus "$_M$".

Observing an Object that Moves at the Speed-of-Light.

$$c*t_S \approx (c*t_B)*\exp(\alpha_M)/2 \qquad ; \qquad x_S \approx (c*t_B)*\exp(\alpha_M)/2$$
$$= (c*t_B)*(\cosh\alpha_M + \sinh\alpha_M)/2 \qquad = (c*t_B)*(\cosh\alpha_M + \sinh\alpha_M)/2$$
$$= (c*t_M + x_M)/2 \qquad = (c*t_M + x_M)/2$$

If "x_M" identifies the back of a baseball and "$x_M + \Delta x_M$" identifies the front of a baseball, then, because "$(c*t_M + x_M)_{back}$" equals "$(c*t_M + x_M)_{front}$", front is earlier than back by "$\Delta x_M/c$".

$$c*t_S = x_S = (c*t_M + x_M)_{back}/2 = (c*t_M + x_M)_{front}/2$$

$$c*t_{Mfront} = c*t_{Mback} - (x_{Mfront} - x_{Mback}) = c*t_{Mback} - \Delta x_M$$

All locations in "$_M$" back to front along the baseball are at one location "x_S" at one time "$c*t_S$" because each location is at a different "t_M" time, per length contraction.

Other Invariants. Other invariants also have large/small-scale cases.

Location (time-like)	$_2r = 1_S*c*t_B*\exp(q_x*\alpha_S)$
Location (space-like)	$_2s = 1_S*q_x*s_{xB}*\exp(q_x*\alpha_S)$
Frequency (time-like)	$_2\omega/c = 1_S*(\omega_B/c)*\exp(q_x*\alpha_S)$
Wave-number (space-like)	$_2k = 1_S*q_x*k_{xB}*\exp(q_x*\alpha_S)$
Energy-Momentum	$_2p = 1_S*m_B*c*\exp(q_x*\alpha_S)$
Charge Density	$_2J = 1_S*\rho_B*\exp(q_x*\alpha_S)$

SPECIAL ALGEBRA FOR SPECIAL RELATIVITY

Wave-number and Frequency Observed at the Speed-of-Light. On the "$_2r$" hypercomplex-plane "$_2\omega/c = 1_B*(\omega_B/c)$" is plotted as stationary horizontal evenly spaced parallel "wave crest" lines that extend left and right for all of "x_B" space. Lines are closer together for higher frequency "ω_B".

As a visualization, two long rods along the floor of the bus, front to back, bounce side to side to make a bang sound everyone seated on the bus hears. The bang sound is the horizontal lines. And people standing along the roadside raise their hands in unison to show they heard each bang.

Then, the bus moves forward with "$\alpha_{S/M} > 0$".

$$_2\omega/c = 1_S*(\omega_B/c)*\exp(q_x*\alpha_{S/M})$$
$$= 1_S*(\omega_B/c)*\cosh(\alpha_{S/M}) + q_{xS}*(\omega_B/c)*\sinh(\alpha_{S/M})$$
$$= 1_S*(\omega_S/c) + q_{xS}*k_{xS}$$

Because the bus is moving, wave crest lines slope up and to the right and are spaced closer together by factor "$\cosh(\alpha_{S/M})$" (because of time dilation and "$\omega_S/c = (\omega_B/c)*\cosh(\alpha_{S/M})$"). People on the roadside each raise their hand when they hear the bang sound of the rods. The speed of the hands is faster than the speed-of-light.

The bus speeds up to (nearly) the speed-of-light so that time dilation is nearly infinite. Hands raise only once and the motion of raising the hand moves at the speed-of-light.

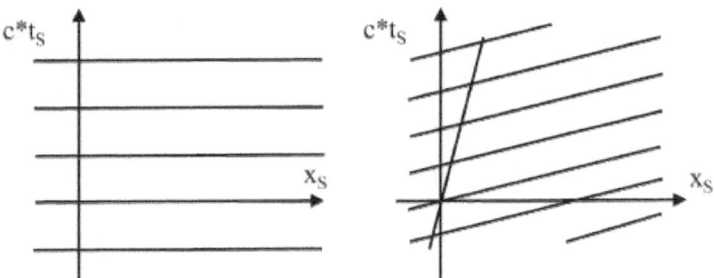

Figure 38a. "$_2\omega/c = 1_S*(\omega_B/c)$" is illustrated on the left. On the right, invariant "$_2\omega/c = 1_S*(\omega_B/c)*\exp(q_x*\alpha_S)$" is illustrated, for "$\alpha_S > 0$".

CHAPTER 5 - PROPOSED THEORY OF NON-FINITE NUMBERS

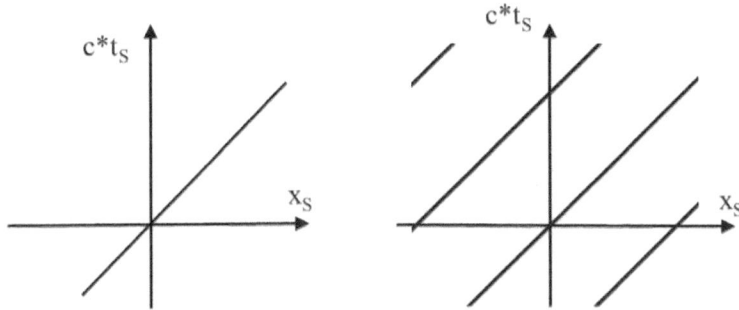

Figure 38b. "$_2\omega/c = 1_S*(\omega_B/c)*\exp(q_x*\alpha_S)$" for "$\alpha_S = +\infty$" is illustrated on left. Right has "$_2\omega/c = 1_S*(\omega_B/c)*\exp(q_x*\alpha_S)$" for the Large-Scale Case "$\alpha_S = (1 - q_x)*(``\Omega\Omega")$".

For the large-scale case, frequency and wavenumber components are finite and equal, "$\omega_S/c \approx k_S/c \approx (\omega_B/c)/2$". Hands raise at the speed-of-light and do so repeatedly.

$$\begin{aligned}\omega_S/c &= (\omega_B/c)*\exp(-``\Omega\Omega")*\cosh(``\Omega\Omega")\\ &= (\omega_B/c)*\exp(-``\Omega\Omega")*(\exp(``\Omega\Omega") + \exp(-``\Omega\Omega"))/2\\ &= (\omega_B/c)*(1 + \exp(-2*``\Omega\Omega"))/2\\ &\approx (\omega_B/c)/2\end{aligned}$$

People on the roadside also hear the small-scale case, for which the bus is stationary. For the small-scale case they all raise their hands simultaneously. Both the large-scale case and small-scale case together are visible in the raising of hands.

Relating Energy-Momentum to Frequency-Wavenumber.

"$_2p$" equals "$_2\omega/c$" times "$\hbar$" (Planck's Constant) per the de Broglie relations.

$$_2p = \hbar*_2\omega/c \quad ; \quad m_B*c = \hbar*\omega_B/c$$

Rest mass and rest frequency are "one in the same" because "$\hbar$" and "c" are measurement unit conversion factors.

The visualization has a person on a bus observing a particle that sits stationary on the bus. Before the actual observation of the particle is made, the particle could be anywhere on the bus, front to back, because the frequency of bang sound occurs everywhere on the bus, front to back. Per that visualization, "$m_B*c = \hbar*\omega_B/c$" pertains to a particle, but only before the particle's location is observed (that is, it pertains when the particle is a wave).

SPECIAL ALGEBRA FOR SPECIAL RELATIVITY

Electric Current Density. Large-scale case electric current density component values are zero because the hyperbolic radius has a zero factor "exp(-"ΩΩ"/2)".

$$(d(Q_B*exp(-"\Omega\Omega"/2))/dx_B)*exp(-"\Omega\Omega")$$

Zero factor "exp(-"ΩΩ"/2)" is derived from classical radius theory, for which "$m_B*c = (1/2)*(Q_B^2/(4*\pi*\ni*c))/r_e$".

$$m_B*c*exp(-"\Omega\Omega") = ((1/2)*(Q_B^2/(4*\pi*\ni*c))/r_e)*exp(-"\Omega\Omega")$$
$$= (1/2)*((Q_B*exp(-"\Omega\Omega"/2))^2/(4*\pi*\ni*c))/r_e$$

Per the above equation the nearly zero value for rest mass requires nearly zero electric charge, accomplished by a "exp(-"ΩΩ"/2)" factor on "Q_B": "$Q_B*exp(-"\Omega\Omega"/2)$" for the large-scale case.

The hyperbolic-radius of "$_2J$" for the large-scale case is modified to include the "exp(-"ΩΩ"/2)" factor. That modification is in addition to the application of the " $\alpha_{S/M} = (1 - q_x)*"\Omega\Omega"$ " large-scale case Lorentz Transformation.

$$J_{tS}/c = (dQ_B/dx_B)*exp(-\Xi)*cosh(\alpha_M + \Xi)$$
$$= (d(Q_B*exp(-"\Omega\Omega"/2))/dx_B)*exp(-"\Omega\Omega")*cosh(\alpha_M + "\Omega\Omega")$$
$$= exp(-"\Omega\Omega"/2)*(dQ_B/dx_B)*exp(-"\Omega\Omega")*(exp(\alpha_M + "\Omega\Omega")$$
$$+ exp(-(\alpha_M + "\Omega\Omega")))/2$$

$$= exp(-"\Omega\Omega"/2)*(dQ_B/dx_B)*exp(-"\Omega\Omega")*(exp(\alpha_M)*exp("\Omega\Omega")$$
$$+ exp(-\alpha_M)*exp(-"\Omega\Omega"))/2$$

$$= exp(-"\Omega\Omega"/2)*(dQ_B/dx_B)*(exp(\alpha_M) + exp(-\alpha_M)*exp^2(-"\Omega\Omega"))/2$$
$$\approx exp(-"\Omega\Omega"/2)*(dQ_B/dx_B)*exp(\alpha_M) \approx 0$$

$$J_{xS}/c = (dQ_B/dx_B)*exp(-\Xi)*sinh(\alpha_M + \Xi)$$
$$= (d(Q_B*exp(-"\Omega\Omega"/2))/dx_B)*exp(-"\Omega\Omega")*sinh(\alpha_M + "\Omega\Omega")$$
$$= exp(-"\Omega\Omega"/2)*(dQ_B/dx_B)*exp(-"\Omega\Omega")*(exp(\alpha_M + "\Omega\Omega")$$
$$- exp(-(\alpha_M + "\Omega\Omega")))/2$$
$$= exp(-"\Omega\Omega"/2)*(dQ_B/dx_B)*exp(-"\Omega\Omega")*(exp(\alpha_M)*exp("\Omega\Omega")$$
$$- exp(-\alpha_M)*exp(-"\Omega\Omega"))/2$$
$$= exp(-"\Omega\Omega"/2)*(dQ_B/dx_B)*(exp(\alpha_M) - exp(-\alpha_M)*exp^2(-"\Omega\Omega"))/2$$
$$\approx exp(-"\Omega\Omega"/2)*(dQ_B/dx_B)*exp(\alpha_M) \approx 0$$

"$J_{tS}/c \approx J_{xS}/c \approx 0$". A photon carries no electric charge.

CHAPTER 5 - PROPOSED THEORY OF NON-FINITE NUMBERS

5.5 Application to Dirac Spinors

Unlike one, four, or six term invariants that represent physical entities in geometric time-space, a Dirac Spinor exists only in mathematics. There is no Dirac Spinor geometric space that exists as an identifiable square-root space. Regardless of this non-geometric but, rather, algebraic, world of Dirac Spinors, they are invariants, and they have a Lorentz Transformation that goes to the speed-of-light with large-scale imprecision.

The Fourth Solution Dirac Spinor is for matter and is the example.

$$_4\Psi\text{-fourth} = e_{1M}*\Psi_{amp}*\exp(q_x*\alpha_M/2)*\exp(-i*(\pm_N k_{xM-m}*x_M - \omega_{M-m}*t_M))$$

Because "$\exp(-i*(\pm_N k_{xM-m}*x_M - \omega_{M-m}*t_M))$" is an invariant, it is combined into wave amplitude "Ψ_{amp}" (to simplify what is written).

$$\Psi_{ampe} = \Psi_{amp}*\exp(-i*(\pm_N k_{xM-m}*x_M - \omega_{M-m}*t_M))$$

<u>General Form</u>.

$$\begin{aligned}
_4\Psi\text{-fourth} &= e_{1M}*\Psi_{ampe}*\exp(q_x*\alpha_M/2) \\
&= e_{1M}*\Psi_{ampe}*\exp(q_x*\alpha_M/2)*1 \\
&= e_{1M}*\Psi_{ampe}*\exp(q_x*\alpha_M/2)*\exp(q_x*\alpha_{S/M}/2)/\exp(q_x*\alpha_{S/M}/2) \\
&= (e_{1M}/\exp(q_x*\alpha_{S/M}/2))*(\Psi_{ampe}*\exp(q_x*\alpha_M/2)*\exp(q_x*\alpha_{S/M}/2)) \\
&= (e_{1M}/\exp(q_x*\alpha_{S/M}/2))*(\Psi_{ampe}*\exp(q_x*(\alpha_M + \alpha_{S/M})/2)) \\
&= e_{1S}*\Psi_{ampe}*\exp(q_x*\alpha_S/2)
\end{aligned}$$

Because the Fourth Solution is matter, "$\alpha_S = \alpha_M + \alpha_{S/M}$".
Continuing with "$\alpha_{S/M} = (1 - q_x)*\Xi$" with "$\Xi$" being either large-scale imprecision "$\Omega\Omega$" or else small-scale imprecision "00":

$$\begin{aligned}
_4\Psi\text{-fourth} &= e_{1S}*(\Psi_{ampe}*\exp(q_x*\alpha_M/2)*\exp(q_x*\alpha_{S/M}/2)) \\
&= e_{1S}*(\Psi_{ampe}*\exp(q_x*\alpha_M/2)*\exp(q_x*(1-q_x)*\Xi/2)) \\
&= e_{1S}*(\Psi_{ampe}*\exp(q_x*\alpha_M/2)*\exp(q_x*\Xi/2)*\exp(-q_x^2*\Xi/2)) \\
&= e_{1S}*(\Psi_{ampe}*\exp(q_x*\alpha_M/2)*\exp(q_x*\Xi/2)*\exp(-\Xi/2)) \\
&= e_{1S}*(\Psi_{ampe}*\exp(q_x*\alpha_M/2)*(\cosh(\Xi/2) + q_x*\sinh(\Xi/2))*\exp(-\Xi/2)) \\
&= e_{1S}*(\Psi_{ampe}*\exp(q_x*\alpha_M/2)*(((\exp(\Xi/2) + \exp(-\Xi/2)) \\
&\quad + q_x*(\exp(\Xi/2) - \exp(-\Xi/2)))*\exp(-\Xi/2))/2) \\
&= e_{1S}*(\Psi_{ampe}*\exp(q_x*\alpha_M/2)*((1 + \exp^2(-\Xi/2)) + q_x*(1 - \exp^2(-\Xi/2)))/2) \\
&= e_{1S}*(\Psi_{ampe}*\exp(q_x*\alpha_M/2)*((1 + q_x)/2 + (1 - q_x)*\exp^2(-\Xi/2)/2))
\end{aligned}$$

SPECIAL ALGEBRA FOR SPECIAL RELATIVITY

Components – Large-Scale Case:

$_4\Psi_{\text{-fourth}} = e_{1S}*\Psi_{\text{ampe}}*\exp(q_x*\alpha_M/2)*((1 + q_x)/2 + (1 - q_x)*\exp^2(-\text{"}\Omega\Omega\text{"}/2)/2)$
$\approx e_{1S}*\Psi_{\text{ampe}}*\exp(q_x*\alpha_M/2)*((1 + q_x)/2 + (1 - q_x)*0/2)$
$= e_{1S}*\Psi_{\text{ampe}}*\exp(q_x*\alpha_M/2)*(1 + q_x)/2$
$= e_{1S}*\Psi_{\text{ampe}}*(\cosh(\alpha_M/2) + q_x*\sinh(\alpha_M/2))*(1 + q_x)/2$
$= e_{1S}*\Psi_{\text{ampe}}*((\cosh(\alpha_M/2) + q_x^2*\sinh(\alpha_M/2)) + q_x*(\cosh(\alpha_M/2) + \sinh(\alpha_M/2)))/2$
$= e_{1S}*\Psi_{\text{ampe}}*((\cosh(\alpha_M/2) + \sinh(\alpha_M/2)) + q_x*(\cosh(\alpha_M/2) + \sinh(\alpha_M/2)))/2$
$= e_{1S}*\Psi_{\text{ampe}}*\exp(\alpha_M/2)*(1 + q_x)/2$

$\Psi_{1S\text{-fourth}} = \Psi_{\text{ampe}}*\cosh((\alpha_M + \alpha_{S/M})/2)$
$\approx \Psi_{\text{ampe}}*\exp(\alpha_M/2)/2$

$\Psi_{4S\text{-fourth}} = -\Psi_{\text{ampe}}*\sinh((\alpha_M + \alpha_{S/M})/2)$
$\approx -\Psi_{\text{ampe}}*\exp(\alpha_M/2)/2$

"$\Psi_{1S\text{-fourth}}$" and "$\Psi_{4S\text{-fourth}}$" are component values for the fourth solution of the Dirac Equation for no external voltage. Per the proposed Lorentz Transformation that adds large-scale imprecision, that moves the observer to the speed-of-light, the two components become negatives of each other. This is different from the energy and momentum components that became equal without the negative.

Label Numbers – Large-Scale Case.

$e_{1S} = e_{1M}/\exp(q_x*\alpha_{S/M}/2)$
$= e_{1M}*\exp(-q_x*(1 - q_x)*\Xi/2)$
$= e_{1M}*\exp(-q_x*\Xi/2)*\exp(\Xi/2)$
$= e_{1M}*(\cosh(\Xi/2) - q_x*\sinh(\Xi/2))*\exp(\Xi/2)$
$= e_{1M}*((\exp(\Xi/2) + \exp(-\Xi/2)) - q_x*(\exp(\Xi/2) - \exp(-\Xi/2)))*(\exp(\Xi/2))/2$
$= e_{1M}*((1 + \exp^2(\Xi/2)) - q_x*(-1 + \exp^2(\Xi/2)))/2$
$= e_{1M}*((1 + q_x)/2 + (1 - q_x)*(\exp^2(\Xi/2))/2)$

Continuing with "$\alpha_{S/M} = (1 - q_x)*\Xi$" with "$\Xi$" being either large-scale imprecision "$\Omega\Omega$" or else small-scale imprecision "00". Again, the label numbers for Dirac Spinors relate per the proposed "$e_{1M} = -e_{4M}*q_x$".

CHAPTER 5 - PROPOSED THEORY OF NON-FINITE NUMBERS

$$e_{1S} = e_{1M}*((1 + q_x)/2 + (1 - q_x)*(\exp^2(\Xi/2))/2)$$
$$= e_{1M}*((1 + q_x)/2 + (1 - q_x)*(\exp^2(\text{``}\Omega\Omega\text{''}/2))/2)$$
$$\approx e_{1M}*(1 + q_x)/2$$
$$= (e_{1M} + e_{1M}*q_x)/2$$
$$= (e_{1M} - e_{4M})/2$$

$$e_{4S} = -e_{1S}*q_x$$
$$= (-(e_{1M} - e_{4M})/2)*q_x$$
$$= -(e_{1M}*q_x - e_{4M}*q_x)/2$$
$$= -(-e_{4M} + e_{1M})/2$$
$$= -(e_{1M} - e_{4M})/2$$
$$= -e_{1S}$$

Label numbers "e_{1S}" and "e_{4S}" are negatives of each other, as could be expected because the components were equal and negatives. Recall, it is the components that are important, and not the label numbers. This was evident in the Dirac Equation which used only component values and not label numbers.

The hyperbolic radius equals zero, as expected for motion at the speed-of-light due to the use of a singular label number.

$$_4\Psi\text{-fourth}^{*j}*_4\Psi\text{-fourth} = (\Psi_{1S\text{-fourth}}*e_{1S}^{*j} + \Psi_{4S\text{-fourth}}*e_{4S}^{*j})*(e_{1S}*\Psi_{1S\text{-fourth}} + e_{4S}*\Psi_{4S\text{-fourth}})$$
$$= ((\Psi_{1S\text{-fourth}} - \Psi_{4S\text{-fourth}})*e_{1S}^{*j})*(e_{1S}*(\Psi_{1S\text{-fourth}} - \Psi_{4S\text{-fourth}}))$$
$$= (2*\Psi_{1S\text{-fourth}}*e_{1S}^{*j})*(e_{1S}*2*\Psi_{1S\text{-fourth}})$$
$$= 4*\Psi_{1S\text{-fourth}}^2*(e_{1S}^{*j}*e_{1S})$$
$$= 4*\Psi_{1S\text{-fourth}}^2*((e_{1M}*(1 + q_x))^{*j}*(e_{1M}*(1 + q_x)))$$
$$= 4*\Psi_{1S\text{-fourth}}^2*(e_{1M}^{*j}*e_{1M})*((1 + q_x)^{*j}*(1 + q_x))$$
$$= 4*\Psi_{1S\text{-fourth}}^2*(1)*(0)$$
$$= 0$$

Small-Scale Case for the Dirac Spinor.

$$_4\Psi\text{-fourth} = e_{1S}*(\Psi_{ampe}*\exp(q_x*\alpha_M/2)*((1 + q_x)/2 + (1 - q_x)*\exp^2(-\Xi/2)/2))$$
$$= e_{1S}*(\Psi_{ampe}*\exp(q_x*\alpha_M/2)*((1 + q_x)/2 + (1 - q_x)*\exp^2(-\text{``}00\text{''}/2)/2))$$
$$\approx e_{1S}*(\Psi_{ampe}*\exp(q_x*\alpha_M/2)*((1 + q_x)/2 + (1 - q_x)/2))$$
$$= e_{1S}*\Psi_{ampe}*\exp(q_x*\alpha_M/2)$$

$$\Psi_{1S\text{-fourth}} \approx \Psi_{ampe}*\cosh(\alpha_M/2) = \Psi_{1M\text{-fourth}}$$
$$\Psi_{4S\text{-fourth}} \approx -\Psi_{ampe}*\sinh(\alpha_M/2) = \Psi_{4M\text{-fourth}}$$

SPECIAL ALGEBRA FOR SPECIAL RELATIVITY

Label Numbers – Small-Scale Case

$e_{1S} = e_{1M}*((1 + q_x)/2 + (1 - q_x)*(\exp^2(\Xi/2))/2)$
$= e_{1M}*((1 + q_x)/2 + (1 - q_x)*(\exp^2(\text{``00''}/2))/2)$
$\approx e_{1M}*((1 + q_x)/2 + (1 - q_x)/2)$
$= e_{1M}$

$e_{4S} = -e_{1M}*q_x$
$= e_{4M}$

The small-scale case has no change in time or in space, as applies to the originally measured motion of the sub-light speed particle.

Proposed Dirac Equation form of Maxwell's Wave Equation.

The large-scale case pertained to Dirac Spinor motion at the speed-of-light relative to an observer "S". Electromagnetic waves travel at the speed-of-light and, per this proposed theory, are the same thing.

For the large-scale case, terms that are only "m_B" and "Q_B" are set to zero because of factors "$\exp(-\text{``}\Omega\Omega\text{''})$" and "$\exp(-3*\text{``}\Omega\Omega\text{''}/2)$", respectively.

As a further simplification, divide each term by "$i*\hbar$". The result is the proposed Dirac Equation form of Maxwell's Wave Equation:

$$\partial/\partial ct_S * \begin{pmatrix} 1&0&0&0 \\ 0&1&0&0 \\ 0&0&1&0 \\ 0&0&0&1 \end{pmatrix} \pm_N (-\partial/\partial x_S) * \begin{pmatrix} 0&0&0&1 \\ 0&0&1&0 \\ 0&1&0&0 \\ 1&0&0&0 \end{pmatrix} \pm_N (-\partial/\partial y_S) * \begin{pmatrix} 0&0&0&k \\ 0&0&-k&0 \\ 0&k&0&0 \\ -k&0&0&0 \end{pmatrix} \pm_N (-\partial/\partial z_S) * \begin{pmatrix} 0&0&1&0 \\ 0&0&0&-1 \\ 1&0&0&0 \\ 0&-1&0&0 \end{pmatrix} * \begin{pmatrix} \Psi_{z\pm 1S} \\ \Psi_{z\pm 2S} \\ \Psi_{z\pm 3S} \\ \Psi_{z\pm 4S} \end{pmatrix} = \begin{pmatrix} 0 \\ 0 \\ 0 \\ 0 \end{pmatrix}$$

Written as a set of linear partial differential equations:

$\partial\Psi_{z\pm 1S}/\partial ct_S \pm_N -\partial\Psi_{z\pm 4S}/\partial x_S \pm_N -k*\partial\Psi_{z\pm 4S}/\partial y_S \pm_N -\partial\Psi_{z\pm 3S}/\partial z_S = 0$
$\partial\Psi_{z\pm 2S}/\partial ct_S \pm_N -\partial\Psi_{z\pm 3S}/\partial x_S \pm_N k*\partial\Psi_{z\pm 3S}/\partial y_S \pm_N \partial\Psi_{z\pm 4S}/\partial z_S = 0$
$\partial\Psi_{z\pm 3S}/\partial ct_S \pm_N -\partial\Psi_{z\pm 2S}/\partial x_S \pm_N -k*\partial\Psi_{z\pm 2S}/\partial y_S \pm_N -\partial\Psi_{z\pm 1S}/\partial z_S = 0$
$\partial\Psi_{z\pm 4S}/\partial ct_S \pm_N -\partial\Psi_{z\pm 1S}/\partial x_S \pm_N k*\partial\Psi_{z\pm 1S}/\partial y_S \pm_N \partial\Psi_{z\pm 2S}/\partial z_S = 0$

Substitute in fourth solution "$_4\Psi_{\text{-fourth}}$" large-scale case components.

$_4\Psi_{\text{-fourth}} = e_{1S}*\Psi_{\text{amp}}*\exp(\alpha_M/2)*\exp(-i*(\pm_N k_{xS\text{-}m}*x_S - \omega_{S\text{-}m}*t_S))*(1 + q_x)/2$

218

CHAPTER 5 - PROPOSED THEORY OF NON-FINITE NUMBERS

"$_4\Psi_{\text{-fourth}}$" applied to the first equation "$\partial\Psi_{z\pm1S}/\partial ct_S \pm_N -\partial\Psi_{z\pm4S}/\partial x_S = 0$" requires "$\partial\Psi_{z\pm1S\text{-fourth}}/\partial ct_S$" and "$\partial\Psi_{z\pm4S\text{-fourth}}/\partial x_S$".

$\Psi_{z\pm1S\text{-fourth}} = \Psi_{amp}*\exp(\alpha_M/2)*\exp(-i*(\pm_N k_{xS-m}*x_S - \omega_{S-m}*t_S))/2$

$\partial\Psi_{z\pm1S\text{-fourth}}/\partial ct_S = i*(\omega_{S-m}/c)*\Psi_{z\pm1S\text{-fourth}}$

$\Psi_{z\pm4S\text{-fourth}} = -\Psi_{amp}*\exp(\alpha_M/2)*\exp(-i*(\pm_N k_{xS-m}*x_S - \omega_{S-m}*t_S))/2$
$= -\Psi_{z\pm1S\text{-fourth}}$

$\partial\Psi_{z\pm4S\text{-fourth}}/\partial x_S = --i*\pm_N k_{S-m}*\Psi_{z\pm1S\text{-fourth}}$

At the speed-of-light "$\omega_{S-m}/c = k_{S-m}$".

$\partial\Psi_{z\pm1S}/\partial ct_S \pm_N -\partial\Psi_{z\pm4S}/\partial x_S \pm_N -k*\partial\Psi_{z\pm4S}/\partial y_S \pm_N -\partial\Psi_{z\pm3S}/\partial z_S$

$= \partial\Psi_{z\pm1S}/\partial ct_S \pm_N -\partial\Psi_{z\pm4S}/\partial x_S$
$= (i*(\omega_{S-m}/c) \pm_N -i*\pm_N k_{S-m})*\Psi_{z\pm1S\text{-fourth}}$
$= i*((\omega_{S-m}/c) - k_{S-m})*\Psi_{z\pm1S\text{-fourth}}$
$= 0$ Okay. Good Check.

Substitute electromagnetic components into the equation ($k = i$).

$\Psi_{z\pm1S} = E_{zS} + \pm_N i*K_{zS}$

$\Psi_{z\pm2S} = (E_{xS} + \pm_N i*K_{xS}) - k*(E_{yS} + \pm_N i*K_{yS})$
$= (E_{xS} + \pm_N K_{yS}) + i*(-E_{yS} + \pm_N K_{xS})$

$\Psi_{z\pm3S} = E_{zS} + \pm_N i*K_{zS}$

$\Psi_{z\pm4S} = (E_{xS} + \pm_N i*K_{xS}) - k*(E_{yS} + \pm_N i*K_{yS})$
$= (E_{xS} + \pm_N K_{yS}) + i*(-E_{yS} + \pm_N K_{xS})$

"$\pm_N$" is on the imaginary portion, which is the magnetic component: "$i*K_{xS}$", "$i*K_{yS}$" and "$i*K_{zS}$". "$\pm_N$" accounts for motion being left ("-") or right ("+") because an electromagnetic wave going left rather than right has the magnetic field reversed in sign for a given sign on the electric field.

SPECIAL ALGEBRA FOR SPECIAL RELATIVITY

(Notice the "Alternative First Dirac Spinor Solution" is incompatible with the above substitutions, due to the different placement of "$\pm_N$".)

The four differential equations become:

$\partial(E_{zS} + \pm_N i^* K_{zS})/\partial ct_S \pm_N -\partial((E_{xS} + \pm_N i^* K_{xS}) - k^*(E_{yS} + \pm_N i^* K_{yS}))/\partial x_S$
$\pm_N -k^* \partial((E_{xS} + \pm_N i^* K_{xS}) - k^*(E_{yS} + \pm_N i^* K_{yS}))/\partial y_S \pm_N -\partial(E_{zS} + \pm_N i^* K_{zS})/\partial z_S = 0$

$\partial((E_{xS} + \pm_N i^* K_{xS}) - k^*(E_{yS} + \pm_N i^* K_{yS}))/\partial ct_S \pm_N -\partial(E_{zS} + \pm_N i^* K_{zS})/\partial x_S$
$\pm_N k^* \partial(E_{zS} + \pm_N i^* K_{zS})/\partial y_S \pm_N \partial((E_{xS} + \pm_N i^* K_{xS}) - k^*(E_{yS} + \pm_N i^* K_{yS}))/\partial z_S = 0$

$\partial(E_{zS} + \pm_N i^* K_{zS})/\partial ct_S \pm_N -\partial((E_{xS} + \pm_N i^* K_{xS}) - k^*(E_{yS} + \pm_N i^* K_{yS}))/\partial x_S$
$\pm_N -k^* \partial((E_{xS} + \pm_N i^* K_{xS}) - k^*(E_{yS} + \pm_N i^* K_{yS}))/\partial y_S \pm_N -\partial(E_{zS} + \pm_N i^* K_{zS})/\partial z_S = 0$

$\partial((E_{xS} + \pm_N i^* K_{xS}) - k^*(E_{yS} + \pm_N i^* K_{yS}))/\partial ct_S \pm_N -\partial(E_{zS} + \pm_N i^* K_{zS})/\partial x_S)$
$\pm_N k^* \partial(E_{zS} + \pm_N i^* K_{zS})/\partial y_S \pm_N \partial((E_{xS} + \pm_N i^* K_{xS}) - k^*(E_{yS} + \pm_N i^* K_{yS}))/\partial z_S = 0$

For motion in the "x" direction:

$\partial \Psi_{z\pm 1S}/\partial ct_S - \pm_N \partial \Psi_{z\pm 4S}/\partial x_S = 0$
$\partial \Psi_{z\pm 2S}/\partial ct_S - \pm_N \partial \Psi_{z\pm 3S}/\partial x_S = 0$
$\partial \Psi_{z\pm 3S}/\partial ct_S - \pm_N \partial \Psi_{z\pm 2S}/\partial x_S = 0$
$\partial \Psi_{z\pm 4S}/\partial ct_S - \pm_N \partial \Psi_{z\pm 1S}/\partial x_S = 0$

Written in terms of electromagnetic field components:

$\partial(E_{zS} + \pm_N i^* K_{zS})/\partial ct_S \pm_N --k^* \partial(E_{yS} + \pm_N i^* K_{yS})/\partial x_S = 0$
$-k^* \partial(E_{yS} + \pm_N i^* K_{yS})/\partial ct_S \pm_N -\partial(E_{zS} + \pm_N i^* K_{zS})/\partial x_S = 0$
$\partial(E_{zS} + \pm_N i^* K_{zS})/\partial ct_S \pm_N --k^* \partial(E_{yS} + \pm_N i^* K_{yS})/\partial x_S = 0$
$-k^* \partial(E_{yS} + \pm_N i^* K_{yS})/\partial ct_S \pm_N -\partial(E_{zS} + \pm_N i^* K_{zS})/\partial x_S = 0$

The above equations were derived from the below substitutions, for which "E_{xS}" and "K_{xS}" have been set to zero:

$\Psi_{z\pm 1S} = E_{zS} + \pm_N i^* K_{zS}$
$\Psi_{z\pm 2S} = -k^*(E_{yS} + \pm_N i^* K_{yS})$
$\Psi_{z\pm 3S} = E_{zS} + \pm_N i^* K_{zS}$
$\Psi_{z\pm 4S} = -k^*(E_{yS} + \pm_N i^* K_{yS})$

CHAPTER 5 - PROPOSED THEORY OF NON-FINITE NUMBERS

The example is the Fourth Spiral Wave. The example shows "$\Psi_{z1S} = -\Psi_{z4S}$". "$\pm_N$" is removed because the Lorentz Transformation assumed motion in the positive "x" direction. (In violation of that assumption, the "-" of "$\pm_N$" represents motion in the negative "x" direction such that its Lorentz Transformation is incorrect.) We don't drop the "$\pm_N$" in general because it is useful later, in the separation of terms for the force density invariant. But, here, we do drop the "$\pm_N$".

Fourth Spiral Wave

$(_6E_{fourth}) = E_{amp}*(-k_{yS} + p_{zS})*\exp(-i*(k_{xS}*(x_S - c*t_S)))$
$(E_{yS} + i*K_{yS})_{fourth} = -i*E_{amp}*\exp(-i*(k_{xS}*(x_S - c*t_S)))$
$(E_{zS} + i*K_{zS})_{fourth} = E_{amp}*\exp(-i*(k_{xS}*(x_S - c*t_S)))$
$E_{ySfourth} = -E_{amp}*\sin(k_{xS}*(x_S - c*t_S))$
$K_{ySfourth} = -E_{amp}*\cos(k_{xS}*(x_S - c*t_S))$
$E_{zSfourth} = E_{amp}*\cos(k_{xS}*(x_S - c*t_S))$
$K_{zSfourth} = -E_{amp}*\sin(k_{xS}*(x_S - c*t_S))$

$_4\Psi_{-fourth} = e_{1S}*\Psi_{amp}*\exp(\alpha_M/2)*\exp(-i*(k_{xS}*x_S - \omega_S*t_S))*(1 + q_x)/2$
$= (e_{1S} - e_{4S})*\Psi_{amp}*\exp(\alpha_M/2)*\exp(-i*(k_{xS}*x_S - \omega_S*t_S))/2$
$= (e_{1S} - e_{4S})*\Psi_{amp}*\exp(\alpha_M/2)*\exp(-i*(k_{xS}*(x_S - c*t_S)))/2$
$= (e_{1S} - e_{4S})*E_{amp}*\exp(-i*(k_{xS}*(x_S - c*t_S)))$
$= (e_{1S} - e_{4S})*E_{amp}*(\cos(k_{xS}*(x_S - c*t_S)) - i*\sin(k_{xS}*(x_S - c*t_S)))$

$= e_{1S}*E_{amp}*(\cos(k_{xS}*(x_S - c*t_S)) - i*\sin(k_{xS}*(x_S - c*t_S)))$
$+ - e_{4S}*E_{amp}*(\cos(k_{xS}*(x_S - c*t_S)) - i*\sin(k_{xS}*(x_S - c*t_S)))$

$= e_{1S}*(E_{amp}*\cos(k_{xS}*(x_S - c*t_S)) + i*-E_{amp}*\sin(k_{xS}*(x_S - c*t_S)))$
$+ e_{4S}*-k*(k*-E_{amp}*\cos(k_{xS}*(x_S - c*t_S)) + -E_{amp}*\sin(k_{xS}*(x_S - c*t_S)))$

$= e_{1S}*(E_{zSfourth} + i*K_{zSfourth}) + e_{4S}*-k*(k*K_{ySfourth} + E_{ySfourth})$

<u>Matrix Isomorph Forms of the Dirac Equation</u>. The "z" arrangement of the Dirac Equation / Maxwell's Wave Equation caused an apparent preference for "z" components because "E_{zS}" and "K_{zS}" are not confounded with other components. Unlike "z" components, "x" components "E_{xS}" and "K_{xS}" are confounded with "y" components "K_{yS}" and "E_{yS}", respectively, because of the below equation.

$(E_{xS} + \pm_N i*K_{xS}) - k*(E_{yS} + \pm_N i*K_{yS}) = (E_{xS} + \pm_N K_{yS}) + i*(\pm_N i*K_{xS} - E_{yS})$

SPECIAL ALGEBRA FOR SPECIAL RELATIVITY

"y" Arrangement.

$$\partial/\partial ct_S * \begin{vmatrix} 1&0&0&0 \\ 0&1&0&0 \\ 0&0&1&0 \\ 0&0&0&1 \end{vmatrix} \pm_N (-\partial/\partial z_S)* \begin{vmatrix} 0&0&0&1 \\ 0&0&1&0 \\ 0&1&0&0 \\ 1&0&0&0 \end{vmatrix} \pm_N (-\partial/\partial x_S)* \begin{vmatrix} 0&0&0&k \\ 0&0&-k&0 \\ 0&k&0&0 \\ -k&0&0&0 \end{vmatrix} \pm_N (-\partial/\partial y_S)* \begin{vmatrix} 0&0&1&0 \\ 0&0&0&-1 \\ 1&0&0&0 \\ 0&-1&0&0 \end{vmatrix} * \begin{vmatrix} \Psi_{y\pm1S} \\ \Psi_{y\pm2S} \\ \Psi_{y\pm3S} \\ \Psi_{y\pm4S} \end{vmatrix} = \begin{vmatrix} 0 \\ 0 \\ 0 \\ 0 \end{vmatrix}$$

Letters "x", "y", "z" were rotated.

$\Psi_{y\pm1S} = E_{yS} + \pm_N i * K_{yS}$
$\Psi_{y\pm2S} = (E_{zS} + \pm_N i * K_{zS}) - k*(E_{xS} + \pm_N i * K_{xS})$
$\Psi_{y\pm3S} = E_{yS} + \pm_N i * K_{yS}$
$\Psi_{y\pm4S} = (E_{zS} + \pm_N i * K_{zS}) - k*(E_{xS} + \pm_N i * K_{xS})$

For motion in the "x" direction, the "$_4\Psi_{y\pm S}$" proposed Dirac Equation form of Maxwell's Wave Equation reduces to

$\partial \Psi_{y\pm1S}/\partial ct_S - \pm_N k * \partial \Psi_{y\pm4S}/\partial x_S = 0$
$\partial \Psi_{y\pm4S}/\partial ct_S + \pm_N k * \partial \Psi_{y\pm1S}/\partial x_S = 0$

$\partial \Psi_{y\pm2S}/\partial ct_S + \pm_N k * \partial \Psi_{y\pm3S}/\partial x_S = 0$
$\partial \Psi_{y\pm3S}/\partial ct_S - \pm_N k * \partial \Psi_{y\pm2S}/\partial x_S = 0$

Specific solutions for "x" direction motion:

$\Psi_{y\pm1S} = E_{yS} + \pm_N i * K_{yS}$
$\Psi_{y\pm2S} = E_{zS} + \pm_N i * K_{zS}$
$\Psi_{y\pm3S} = E_{yS} + \pm_N i * K_{yS}$
$\Psi_{y\pm4S} = E_{zS} + \pm_N i * K_{zS}$

"x" Arrangement.

$$\partial/\partial ct_S * \begin{vmatrix} 1&0&0&0 \\ 0&1&0&0 \\ 0&0&1&0 \\ 0&0&0&1 \end{vmatrix} \pm_N (-\partial/\partial y_S)* \begin{vmatrix} 0&0&0&1 \\ 0&0&1&0 \\ 0&1&0&0 \\ 1&0&0&0 \end{vmatrix} \pm_N (-\partial/\partial z_S)* \begin{vmatrix} 0&0&0&k \\ 0&0&-k&0 \\ 0&k&0&0 \\ -k&0&0&0 \end{vmatrix} \pm_N (-\partial/\partial x_S)* \begin{vmatrix} 0&0&1&0 \\ 0&0&0&-1 \\ 1&0&0&0 \\ 0&-1&0&0 \end{vmatrix} * \begin{vmatrix} \Psi_{x\pm1S} \\ \Psi_{x\pm2S} \\ \Psi_{x\pm3S} \\ \Psi_{x\pm4S} \end{vmatrix} = \begin{vmatrix} 0 \\ 0 \\ 0 \\ 0 \end{vmatrix}$$

Chapter 5 - Proposed Theory of Non-Finite Numbers

$\Psi_{x\pm 1S} = E_{xS} + \pm_N i^* K_{xS}$
$\Psi_{x\pm 2S} = (E_{yS} + \pm_N i^* K_{yS}) - k^*(E_{zS} + \pm_N i^* K_{zS})$
$\Psi_{x\pm 3S} = E_{xS} + \pm_N i^* K_{xS}$
$\Psi_{x\pm 4S} = (E_{yS} + \pm_N i^* K_{yS}) - k^*(E_{zS} + \pm_N i^* K_{zS})$

For motion in the "x" direction, the "$\Psi_{x\pm S}$" proposed Dirac Equation form of Maxwell's Wave Equation reduces to:

$\partial \Psi_{x\pm 1S}/\partial ct_S - \pm_N \partial \Psi_{x\pm 3S}/\partial x_S = 0$
$\partial \Psi_{x\pm 4S}/\partial ct_S + \pm_N \partial \Psi_{x\pm 2S}/\partial x_S = 0$
$\partial \Psi_{x\pm 2S}/\partial ct_S + \pm_N \partial \Psi_{x\pm 4S}/\partial x_S = 0$
$\partial \Psi_{x\pm 3S}/\partial ct_S - \pm_N \partial \Psi_{x\pm 1S}/\partial x_S = 0$

Specific solutions for "x" direction motion:

$\Psi_{x\pm 1S} = 0$
$\Psi_{x\pm 2S} = (E_{yS} + \pm_N i^* K_{yS}) - k^*(E_{zS} + \pm_N i^* K_{zS}) = 0$
$\Psi_{x\pm 3S} = 0$
$\Psi_{x\pm 4S} = (E_{yS} + \pm_N i^* K_{yS}) - k^*(E_{zS} + \pm_N i^* K_{zS}) = 0$

Anti-Commute Label-Numbers. We now propose a justification for why anti-commutative property "$e_{z1S} = q_x{}^* e_{z4S} = -e_{z4S}{}^* q_x$" was appropriate. The clue is that "$E_{zS} + \pm_N i^* K_{zS}$" has compound-label-number "e_{z1S}" in the Dirac Equation and has "p_{zS}" in Maxwell's Equations. The implication is that "e_{z1S}" replaces "p_{zS}", but there is no formal theory for that, not yet.

As a proto-theory, first substitute "q_z" for "e_{z1S}" so that "e_{z1S}" is a column of "q_z" over top of three zeros. Secondly, because "$E_{xS} + \pm_N i^* K_{xS} = 0$", "$e_{z4S}$" is replaced by a column-vector of three zeros on top of "$-k^* q_y = -j_y$" on the bottom. Lastly, replace the upper right and lower left "1"'s inside the matrix for "q_x" with the simple-label-number "q_x".

These three substitutions for "e_{z1S}", "e_{z4S}", and "q_x" mean "$e_{z1S} = q_x{}^* e_{z4S}$" includes "$q_z = -j_y{}^* q_x$". Anti-commutative property for "$q_z = -(q_x{}^* -j_y)$" implies "$e_{z1S} = -e_{z4S}{}^* q_x$". For reference, see the front and back multiplication operations in the Appendix on Octonions.

SPECIAL ALGEBRA FOR SPECIAL RELATIVITY

Large-Scale Case of the Gradient Differential Operator. "$_4\nabla^{sn}$" is normally a space-negative. Therefore, space-negative of "$\exp(q_x * \alpha_{S/M})$" is used in the Lorentz Transformation.

$$_4\nabla^{sn} = 1_M^{sn} * \nabla_{tM} + q_{xM}^{sn} * \nabla_{xM} + q_{yM}^{sn} * \nabla_{yM} + q_{zM}^{sn} * \nabla_{zM}$$

$$= \exp(-\kappa^{sn} * \varsigma/2) * 1 * (1^{sn} * \nabla_{tM} + q_x^{sn} * \nabla_{xM}$$
$$+ q_y^{sn} * \nabla_{yM} + q_z^{sn} * \nabla_{zM}) * 1 * \exp(-\kappa^{sn} * \varsigma/2)$$

$$= \exp(-\kappa^{sn} * \varsigma/2) * \exp(-q^{sn} * \alpha_{S/M}/2) * \exp(q^{sn} * \alpha_{S/M}/2) * (1^{sn} * \nabla_{tM} + q_x^{sn} * \nabla_{xM}$$
$$+ q_y^{sn} * \nabla_{yM} + q_z^{sn} * \nabla_{zM}) * \exp(q^{sn} * \alpha_{S/M}/2) * \exp(-q^{sn} * \alpha_{S/M}/2) * \exp(-\kappa^{sn} * \varsigma/2)$$

The above equation requires "q" be only one simple-label-number "q_x", "q_y", or "q_z", and is why that restriction exists.

Left and right "$\exp(-\kappa^{sn} * \varsigma/2)$" with "$\exp(-q^{sn} * \alpha_{S/M}/2)$" form "$1_S^{sn}$".
The two "$\exp(q^{sn} * \alpha_{S/M}/2)$" pertain to the four components of "$_4\nabla^{sn}$".
For "$\alpha_{S/M} = (1 - q^{sn}) * \Xi$", the "$\exp(q^{sn} * \alpha_{S/M}/2)$" factor is given below.

$$\exp(q^{sn} * \alpha_{S/M}/2) = \exp(q^{sn} * ((1 - q^{sn}) * \Xi)/2)$$
$$= \exp(-"\Omega\Omega"/2) * \exp(q^{sn} * "\Omega\Omega"/2) \quad \text{Large-scale case}$$
or
$$= \exp(-"00"/2) * \exp(q^{sn} * "00"/2) \approx 1 \quad \text{Small-scale case}$$

To visualize the large-scale case, replace "q^{sn}" with "q_x^{sn}".

$$\nabla_{tS} = \exp(-"\Omega\Omega"/2) * (\cosh("\Omega\Omega"/2) * \nabla_{tM} - \sinh("\Omega\Omega"/2) * \nabla_{xM})$$

$$= \exp(-"\Omega\Omega"/2) * (\exp("\Omega\Omega"/2) * \nabla_{tM}$$
$$- \sinh("\Omega\Omega"/2) * \nabla_{tM} - \sinh("\Omega\Omega"/2) * \nabla_{xM})$$

$$= \nabla_{tM} - \exp(-"\Omega\Omega"/2) * \sinh("\Omega\Omega"/2) * (\nabla_{tM} + \nabla_{xM})$$
$$= \nabla_{tM} - \exp(-"\Omega\Omega"/2) * (\exp("\Omega\Omega"/2) - \exp(-"\Omega\Omega"/2)) * (\nabla_{tM} + \nabla_{xM})/2$$
$$= \nabla_{tM} - (\nabla_{tM} + \nabla_{xM})/2 + \exp^2(-"\Omega\Omega"/2) * (\nabla_{tM} + \nabla_{xM})/2$$
$$= (\nabla_{tM} - \nabla_{xM})/2 + \exp^2(-"\Omega\Omega"/2) * (\nabla_{tM} + \nabla_{xM})/2$$
$$\approx (\nabla_{tM} - \nabla_{xM})/2$$

$$\nabla_{xS} = \exp(-"\Omega\Omega"/2) * (-\sinh("\Omega\Omega"/2) * \nabla_{tM} + \cosh("\Omega\Omega"/2) * \nabla_{xM})$$
$$\approx (-\nabla_{tM} + \nabla_{xM})/2$$

$$\nabla_{yS} = \nabla_{yM} \quad ; \quad \nabla_{zS} = \nabla_{zM}$$

CHAPTER 5 - PROPOSED THEORY OF NON-FINITE NUMBERS

"$\nabla_{xS} = -\nabla_{tS}$" states a field moving at the speed-of-light has the space gradient equal to the negative of the time gradient. For example:

$$_6E_{second} = -E_{amp}*(k_{yM} - p_{zM})*\exp(i*(k_{xM}*(x_M - c*t_M)))$$

$$\nabla_{xS}*_6E_{second} = i*k_{xM}*_6E_{second} = --i*k_{xM}*c*_6E_{second}/c = -\nabla_{tS}*_6E_{second}$$

5.6 Dirac Equation Form - Development

<u>Mechanical Energy-Momentum Components</u>. "$_4p$" of a particle:

$$_4p = \exp(-\kappa*\varsigma/2)*m_B*c*\exp(q*\alpha_M)*\exp(-\kappa*\varsigma/2)$$

"q" is restricted to being only one of "q_x", "q_y", or "q_z" with the choice known or knowable, and that contrasts with "κ", for which the choice of "q_x", "q_y", or "q_z" is unknown and unknowable. General form:

$$\begin{aligned}_4p &= \exp(-\kappa*\varsigma/2)*1*m_B*c*\exp(q*\alpha_M)*1*\exp(-\kappa*\varsigma/2)\\ &= \exp(-\kappa*\varsigma/2)*\exp(-q*\alpha_{S/M}/2)\\ &\quad *\exp(q*\alpha_{S/M}/2)*m_B*c*\exp(q*\alpha_M)*\exp(q*\alpha_{S/M}/2)\\ &\quad *\exp(-q*\alpha_{S/M}/2)*\exp(-\kappa*\varsigma/2)\end{aligned}$$

"α_M" is a rational truncated number calculated from speed "$v_M = c*\tanh\alpha_M$". "α_M" becomes a local-real number by the addition of imprecision term "Ξ" (xi) in a Lorentz Transformation. "Ξ" also applies to the hyperbolic-radius per the "$1 - q$" factor in "$\alpha_{S/M} = (1 - q)*\Xi$".

$$\begin{aligned}_4p &= \exp(-\kappa*\varsigma/2)*\exp(-q*(1 - q)*\Xi/2)\\ &\quad *\exp(q*(1 - q)*\Xi/2)*m_B*c*\exp(q*\alpha_M)*\exp(q*(1 - q)*\Xi/2)\\ &\quad *\exp(-q*(1 - q)*\Xi/2)*\exp(-\kappa*\varsigma/2)\end{aligned}$$

$$\begin{aligned}&= \exp(-\kappa*\varsigma/2)*\exp(-q*(1 - q)*"\Omega\Omega"/2) \quad \text{Large-scale case}\\ &\quad *\exp(q*(1 - q)*"\Omega\Omega"/2)*m_B*c*\exp(q*\alpha_M)*\exp(q*(1 - q)*"\Omega\Omega"/2)\\ &\quad *\exp(-q*(1 - q)*"\Omega\Omega"/2)*\exp(-\kappa*\varsigma/2)\end{aligned}$$

$$\begin{aligned}&= \exp(-\kappa*\varsigma/2)*\exp(-q*"\Omega\Omega"/2)*\exp("\Omega\Omega"/2)\\ &\quad *m_B*c*\exp(-"\Omega\Omega")*\exp(q*(\alpha_M + "\Omega\Omega"))\\ &\quad *\exp("\Omega\Omega"/2)*\exp(-q*"\Omega\Omega"/2)*\exp(-\kappa*\varsigma/2)\end{aligned}$$

SPECIAL ALGEBRA FOR SPECIAL RELATIVITY

The above expression for Lorentz Transformed "$_4$p" has the compound-label-number in the top and bottom lines and has components in the middle line. Large-scale case components:

energy	$m_B*c*\exp(\text{-"}\Omega\Omega\text{"})*\cosh(\alpha_M + \text{"}\Omega\Omega\text{"})$
momentum	$q*m_B*c*\exp(\text{-"}\Omega\Omega\text{"})*\sinh(\alpha_M + \text{"}\Omega\Omega\text{"})$
rest mass	$m_B*c*\exp(\text{-"}\Omega\Omega\text{"})$

"$m_B*c*\exp(\text{-"}\Omega\Omega\text{"})$" includes the "$\exp(\text{-"}\Omega\Omega\text{"})$" nearly-zero and therefore the rest mass term is dropped from the Pythagorean theorem.

$$0 \approx (m_B*c*\exp(\text{-"}\Omega\Omega\text{"})*\cosh(\alpha_M + \text{"}\Omega\Omega\text{"}))^2$$
$$- (\pm q)^2*(m_B*c*\exp(\text{-"}\Omega\Omega\text{"})*\sinh(\alpha_M + \text{"}\Omega\Omega\text{"}))^2$$

Enabler functions "PP$_S$" and "QQ$_S$" are applied, same as in the previous chapter.

$$0 \approx (m_B*c*\exp(\text{-"}\Omega\Omega\text{"})*\cosh(\alpha_M + \text{"}\Omega\Omega\text{"}))^2*(PP_S)*(-QQ_S)$$
$$- (\pm q)^2*(m_B*c*\exp(\text{-"}\Omega\Omega\text{"})*\sinh(\alpha_M + \text{"}\Omega\Omega\text{"}))^2*(-QQ_S)*(PP_S)$$

$$(m_B*c*\exp(\text{-"}\Omega\Omega\text{"})*\cosh(\alpha_M + \text{"}\Omega\Omega\text{"}))*PP_S$$
$$*(m_B*c*\exp(\text{-"}\Omega\Omega\text{"})*\cosh(\alpha_M + \text{"}\Omega\Omega\text{"}))*(-QQ_S)$$
$$= \pm q*(m_B*c*\exp(\text{-"}\Omega\Omega\text{"})*\sinh(\alpha_M + \text{"}\Omega\Omega\text{"}))*(-QQ_S)$$
$$*(\pm q)*(m_B*c*\exp(\text{-"}\Omega\Omega\text{"})*\sinh(\alpha_M + \text{"}\Omega\Omega\text{"}))*PP_S$$

Enabler functions "PP$_S$" and "QQ$_S$" split the Pythagorean Theorem into two separate equations. The first line in the above equation is set equal to the third, and second to fourth. After the split, the two equations are placed into a matrix equation.

$r*\cosh(s)$	$\pm q*r*\sinh(s)$		PP$_S$		0
		*		=	
$\pm q*r*\sinh(s)$	$r*\cosh(s)$		QQ$_S$		0

"$r = m_B*c*\exp(\text{-"}\Omega\Omega\text{"})$" and "$s = \alpha_M + \text{"}\Omega\Omega\text{"}$" (large-scale case). For the above algebraic equation to be valid, "QQ$_S$ = $\pm q*$PP$_S$" and "PP$_S$ = $\pm q*$QQ$_S$", and "$r*\cosh(s) \approx (m_B*c)*\exp(\alpha_M)/2$" and "$r*\sinh(s) \approx (m_B*c)*\exp(\alpha_M)/2$". Energy equals momentum, as applies to a photon.

CHAPTER 5 - PROPOSED THEORY OF NON-FINITE NUMBERS

Differential Operator. Relative to observer "S" are gradients in time and space. Gradients, along with space-negative of the compound label numbers ("1_S^{sn}", "q_{xS}^{sn}", "q_{yS}^{sn}", and "q_{zS}^{sn}"), comprise the time-space invariant gradient operator "$_4\nabla^{sn}$". A person on the roadside has only one sense for gradients, because there is only one roadside "S". In contrast, there are two observations for the one bus "M":

- A large-scale case for a speed-of-light bus, for which "$v_{S/M\text{-LargeCase}} \approx c$"
- A small-scale case for a stationary bus, for which "$v_{S/M\text{-SmallCase}} \approx 0$"

$$_4p = 1_M * m_B * c * \cosh\alpha_M + q_M * m_B * c * \sinh\alpha_M$$
$$= 1_S * r * \cosh(s) + q_S * r * \sinh(s)$$

$$= 1_S * m_B * c * \exp(\text{-}``\Omega\Omega\text{''}) * \cosh(\alpha_M + ``\Omega\Omega\text{''}) \quad \text{large-scale case}$$
$$+ q_S * m_B * c * \exp(\text{-}``\Omega\Omega\text{''}) * \sinh(\alpha_M + ``\Omega\Omega\text{''})$$

Total energy-momentum "$\hbar *_4 k$" equals mechanical plus electrical.

$$\hbar *_4 k = {_4}p + {_4}q$$
$$\hbar *_4 k = i * \hbar *_4 \nabla^{sn}$$

$$_4\nabla^{sn} = {_1}\nabla^{sn} + {_3}\nabla^{sn}$$
$$= {_1}\nabla - {_3}\nabla$$
$$= 1_S * \nabla_{tS} - (q_{xS} * \nabla_{xS} + q_{yS} * \nabla_{yS} + q_{zS} * \nabla_{zS})$$
$$= 1_S * \partial/\partial ct_S - (q_{xS} * \partial/\partial x_S + q_{yS} * \partial/\partial y_S + q_{zS} * \partial/\partial z_S)$$

$$_4p = \hbar *_4 k - {_4}q = i * \hbar *_4 \nabla^{sn} - {_4}q$$

$$1_S * r * \cosh(s) = \hbar *_1 k - {_1}q = i * \hbar * ({_1}\nabla) - {_1}q$$
$$q_S * r * \sinh(s) = \hbar *_3 k - {_3}q = -i * \hbar * ({_3}\nabla) - {_3}q$$

Remove the reference "S" by dividing "$\exp(-\kappa * \varsigma/2)$" from left and right, and then "$\exp(-q * \alpha_{S/M}/2)$".

$$1 * r * \cosh(s) = i * \hbar * \nabla_{tS} - Q_B * V_{tS}$$

$$q * r * \sinh(s) = q_x * (-i * \hbar * \nabla_{xS} - Q_B * V_{xS})$$
$$+ q_y * (-i * \hbar * \nabla_{yS} - Q_B * V_{yS}) + q_z * (-i * \hbar * \nabla_{zS} - Q_B * V_{zS})$$

SPECIAL ALGEBRA FOR SPECIAL RELATIVITY

"Q_B" includes "$\exp(-\Omega\Omega''/2)$" and so is dropped in the large-scale case, as presented earlier in the discussion on electric charge of a photon.

Large-scale case:

$$1*r*\cosh(s) = i*\hbar*\nabla_{tS}$$

$$q*r*\sinh(s) = q_x*(-i*\hbar*\nabla_{xS}) + q_y*(-i*\hbar*\nabla_{yS}) + q_z*(-i*\hbar*\nabla_{zS})$$

Small-scale case:

$$1*r*\cosh(s) = i*\hbar*\nabla_{tS} - Q_B*V_{tS}$$

$$q*r*\sinh(s) = q_x*(-i*\hbar*\nabla_{xS} - Q_B*V_{xS})$$
$$+ q_y*(-i*\hbar*\nabla_{yS} - Q_B*V_{yS}) + q_z*(-i*\hbar*\nabla_{zS} - Q_B*V_{zS})$$

The small-scale case is redundant to what was previously developed as the Dirac Equation for modeling dynamics of an electron.

The large-scale case is new.

Proposed Dirac Equation Form for Maxwell's Equations.

$$\begin{vmatrix} r*\cosh(s) & \pm q*r*\sinh(s) \\ \pm q*r*\sinh(s) & r*\cosh(s) \end{vmatrix} * \begin{vmatrix} PP_S \\ QQ_S \end{vmatrix} = \begin{vmatrix} 0 \\ 0 \end{vmatrix}$$

"$\pm$" was applicable for particles, and it becomes "$\pm_N$" for waves.

$$\begin{vmatrix} i*\hbar*\nabla_{tS} & \pm_N(q_x*(-i*\hbar*\nabla_{xS}) + q_y*(-i*\hbar*\nabla_{yS}) + q_z*(-i*\hbar*\nabla_{zS})) \\ \pm_N(q_x*(-i*\hbar*\nabla_{xS}) + q_y*(-i*\hbar*\nabla_{yS}) + q_z*(-i*\hbar*\nabla_{zS})) & i*\hbar*\nabla_{tS} \end{vmatrix} * \begin{vmatrix} PP_S \\ QQ_S \end{vmatrix} = \begin{vmatrix} 0 \\ 0 \end{vmatrix}$$

Divide "$i*\hbar$" from both sides.

$$\begin{vmatrix} \nabla_{tS} & -\pm_N(q_x*\nabla_{xS} + q_y*\nabla_{yS} + q_z*\nabla_{zS}) \\ -\pm_N(q_x*\nabla_{xS} + q_y*\nabla_{yS} + q_z*\nabla_{zS}) & \nabla_{tS} \end{vmatrix} * \begin{vmatrix} PP_S \\ QQ_S \end{vmatrix} = \begin{vmatrix} 0 \\ 0 \end{vmatrix}$$

CHAPTER 5 - PROPOSED THEORY OF NON-FINITE NUMBERS

The matrix equation above has two identical rows. If motion is only in the "x" direction, then:

$$\nabla_{tS}*PP_S + -\pm_N(q_x*\nabla_{xS})*QQ_S = 0$$

$$-\pm_N(q_x*\nabla_{xS})*PP_S + \nabla_{tS}*QQ_S = 0$$

It requires "$PP_S = f(x_S - c*t_S)$" and "$QQ_S = -\pm_N q_x*PP_S$".

$$\nabla_{tS}*PP_S = \partial PP_S/\partial c t_S = \partial f/\partial c t_S = (\partial f/\partial(x_S - c*t_S))*(\partial(x_S - c*t_S)/\partial c = -\partial f/\partial(x_S - c*t_S)$$

$$-\pm_N(q_x*\nabla_{xS})*QQ_S = -\pm_N(q_x*\nabla_{xS})*(-\pm_N q_x*PP_S) = \nabla_{xS}*PP_S = \partial f/\partial x_S$$
$$= (\partial f/\partial(x_S - c*t_S))*(\partial(x_S - c*t_S)/\partial x_S = \partial f/\partial(x_S - c*t_S)$$

"$f(x_S - c*t_S)$" describes motion at the speed-of-light in the positive "x_S" direction. The example is an electromagnetic wave.

The equation applies to a field at a speed other than the speed-of-light after replacing right-side zeros with non-zero value "a".

$$\begin{vmatrix} \nabla_{tS} & -\pm_N(q_x*\nabla_{xS} + q_y*\nabla_{yS} + q_z*\nabla_{zS}) \\ -\pm_N(q_x*\nabla_{xS} + q_y*\nabla_{yS} + q_z*\nabla_{zS}) & \nabla_{tS} \end{vmatrix} * \begin{vmatrix} PP_S \\ -\pm_N q*PP_S \end{vmatrix} = \begin{vmatrix} a \\ -\pm_N q*a \end{vmatrix}$$

To substitute 2x2 matrices for "q_x", "q_y", and "q_z", "PP_S" and "a" must each be split into two-member column-vectors, written in terms of electromagnetic field components and electric current density components. Proposed Dirac Equation form for Maxwell's Equations:

$$\partial/\partial c t_S \begin{vmatrix} 1 & 0 & 0 & 0 \\ 0 & 1 & 0 & 0 \\ 0 & 0 & 1 & 0 \\ 0 & 0 & 0 & 1 \end{vmatrix} \pm_N -\partial/\partial x_S \begin{vmatrix} 0 & 0 & 0 & 1 \\ 0 & 0 & 1 & 0 \\ 0 & 1 & 0 & 0 \\ 1 & 0 & 0 & 0 \end{vmatrix} \pm_N -\partial/\partial y_S \begin{vmatrix} 0 & 0 & 0 & k \\ 0 & 0 & -k & 0 \\ 0 & k & 0 & 0 \\ -k & 0 & 0 & 0 \end{vmatrix}$$

$$\pm_N -\partial/\partial z_S \begin{vmatrix} 0 & 0 & 1 & 0 \\ 0 & 0 & 0 & -1 \\ 1 & 0 & 0 & 0 \\ 0 & -1 & 0 & 0 \end{vmatrix} * \begin{vmatrix} (E_{zS} \pm_N i*K_{zS}) \\ (E_{xS} \pm_N i*K_{xS}) - k*(E_{yS} \pm_N i*K_{yS}) \\ (E_{zS} \pm_N i*K_{zS}) \\ (E_{xS} \pm_N i*K_{xS}) - k*(E_{yS} \pm_N i*K_{yS}) \end{vmatrix} = \begin{vmatrix} (-\pm_N J_{tS} - J_{zS}) \\ (-J_{xS} + k*J_{yS}) \\ (-\pm_N J_{tS} - J_{zS}) \\ (-J_{xS} + k*J_{yS}) \end{vmatrix}$$

SPECIAL ALGEBRA FOR SPECIAL RELATIVITY

Rewritten as two differential equations:

$\partial(E_{zS} \pm_N i^*K_{zS})/\partial ct_S$
$\pm_N -\partial(E_{xS} \pm_N i^*K_{xS})/\partial x_S \pm_N -\partial(-k^*(E_{yS} \pm_N i^*K_{yS}))/\partial x_S$
$\pm_N -k^*\partial(E_{xS} \pm_N i^*K_{xS})/\partial y_S \pm_N -k^*\partial(-k^*(E_{yS} \pm_N i^*K_{yS}))/\partial y_S$
$\pm_N -\partial(E_{zS} \pm_N i^*K_{zS})/\partial z_S$
$= (-\pm_N J_{tS} - J_{zS})$

$\partial(E_{xS} \pm_N i^*K_{xS})/\partial ct_S + \partial(-k^*(E_{yS} \pm_N i^*K_{yS}))/\partial ct_S$
$\pm_N -\partial(E_{zS} \pm_N i^*K_{zS})/\partial x_S$
$\pm_N -(-k)^*\partial(E_{zS} \pm_N i^*K_{zS})/\partial y_S$
$\pm_N -(-1)\partial(E_{xS} \pm_N i^*K_{xS})/\partial z_S \pm_N -(-1)\partial(-k^*(E_{yS} \pm_N i^*K_{yS}))/\partial z_S$
$= (-J_{xS} + k^*J_{yS})$

Validity is established by two tests.

- The first test (the wave test) ensures the above equation does not violate Maxwell's Equations

- The second test (the particle test) ensures the complex-conjugate addition and subtraction equations result in the correct force density invariant "$_4f$" measured for an electromagnetic field

Equivalence to Maxwell's Equations. The first test is satisfied by breaking the above equation into components, and then showing component equations match component equations from Maxwell's Equations. ("s" has been dropped from the components to not clutter what is written.) Component equations from Maxwell's Equations are:

$-\partial K_x/\partial ct - \partial E_z/\partial y + \partial E_y/\partial z = 0$; $-\partial E_x/\partial ct + \partial K_z/\partial y - \partial K_y/\partial z = J_x$
$-\partial K_y/\partial ct - \partial E_x/\partial z + \partial E_z/\partial x = 0$; $-\partial E_y/\partial ct + \partial K_x/\partial z - \partial K_z/\partial x = J_y$
$-\partial K_z/\partial ct - \partial E_y/\partial x + \partial E_x/\partial y = 0$; $-\partial E_z/\partial ct + \partial K_y/\partial x - \partial K_x/\partial y = J_z$
$\partial K_x/\partial x + \partial K_y/\partial y + \partial K_z/\partial z = 0$; $\partial E_x/\partial x + \partial E_y/\partial y + \partial E_z/\partial z = J_t$

First, split real components of the first row per "$\pm_N$" or not:

$\partial(E_{zS})/\partial ct_S \pm_N -\partial(E_{xS})/\partial x_S \pm_N -\partial(-k^*(\pm_N i^*K_{yS}))/\partial x_S$
$\pm_N -k^*\partial(\pm_N i^*K_{xS})/\partial y_S \pm_N -k^*\partial(-k^*(E_{yS}))/\partial y_S \pm_N -\partial(E_{zS})/\partial z_S$
$= (-\pm_N J_{tS} - J_{zS})$

CHAPTER 5 - PROPOSED THEORY OF NON-FINITE NUMBERS

$\partial E_z/\partial ct - \pm_N \partial E_x/\partial x - -\pm_N \pm_N k^* i^* \partial K_y/\partial x - \pm_N \pm_N k^* i^* \partial K_x/\partial y - -\pm_N k^* k^* \partial E_y/\partial y$
$- \pm_N \partial E_z/\partial z = -\pm_N J_t - J_z$

$\partial E_z/\partial ct - \pm_N \partial E_x/\partial x - \partial K_y/\partial x + \partial K_x/\partial y - \pm_N \partial E_y/\partial y - \pm_N \partial E_z/\partial z = -\pm_N J_t - J_z$

$\partial E_z/\partial ct - \partial K_y/\partial x + \partial K_x/\partial y = -J_z$; $-\pm_N \partial E_x/\partial x - \pm_N \partial E_y/\partial y - \pm_N \partial E_z/\partial z = -\pm_N J_t$

$-\partial E_x/\partial x - \partial E_y/\partial y - \partial E_z/\partial z = -J_t$

Second, split imaginary components of the first row:

$\partial(\pm_N i^* K_{zS})/\partial ct_S \pm_N -\partial(\pm_N i^* K_{xS})/\partial x_S \pm_N -\partial(-k^*(E_{yS}))/\partial x_S$
$\pm_N -k^* \partial(E_{xS})/\partial y_S \pm_N -k^* \partial(-k^*(\pm_N i^* K_{yS}))/\partial y_S \pm_N -\partial(\pm_N i^* K_{zS})/\partial z_S$
$= 0$

$\pm_N \partial K_z/\partial ct - \partial K_x/\partial x + \pm_N \partial E_y/\partial x - \pm_N \partial E_x/\partial y - \partial K_y/\partial y - \partial K_z/\partial z = 0$

$\pm_N \partial K_z/\partial ct + \pm_N \partial E_y/\partial x - \pm_N \partial E_x/\partial y = 0$; $-\partial K_x/\partial x - \partial K_y/\partial y - \partial K_z/\partial z = 0$

$-\partial K_z/\partial ct - \partial E_y/\partial x + \partial E_x/\partial y = 0$

Third, split real components of the second row:

$\partial(E_{xS})/\partial ct_S + \partial(-k^*(\pm_N i^* K_{yS}))/\partial ct_S \pm_N -\partial(E_{zS})/\partial x_S$
$\pm_N -(-k)^* \partial(\pm_N i^* K_{zS})/\partial y_S \pm_N -(-1)\partial(E_{xS})/\partial z_S \pm_N -(-1)\partial(-k^*(\pm_N i^* K_{yS}))/\partial z_S$
$= (-J_{xS})$

$\partial E_x/\partial ct + -\pm_N k^* i^* \partial K_y/\partial ct - \pm_N \partial E_z/\partial x - -\pm_N \pm_N k^* i^* \partial K_z/\partial y$
$- -\pm_N \partial E_x/\partial z - -\pm_N \pm_N -k^* i^* \partial K_y/\partial z = -J_x$

$\partial E_x/\partial ct + \pm_N \partial K_y/\partial ct - \pm_N \partial E_z/\partial x - \partial K_z/\partial y + \pm_N \partial E_x/\partial z + \partial K_y/\partial z = -J_x$

$\partial E_x/\partial ct - \partial K_z/\partial y + \partial K_y/\partial z = -J_x$; $\pm_N \partial K_y/\partial ct - \pm_N \partial E_z/\partial x + \pm_N \partial E_x/\partial z = 0$

$-\partial K_y/\partial ct + \partial E_z/\partial x - \partial E_x/\partial z = 0$

Fourth, split imaginary components of the second row:

SPECIAL ALGEBRA FOR SPECIAL RELATIVITY

$\partial(\pm_N i^* K_{xS})/\partial ct_S + \partial(-k^*(E_{yS}))/\partial ct_S \pm_N -\partial(\pm_N i^* K_{zS})/\partial x_S$
$\pm_N -(-k)^*\partial(E_{zS})/\partial y_S \pm_N -(-1)\partial(\pm_N i^* K_{xS})/\partial z_S \pm_N -(-1)\partial(-k^*(E_{yS}))/\partial z_S$
$= (k^* J_{yS})$

$\pm_N i^* \partial K_x/\partial ct - k^* \partial E_y/\partial ct - \pm_N \pm_N i^* \partial K_z/\partial x - -\pm_N k^* \partial E_z/\partial y - -\pm_N \pm_N i^* \partial K_x/\partial z$
$- -\pm_N -k^* \partial E_y/\partial z = k^* J_y$

$\pm_N \partial K_x/\partial ct - \partial E_y/\partial ct - \partial K_z/\partial x + \pm_N \partial E_z/\partial y + \partial K_x/\partial z - \pm_N \partial E_y/\partial z = J_y$

$\pm_N \partial K_x/\partial ct + \pm_N \partial E_z/\partial y - \pm_N \partial E_y/\partial z = 0 \quad ; \quad -\partial E_y/\partial ct - \partial K_z/\partial x + \partial K_x/\partial z = J_y$

$\partial K_x/\partial ct + \partial E_z/\partial y - \partial E_y/\partial z = 0$

Example electromagnetic field:

$E_{yS} = -E_{amp}*\sin(k_{xS}*(x_S + c^*t_S)) \quad ; \quad K_{zS} = E_{amp}*\sin(k_{xS}*(x_S + c^*t_S))$

Maxwell's Equations are satisfied by this field:

$-\partial E_y/\partial ct + \partial K_x/\partial z - \partial K_z/\partial x = J_y \quad ; \quad -\partial K_z/\partial ct - \partial E_y/\partial x + \partial E_x/\partial y = 0$
$-\partial E_y/\partial ct - \partial K_z/\partial x = 0 \quad ; \quad -\partial K_z/\partial ct - \partial E_y/\partial x = 0$
$-(k_{xS})*(-E_{amp}) - (k_{xS})*(E_{amp}) = 0 \quad ; \quad -(k_{xS})*(E_{amp}) - (k_{xS})*(-E_{amp}) = 0$

Checked against the differential equations:

$\partial(E_{yS})/\partial ct_S = (k_{xS})*-E_{amplitude}*\cos(k_{xS}*(x_S + c^*t_S))$

$\partial(E_{yS})/\partial x_S = k_{xS}*-E_{amplitude}*\cos(k_{xS}*(x_S + c^*t_S))$

$\partial(i^* K_{zS})/\partial ct_S = (i^* k_{xS})*E_{amplitude}*\cos(k_{xS}*(x_S + c^*t_S))$

$\partial(i^* K_{zS})/\partial x_S = (i^* k_{xS})*E_{amplitude}*\cos(k_{xS}*(x_S + c^*t_S))$

$\partial(E_{zS} \pm_N i^* K_{zS})/\partial ct_S - \pm_N(-k)^* \partial(E_{yS} \pm_N i^* K_{yS})/\partial x_S$
$= \partial(i^* K_{zS})/\partial ct_S - (-k)^*\partial(E_{yS})/\partial x_S$
$= (i^* k_{xS})*E_{amplitude}*\cos(k_{xS}*(x_S + c^*t_S)) - (-k)^* k_{xS}*-E_{amplitude}*\cos(k_{xS}*(x_S + c^*t_S))$
$= (i^* k_{xS}) - (-k)^* k_{xS}*(-1) = 0$

CHAPTER 5 - PROPOSED THEORY OF NON-FINITE NUMBERS

$$-k*\partial(E_{yS} \pm_N i*K_{yS})/\partial ct_S - \pm_N \partial(E_{zS} \pm_N i*K_{zS})/\partial x_S$$
$$-k*\partial(E_{yS})/\partial ct_S - \partial(i*K_{zS})/\partial x_S$$
$$= -k*(k_{xS})*-E_{amplitude}*\cos(k_{xS}*(x_S + c*t_S)) - (i*k_{xS})*E_{amplitude}*\cos(k_{xS}*(x_S + c*t_S))$$
$$= -k*(k_{xS})*(-1) - (i*k_{xS}) = 0$$

As another check, a more general example is also satisfied:

$$E_{yS} \pm_N i*K_{yS} = \pm_N k*E_{amplitude}*\exp(\pm_N i*(k_{xS}*(x_S + c*t_S)))$$
$$E_{zS} \pm_N i*K_{zS} = E_{amplitude}*\exp(\pm_N i*(k_{xS}*(x_S + c*t_S)))$$

$$\partial(E_{yS} \pm_N i*K_{yS})/\partial ct_S = -k_{xS}*E_{amplitude}*\exp(\pm_N i*(k_{xS}*(x_S + c*t_S)))$$
$$\partial(E_{yS} \pm_N i*K_{yS})/\partial x_S = -k_{xS}*E_{amplitude}*\exp(\pm_N i*(k_{xS}*(x_S + c*t_S)))$$

$$\partial(E_{zS} \pm_N i*K_{zS})/\partial ct_S = (\pm_N i*k_{xS})*E_{amplitude}*\exp(\pm_N i*(k_{xS}*(x_S + c*t_S)))$$
$$\partial(E_{zS} \pm_N i*K_{zS})/\partial x_S = (\pm_N i*k_{xS})*E_{amplitude}*\exp(\pm_N i*(k_{xS}*(x_S + c*t_S)))$$

5.7 Force Density Using the Complex-Conjugate

Force Density Calculation. The below equation is the addition equation, formed by multiplying the complex-conjugate of the Dirac Spinor by the large-scale case equation and adding it to the product of the complex-conjugate of the large-scale case equation and the Dirac Spinor. The addition operation removes imaginary terms.

$$4*(E_z*\partial E_z/\partial ct + K_z*\partial K_z/\partial ct - \pm_N E_z*\partial E_x/\partial x \pm_N -K_z*\partial K_x/\partial x$$
$$- E_z*\partial K_y/\partial x + K_z*\partial E_y/\partial x + E_z*\partial K_x/\partial y - K_z*\partial E_x/\partial y$$
$$-\pm_N E_z*\partial E_y/\partial y \pm_N -K_z*\partial K_y/\partial y -\pm_N E_z*\partial E_z/\partial z \pm_N -K_z*\partial K_z/\partial z$$
$$+ E_x*\partial E_x/\partial ct + K_x*\partial K_x/\partial ct \pm_N E_x*\partial K_y/\partial ct \pm_N -K_x*\partial E_y/\partial ct$$
$$-\pm_N E_x*\partial E_z/\partial x \pm_N -K_x*\partial K_z/\partial x - E_x*\partial K_z/\partial y + K_x*\partial E_z/\partial y$$
$$\pm_N E_x*\partial E_x/\partial z \pm_N K_x*\partial K_x/\partial z + E_x*\partial K_y/\partial z - K_x*\partial E_y/\partial z$$
$$-\pm_N E_y*\partial K_x/\partial ct \pm_N K_y*\partial E_x/\partial ct + E_y*\partial E_y/\partial ct + K_y*\partial K_y/\partial ct$$
$$+ E_y*\partial K_z/\partial x - K_y*\partial E_z/\partial x - \pm_N E_y*\partial E_z/\partial y \pm_N -K_y*\partial K_z/\partial y$$
$$- E_y*\partial K_x/\partial z + K_y*\partial E_x/\partial z \pm_N E_y*\partial E_y/\partial z \pm_N K_y*\partial K_y/\partial z)$$

$$= 4*(-\pm_N E_z*J_t - E_z*J_z - E_x*J_x \pm_N K_x*J_y - E_y*J_y \pm_N -K_y*J_x)$$

Divide it by four and separate away "$\pm_N$" terms.

SPECIAL ALGEBRA FOR SPECIAL RELATIVITY

First Equation:

$-((\nabla_y * K_z - \nabla_z * K_y) * E_x + (\nabla_z * K_x - \nabla_x * K_z) * E_y + (\nabla_x * K_y - \nabla_y * K_x) * E_z)$
$+ (\nabla_y * E_z - \nabla_z * E_y) * K_x + (\nabla_z * E_x - \nabla_x * E_z) * K_y + (\nabla_x * E_y - \nabla_y * E_x) * K_z)$
$+ (\nabla_t * K_x) * K_x + (\nabla_t * K_y) * K_y + (\nabla_t * K_z) * K_z$
$+ (\nabla_t * E_x) * E_x + (\nabla_t * E_y) * E_y + (\nabla_t * E_z) * E_z$
$= -(J_x * E_x + J_y * E_y + J_z * E_z)$
$= -f_{tr}$

Second Equation:

$\pm_N * (((\nabla_t * K_y) * E_x - (\nabla_t * K_x) * E_y) + ((\nabla_t * E_x) * K_y - (\nabla_t * E_y) * K_x)$
$- (\nabla_y * K_z - \nabla_z * K_y) * K_y + (\nabla_z * K_x - \nabla_x * K_z) * K_x$
$- (\nabla_y * E_z - \nabla_z * E_y) * E_y + (\nabla_z * E_x - \nabla_x * E_z) * E_x$
$- (\nabla_x * K_x + \nabla_y * K_y + \nabla_z * K_z) * K_z - (\nabla_x * E_x + \nabla_y * E_y + \nabla_z * E_z) * E_z)$
$= \pm_N * ((-J_x * K_y + J_y * K_x) - J_t * E_z)$
$= \pm_N * -f_{zr}$

In the Second Equation are six of Maxwell's Equations.

- "$-J_t * E_z$" equates to "$-(\nabla_x * E_x + \nabla_y * E_y + \nabla_z * E_z) * E_z$"
- "$J_y * K_x$" equates to "$-(\nabla_t * E_y) * K_x + (\nabla_z * K_x - \nabla_x * K_z) * K_x$"
- "$-J_x * K_y$" equates to "$(\nabla_t * E_x) * K_y - (\nabla_y * K_z - \nabla_z * K_y) * K_y$"
- "0" equates to "$-(\nabla_x * K_x + \nabla_y * K_y + \nabla_z * K_z) * K_z$"
- "0" equates to "$(\nabla_t * K_y) * E_x + (\nabla_z * E_x - \nabla_x * E_z) * E_x$"
- "0" equates to "$-(\nabla_t * K_x) * E_y - (\nabla_y * E_z - \nabla_z * E_y) * E_y$"

These match Maxwell's Equations found in the "$-f_{zr}$" component equation in the chapter on electromagnetic fields:

- "$-J_t * E_z$" equates to "$-(\partial E_x/\partial x + \partial E_y/\partial y + \partial E_z/\partial z) * E_z$"
- "$J_y * K_x$" equates to "$-(\partial E_y/\partial ct) * K_x - (\partial K_z/\partial x - \partial K_x/\partial z) * K_x$"
- "$-J_x * K_y$" equates to "$(\partial E_x/\partial ct) * K_y + (-\partial K_z/\partial y + \partial K_y/\partial z) * K_y$"
- "0" equates to "$-(\partial K_x/\partial x + \partial K_y/\partial y + \partial K_z/\partial z) * K_z$"
- "0" equates to "$(\partial K_y/\partial ct) * E_x - (\partial E_z/\partial x - \partial E_x/\partial z) * E_x$"
- "0" equates to "$-(\partial K_x/\partial ct) * E_y + (-\partial E_z/\partial y + \partial E_y/\partial z) * E_y$"

The other two 2x2 matrix selections for "q_x", "q_y", and "q_z" have "x-y-z" rotated to "z-x-y" and then to "y-z-x". All three selections are valid, and all three selections apply simultaneously. The other two selections lead to the below two equations, as well as to repeats of the first equation above.

CHAPTER 5 - PROPOSED THEORY OF NON-FINITE NUMBERS

Third Equation:

$\pm_N * (((\nabla_t * K_z) * E_y - (\nabla_t * K_y) * E_z) + ((\nabla_t * E_y) * K_z - (\nabla_t * E_z) * K_y)$
$- (\nabla_z * K_x - \nabla_x * K_z) * K_z + (\nabla_x * K_y - \nabla_y * K_x) * K_y$
$- (\nabla_z * E_x - \nabla_x * E_z) * E_z + (\nabla_x * E_y - \nabla_y * E_x) * E_y$
$- (\nabla_y * K_y + \nabla_z * K_z + \nabla_x * K_x) * K_x - (\nabla_y * E_y + \nabla_z * E_z + \nabla_x * E_x) * E_x)$
$= \pm_N * ((-J_y * K_z + J_z * K_y) - J_t * E_x)$
$= \pm_N * -f_{xr}$

Fourth Equation:

$\pm_N * (((\nabla_t * K_x) * E_z - (\nabla_t * K_z) * E_x) + ((\nabla_t * E_z) * K_x - (\nabla_t * E_x) * K_z)$
$- (\nabla_x * K_y - \nabla_y * K_x) * K_x + (\nabla_y * K_z - \nabla_z * K_y) * K_z$
$- (\nabla_x * E_y - \nabla_y * E_x) * E_x + (\nabla_y * E_z - \nabla_z * E_y) * E_z$
$- (\nabla_z * K_z + \nabla_x * K_x + \nabla_y * K_y) * K_y - (\nabla_z * E_z + \nabla_x * E_x + \nabla_y * E_y) * E_y)$
$= \pm_N * ((-J_z * K_x + J_x * K_z) - J_t * E_y)$
$= \pm_N * -f_{yr}$

The subtraction equation retains only imaginary terms. "i" is replaced by "$\pm_W i$" to account for ambiguity of which term is subtracted from the other.

$\pm_W 4 * i * (\pm_N E_z * \partial K_z / \partial ct \pm_N -K_z * \partial E_z / \partial ct - E_z * \partial K_x / \partial x + K_z * \partial E_x / \partial x$
$\pm_N E_z * \partial E_y / \partial x \pm_N K_z * \partial K_y / \partial x - \pm_N E_z * \partial E_x / \partial y \pm_N -K_z * \partial K_x / \partial y$
$+ -E_z * \partial K_y / \partial y + K_z * \partial E_y / \partial y + -E_z * \partial K_z / \partial z + K_z * \partial E_z / \partial z$
$\pm_N E_x * \partial K_x / \partial ct \pm_N -K_x * \partial E_x / \partial ct - E_x * \partial E_y / \partial ct - K_x * \partial K_y / \partial ct$
$+ -E_x * \partial K_z / \partial x + K_x * \partial E_z / \partial x \pm_N E_x * \partial E_z / \partial y \pm_N K_x * \partial K_z / \partial y$
$+ E_x * \partial K_x / \partial z - K_x * \partial E_x / \partial z - \pm_N E_x * \partial E_y / \partial z \pm_N -K_x * \partial K_y / \partial z$
$+ E_y * \partial E_x / \partial ct + K_y * \partial K_x / \partial ct \pm_N E_y * \partial K_y / \partial ct \pm_N -K_y * \partial E_y / \partial ct$
$+ -\pm_N E_y * \partial E_z / \partial x \pm_N -K_y * \partial K_z / \partial x - E_y * \partial K_z / \partial y + K_y * \partial E_z / \partial y$
$\pm_N E_y * \partial E_x / \partial z \pm_N K_y * \partial K_x / \partial z + E_y * \partial K_y / \partial z - K_y * \partial E_y / \partial z)$

$= \pm_W 4 * i * (K_z * J_t \pm_N K_z * J_z + E_x * J_y \pm_N K_x * J_x - E_y * J_x \pm_N K_y * J_y)$

Below are the two equations formed by separating terms.

SPECIAL ALGEBRA FOR SPECIAL RELATIVITY

Fifth Equation:

$\pm_w \pm_N i * ((\nabla_y * K_z - \nabla_z * K_y) * K_x + (\nabla_z * K_x - \nabla_x * K_z) * K_y + (\nabla_x * K_y - \nabla_y * K_x) * K_z$
$- (\nabla_z * E_y - \nabla_y * E_z) * E_x - (\nabla_x * E_z - \nabla_z * K_x) * E_y - (\nabla_y * E_x - \nabla_x * E_y) * E_z$
$+ (\nabla_t * K_x) * E_x + (\nabla_t * K_y) * E_y + (\nabla_t * K_z) * E_z$
$- (\nabla_t * E_x) * K_x - (\nabla_t * E_y) * K_y - (\nabla_t * E_z) * K_z)$
$= \pm_w \pm_N i * (J_x * K_x + J_y * K_y + J_z * K_z)$
$= \pm_N * -f_{ti}$

Sixth Equation:

$\pm_w i * (-((\nabla_t * K_y) * K_x - (\nabla_t * K_x) * K_y) + ((\nabla_t * E_x) * E_y - (\nabla_t * E_y) * E_x)$
$- (\nabla_y * K_z - \nabla_z * K_y) * E_y + (\nabla_z * K_x - \nabla_x * K_z) * E_x$
$+ (\nabla_y * E_z - \nabla_z * E_y) * K_y - (\nabla_z * E_x - \nabla_x * E_z) * K_x$
$- (\nabla_x * K_x + \nabla_y * K_y + \nabla_z * K_z) * E_z + (\nabla_x * E_x + \nabla_y * E_y + \nabla_z * E_z) * K_z$
$= \pm_w i * (-(J_x * E_y - J_y * E_x) + J_t * K_z)$
$= -f_{zi}$

In the Sixth Equation are six of Maxwell's Equations.

- "$J_t * K_z$" equates to "$(\nabla_x * E_x + \nabla_y * E_y + \nabla_z * E_z) * K_z$"
- "$J_y * E_x$" equates to "$-(\nabla_t * E_y) * E_x + (\nabla_z * K_x - \nabla_x * K_z) * E_x$"
- "$-J_x * E_y$" equates to "$(\nabla_t * E_x) * E_y - (\nabla_y * K_z - \nabla_z * K_y) * E_y$"
- "0" equates to "$-(\nabla_x * K_x + \nabla_y * K_y + \nabla_z * K_z) * E_z$"
- "0" equates to "$-(\nabla_t * K_y) * K_x - (\nabla_z * E_x - \nabla_x * E_z) * K_x$"
- "0" equates to "$-(\nabla_t * K_x) * K_y + (\nabla_y * E_z - \nabla_z * E_y) * K_y$"

These match Maxwell's Equations found in the "$-f_{zi}$" component equation in the chapter on electromagnetic fields.

- "$\pm J_t * K_z$" equates to "$\pm(\partial E_x/\partial x + \partial E_y/\partial y + \partial E_z/\partial z) * K_z$"
- "$\pm J_y * E_x$" equates to "$\pm -(\partial E_y/\partial ct) * E_x \pm -(\partial K_z/\partial x - \partial K_x/\partial z) * E_x$"
- "$\pm -J_x * E_y$" equates to "$\pm(\partial E_x/\partial ct) * E_y \pm (-\partial K_z/\partial y + \partial K_y/\partial z) * E_y$"
- "0" equates to "$\pm -(\partial K_x/\partial x + \partial K_y/\partial y + \partial K_z/\partial z) * E_z$"
- "0" equates to "$\pm -(\partial K_y/\partial ct) * K_x \pm (\partial E_z/\partial x - \partial E_x/\partial z) * K_x$"
- "0" equates to "$\pm(\partial K_x/\partial ct) * K_y \pm (\partial E_z/\partial y - \partial E_y/\partial z) * K_y$"

CHAPTER 5 - PROPOSED THEORY OF NON-FINITE NUMBERS

Seventh Equation:
$\pm_w i*(-((\nabla_t*K_z)*K_y - (\nabla_t*K_y)*K_z) + ((\nabla_t*E_y)*E_z - (\nabla_t*E_z)*E_y)$
$- (\nabla_z*K_x - \nabla_x*K_z)*E_z + (\nabla_x*K_y - \nabla_y*K_x)*E_y$
$+ (\nabla_z*E_x - \nabla_x*E_z)*K_z - (\nabla_x*E_y - \nabla_y*E_x)*K_y$
$- (\nabla_y*K_y + \nabla_z*K_z + \nabla_x*K_x)*E_x + (\nabla_y*E_y + \nabla_z*E_z + \nabla_x*E_x)*K_x$
$= \pm_w i*(-(J_y*E_z - J_z*E_y) + J_t*K_x)$
$= -f_{xi}$

Eighth Equation:
$\pm_w i*(-((\nabla_t*K_x)*K_z - (\nabla_t*K_z)*K_x) + ((\nabla_t*E_z)*E_x - (\nabla_t*E_x)*E_z)$
$- (\nabla_x*K_y - \nabla_y*K_x)*E_x + (\nabla_y*K_z - \nabla_z*K_y)*E_z$
$+ (\nabla_x*E_y - \nabla_y*E_x)*K_x - (\nabla_y*E_z - \nabla_z*E_y)*K_z$
$- (\nabla_z*K_z + \nabla_x*K_x + \nabla_y*K_y)*E_y + (\nabla_z*E_z + \nabla_x*E_x + \nabla_y*E_y)*K_y$
$= \pm_w i*(-(J_z*E_x - J_x*E_z) + J_t*K_y)$
$= -f_{yi}$

Energy Density. Inside the First Equation is the time gradient of energy, "$+(\nabla_t*K_x)*K_x+(\nabla_t*K_y)*K_y+(\nabla_t*K_z)*K_z+(\nabla_t*E_x)*E_x+(\nabla_t*E_y)*E_y+(\nabla_t*E_z)*E_z$". It includes a negative when combined with compound label numbers.

$$(_1\nabla^{sn}*_3E)\bullet_3E - (_1\nabla^{sn}*_3K)\bullet_3K$$

$$= (\nabla_t*E_x)*E_x*1_M{}^{sn}*p_{xM}*p_{xM} + (\nabla_t*E_y)*E_y*1_M{}^{sn}*p_{yM}*p_{yM}$$
$$+ (\nabla_t*E_z)*E_z*1_M{}^{sn}*p_{zM}*p_{zM} - (\nabla_t*K_x)*K_x*1_M{}^{sn}*k_{xM}*k_{xM}$$
$$- (\nabla_t*K_y)*K_y*1_M{}^{sn}*k_{yM}*k_{yM} - (\nabla_t*K_z)*K_z*1_M{}^{sn}*k_{zM}*k_{zM}$$

$$= 1_M*((\nabla_t*E_x)*E_x + (\nabla_t*E_y)*E_y + (\nabla_t*E_z)*E_z$$
$$+ (\nabla_t*K_x)*K_x + (\nabla_t*K_y)*K_y + (\nabla_t*K_z)*K_z)$$

$$= 1_M*\nabla_t*(E_x*E_x + E_y*E_y + E_z*E_z + K_x*K_x + K_y*K_y + K_z*K_z)/2$$

SPECIAL ALGEBRA FOR SPECIAL RELATIVITY

<u>Including Compound-Label-Numbers</u>. First of eight equations combined with compound label numbers written using pieces of invariants:

$$(_3\nabla^{sn}\mathbf{x}_3K)\bullet_3E - (_3\nabla^{sn}\mathbf{x}_3E)\bullet_3K + (_1\nabla^{sn}*_3E)\bullet_3E - (_1\nabla^{sn}*_3K)\bullet_3K = -_3J\bullet_3E = 1_M*-f_{tr}$$

"$-_3J\bullet_3E$" has the example of electric direct current in a wire, for which both current and electric field are positive. Current in the wire creates heat as a loss of energy from the system. The rate of that loss of energy from the system is the real time component "$-_1f_r = 1_M*-f_{tr}$" of the four-component force density "$-_4f$".

$$-_1f_r = (\,(_1\nabla^{sn}*_3E + _3\nabla^{sn}\mathbf{x}_3K)\bullet_3E\,) - (\,(_1\nabla^{sn}*_3K + _3\nabla^{sn}\mathbf{x}_3E)\bullet_3K\,)$$
$$= (-_3J)\bullet_3E - (_30)\bullet_3K$$

Second, third, and fourth equations written using pieces of invariants:
$$\pm_N(-(_1\nabla^{sn}*_3K)\mathbf{x}_3E + (_1\nabla^{sn}*_3E)\mathbf{x}_3K - (_3\nabla^{sn}\mathbf{x}_3E)\mathbf{x}_3E + (_3\nabla^{sn}\mathbf{x}_3K)\mathbf{x}_3K$$
$$+ (_3\nabla^{sn}\bullet_3E)*_3E - (_3\nabla^{sn}\bullet_3K)*_3K) = \pm_N(-_3J\mathbf{x}_3K - _1J*_3E) = \pm_N-_3f_r$$

"$-_3J\mathbf{x}_3K - _1J*_3E$" is the force due to charge in an electromagnetic field and "$-_3f_r$" has the real space components of "$-_4f$". Because "$-_3f_r$" is measurable as a particle (not wave) property, it does not have the "$\pm_N$" factor.

The fifth equation is also written with pieces of invariants.

$$\pm_N(-(_3\nabla^{sn}\mathbf{x}_3K)\bullet_3K + (_3\nabla^{sn}\mathbf{x}_3E)\bullet_3E - (_1\nabla^{sn}*_3E)\bullet_3K + (_1\nabla^{sn}*_3K)\bullet_3E)$$
$$= \pm_N(-(-_3J\bullet_3K)) = \pm_N-_1f_i$$

$-_1f_i$" of "$-_4f$" is not directly measurable as an energy.

$$-\pm_N\pm_W_1f_i = \pm_N(\pm_W(\,(_1\nabla^{sn}*_3K + _3\nabla^{sn}\mathbf{x}_3E)\bullet_3E\,) \pm_W -(\,(_1\nabla^{sn}*_3E + _3\nabla^{sn}\mathbf{x}_3K)\bullet_3K\,))$$
$$= \pm_N(\pm_W(_30)\bullet_3E \pm_W -(-_3J)\bullet_3K)$$

"$\pm_W$" applies to imaginary terms of energy density and is appropriate because imaginary energy density is not measurable, and, therefore, is both/neither positive and/nor negative. "$\pm_N$" is dropped because force pertains to particles.

$$-_1f = -_1f_r + -\pm_W_1f_i$$
$$= (-_3J)\bullet_3E \pm_W (_30)\bullet_3E \pm_W -(-_3J)\bullet_3K - (_30)\bullet_3K$$
$$= (\,(_1\nabla^{sn}*_3E + _3\nabla^{sn}\mathbf{x}_3K)\bullet_3E\,) \pm_W (\,(_1\nabla^{sn}*_3K + _3\nabla^{sn}\mathbf{x}_3E)\bullet_3E\,)$$
$$\pm_W -(\,(_1\nabla^{sn}*_3E + _3\nabla^{sn}\mathbf{x}_3K)\bullet_3K\,) - (\,(_1\nabla^{sn}*_3K + _3\nabla^{sn}\mathbf{x}_3E)\bullet_3K\,)$$

CHAPTER 5 - PROPOSED THEORY OF NON-FINITE NUMBERS

Sixth, seventh, and eighth equations written with pieces of invariants:

$$-\pm_w{}_3f_i = \pm_w(\ (_3\nabla^{sn}\bullet_3K)*_3E + (_1\nabla^{sn}*_3K + _3\nabla^{sn}\mathbf{x}_3E)\mathbf{x}_3K\ )$$
$$-\pm_w(\ (_3\nabla^{sn}\bullet_3E)*_3K + (_1\nabla^{sn}*_3E + _3\nabla^{sn}\mathbf{x}_3K)\mathbf{x}_3E\ )$$

$$= \pm_w\ (\ (0)*_3E + (_30)\mathbf{x}_3K\ ) \pm_w -(\ (-_1J)*_3K + (-_3J)\mathbf{x}_3E\ )$$

$$-_3f = -_3f_r - \pm_w{}_3f_i$$
$$= ((\ (-_1J)*_3E + (-_3J)\mathbf{x}_3K\ ) \pm_w\ (\ (0)*_3E + (_30)\mathbf{x}_3K\ )$$
$$\pm_w -(\ (-_1J)*_3K + (-_3J)\mathbf{x}_3E\ ) - (\ (0)*_3K + (_30)\mathbf{x}_3E\ ))$$

$$= ((\ (_3\nabla^{sn}\bullet_3E)*_3E + (_1\nabla^{sn}*_3E + _3\nabla^{sn}\mathbf{x}_3K)\mathbf{x}_3K\ )$$
$$\pm_w\ (\ (_3\nabla^{sn}\bullet_3K)*_3E + (_1\nabla^{sn}*_3K + _3\nabla^{sn}\mathbf{x}_3E)\mathbf{x}_3K\ )$$
$$\pm_w -(\ (_3\nabla^{sn}\bullet_3E)*_3K + (_1\nabla^{sn}*_3E + _3\nabla^{sn}\mathbf{x}_3K)\mathbf{x}_3E\ )$$
$$- (\ (_3\nabla^{sn}\bullet_3K)*_3K + (_1\nabla^{sn}*_3K + _3\nabla^{sn}\mathbf{x}_3E)\mathbf{x}_3E\ ))$$

These four equations are combined below.

$$-_4f = -_1f + -_3f = -_1f_r + -\pm_w{}_1f_i + -_3f_r + -\pm_w{}_3f_i$$

$$= (\ (_1\nabla^{sn}*_3E + _3\nabla^{sn}\mathbf{x}_3K)\bullet_3E\ ) \pm_w\ (\ (_1\nabla^{sn}*_3K + _3\nabla^{sn}\mathbf{x}_3E)\bullet_3E\ )$$
$$\pm_w -(\ (_1\nabla^{sn}*_3E + _3\nabla^{sn}\mathbf{x}_3K)\bullet_3K\ ) - (\ (_1\nabla^{sn}*_3K + _3\nabla^{sn}\mathbf{x}_3E)\bullet_3K\ )$$
$$+ ((\ (_3\nabla^{sn}\bullet_3E)*_3E + (_1\nabla^{sn}*_3E + _3\nabla^{sn}\mathbf{x}_3K)\mathbf{x}_3K\ )$$
$$\pm_w\ (\ (_3\nabla^{sn}\bullet_3K)*_3E + (_1\nabla^{sn}*_3K + _3\nabla^{sn}\mathbf{x}_3E)\mathbf{x}_3K\ )$$
$$\pm_w -(\ (_3\nabla^{sn}\bullet_3E)*_3K + (_1\nabla^{sn}*_3E + _3\nabla^{sn}\mathbf{x}_3K)\mathbf{x}_3E\ )$$
$$- (\ (_3\nabla^{sn}\bullet_3K)*_3K + (_1\nabla^{sn}*_3K + _3\nabla^{sn}\mathbf{x}_3E)\mathbf{x}_3E\ ))$$

It is the same force density equation developed in the chapter on electromagnetism.

$$-_4f = ((_3\nabla^{sn}\bullet_3E)*_3E + (_1\nabla^{sn}*_3E + _3\nabla^{sn}\mathbf{x}_3K)\mathbf{x}_3K + (_1\nabla^{sn}*_3E + _3\nabla^{sn}\mathbf{x}_3K)\bullet_3E)$$
$$\pm (\ (_3\nabla^{sn}\bullet_3K)*_3E + (_1\nabla^{sn}*_3K + _3\nabla^{sn}\mathbf{x}_3E)\mathbf{x}_3K + (_1\nabla^{sn}*_3K + _3\nabla^{sn}\mathbf{x}_3E)\bullet_3E\ )$$
$$\pm -(\ (_3\nabla^{sn}\bullet_3E)*_3K + (_1\nabla^{sn}*_3E + _3\nabla^{sn}\mathbf{x}_3K)\mathbf{x}_3E + (_1\nabla^{sn}*_3E + _3\nabla^{sn}\mathbf{x}_3K)\bullet_3K\ )$$
$$- (\ (_3\nabla^{sn}\bullet_3K)*_3K + (_1\nabla^{sn}*_3K + _3\nabla^{sn}\mathbf{x}_3E)\mathbf{x}_3E + (_1\nabla^{sn}*_3K + _3\nabla^{sn}\mathbf{x}_3E)\bullet_3K\ )$$

Other terms of the force density equation are:

SPECIAL ALGEBRA FOR SPECIAL RELATIVITY

$$-_4f = -_1f + -_3f = -_1f_r + -\pm_{w1}f_i + -_3f_r + -\pm_{w3}f_i$$

$$= (-_3J)\bullet_3E \pm_w (_30)\bullet_3E$$
$$\pm_w -(-_3J)\bullet_3K - (_30)\bullet_3K$$
$$+ (((-_1J)*_3E + (-_3J)\mathbf{x}_3K) \pm_w ((0)*_3E + (_30)\mathbf{x}_3K)$$
$$\pm_w -((-_1J)*_3K + (-_3J)\mathbf{x}_3E) - ((0)*_3K + (_30)\mathbf{x}_3E))$$

This was a long derivation in which components had to have compound label-numbers tagged onto them as factors, and it might have felt ugly, but regardless of all that pain, "$-_4f$" is mathematically beautiful because it follows directly from the large-scale case Dirac Equation without artificially introducing negatives.

Concluding Remarks. The large-scale case Dirac Equation (to model dynamics of a photon) and the small-scale case Dirac Equation (to model dynamics of an electron) were developed simultaneously using the same steps. Co-development combines the photon's electromagnetic wave with the electron's matter-wave in a model in which both sets of waves originate from the same electron.

The large-scale case Dirac Equation form for Maxwell's Equations is fundamental to Maxwell's Equations because there are no ugly artificially introduced negatives in the calculation of the measurable force density invariant. The success suggests continuum/real numbers in applied mathematics actually do have finite imprecision. And it suggests the derivation of the proposed Theory of Non-Finite Numbers is valid, and that Axiomatic Set Theory was modified correctly. The modifications were: drop the *axiom of choice*, add the *reciprocal-of-zero axiom* to create an algebra for "1/0" and make "1/0" the quantity for a set, derive the Modified Continuum Hypothesis by which there are no actual infinities as quantities of sets, and define practical real numbers as rational numbers with either large-scale imprecision or else small-scale imprecision.

Electromagnetic theory was used as the testing environment for the transitioning of Axiomatic Set Theory from pure mathematics into applied mathematics. E-m theory was chosen because e-m theory is over a hundred years old and is simple because it is linear. With validity of the proposed Theory of Non-Finite Numbers now established, the proposed modifications to Axiomatic Set Theory on which it is based can now be applied in more modern theories of physics.

CHAPTER 5 - PROPOSED THEORY OF NON-FINITE NUMBERS

5.8 Spin of a Photon

A spin of one unit is appropriate for a photon.

Mathematics for spin of an electron suspended in a magnetic field is found in *quantum physics* by Stephen Gasiorowicz (Wiley, 1974), Pages 234 to 237. The electron's magnetic field (caused by rotating electric charge) is perpendicular to external field "B". A force attempts to align the electron's magnetic field with the external magnetic field by forcing the axis of rotation to become parallel, rather than perpendicular, to the external magnetic field. Mechanical angular momentum of the electron prevents alignment from happening and, instead, causes "precessing motion" in which the electron's axis of rotation rotates in the plane perpendicular to the external magnetic field.

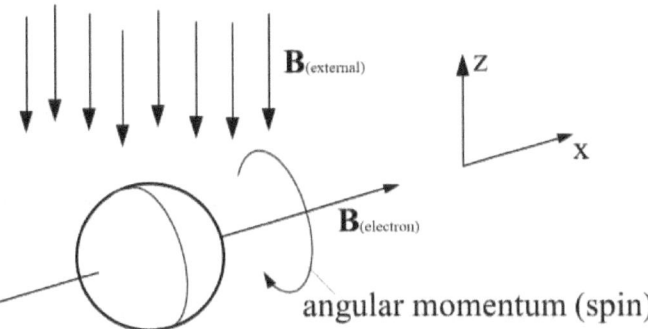

Figure 39. Electron in a magnetic field, if the electron were a ball.

The electron has frequency "ω" proportional to its total energy, "$E = \hbar*\omega$". (This frequency "ω" is not the rate of rotation of the spinning electron around its axis.)

$$\omega = B*(2*g)*(Q_B/m_B)/(4*c)$$

"Q_B" is electric charge. "m_B" is rest mass. "c" is speed-of-light. Magnetic moment "g" of an electron equals approximately "1.0011596520991...", almost one (To understand what "g" represents, please read the simple to read book: *QED, The Strange Theory of Light and Matter* by Richard P. Feynman, Princeton University Press, 1985.)

A similar system is an object in orbit around a planet. Kinetic energy perpendicular to gravity keeps the object in orbit. For both the object in orbit and the precessing electron, both potential energy and kinetic energy remain

SPECIAL ALGEBRA FOR SPECIAL RELATIVITY

constant. Another analogy for the electron is a bicycle wheel with one end of the axle on the edge of a table. Instead of the wheel falling to the ground, the wheel axle rotates with precessing motion in the plane perpendicular to gravity.

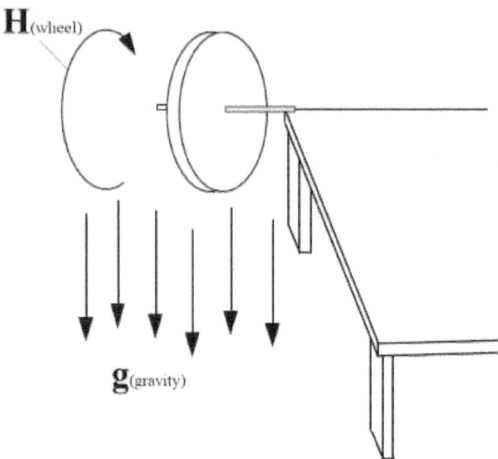

Figure 40. Bicycle wheel suspended in a gravitational field.

The electron's potential energy "$E = \hbar*\omega$" was the energy required to change the electron axis-of-rotation from parallel with "B" to perpendicular. It is proportional to "B", to "Q_B" and to speed of rotation. Speed of rotation is inversely proportional to rest mass "m_B" (assuming size stays the same and only density changes) for constant angular momentum.

Potential energy does not include "$m*c^2$" because "$_4p$" is not considered.

Gasiorowicz's textbook concludes that the electron's axis of rotation rotates in the plane perpendicular to the external magnetic field at rate "$2*\omega$". Factor "2" is due to the square operation for a particle property calculated as a spinor times its complex conjugate.

precession frequency = $2*B*Q_B*(2*g)/(4*m_B*c)) \approx B*(Q_B*/(m_B*c)))$

An electron (that is, the projection of an electron proposed to be a photon) moving at the speed-of-light may be visualized as having its axis of rotation perpendicular to the direction of motion. To justify this idealized visualization: Per quantum mechanics effects, the axis of rotation cannot be exclusively in the direction of motion. And, due to complete length contraction at the speed-of-

CHAPTER 5 - PROPOSED THEORY OF NON-FINITE NUMBERS

light, the axis of rotation of the electron projection (that is, of the photon) must be completely perpendicular to the direction of motion.

For the large-scale case (for the photon), both "m_B" and "Q_B" have a factor smaller in magnitude than the smallest positive truncated number, "< 2^-Lmax". "B", too, has a similarly small factor "< 2^-Lmax". The three similar small factors (two in the numerator and one in the denominator) form the finished-calculation of the hyperbolic radius and that hyperbolic radius is the "precession frequency $\approx B*(Q_B*/(m_B*c))$".

For the large-scale case, small factor "< 2^-Lmax" on the hyperbolic-radius is countered by a large factor (larger in magnitude than a truncated number, "> 2^Lmax") from the large-scale imprecision hyperbolic-angle, for a product that is a finite finished-calculation result. Therefore, final-result components "ω_S/c" and "k_{xS}" in "$_4\omega = 1_S*\omega_S/c + q_{xS}*k_{xS}$" are finite, so that the photon's precession frequency "ω_S" observed through measurements is finite and is not zero (regardless of the hyperbolic-radius "$B*(Q_B*/(m_B*c))$" being so small it may be considered to be a zero).

Precession frequency "ω_S" is the frequency of the electromagnetic wave. Imagine a bar magnet with its magnetic field spiraling through space pointing radially outward to create the magnetic field component of a spiral electromagnetic wave. Induction adds the electric field and increases the amplitude from infinitesimal to finite.

Per this proposed model, spin of a photon due to precessing motion has a factor of two compared to frequency of an electron. That appears to be why a photon has a spin (angular momentum) of one unit (of Planck's constant) and an electron has a spin half of one unit.

More mathematical rigor is needed, but the point is made that photon spin is not required to be zero (or a half) by the proposed Theory of Special Relativity with Non-Finite Numbers.

SPECIAL ALGEBRA FOR SPECIAL RELATIVITY

5.9 Exercises

Exercises for Text Comprehension

1) Prove square-root-of-two "$\sqrt{2}$" is irrational. Apply the algebra for "1/0" to prove "$1/0 * \sqrt{2} = 1/0$" so that "1/0" is, effectively, numerator and denominator.

2) Find the rational approximation for "$\log_2 3$" in base two for thirty-two place-value digits after the decimal point.

3) What are truncated numbers for "Lmax = 0" and "Lmax = 1"? How many truncated numbers are for "Lmax = 9", and what are the largest and smallest?

4) Write local-zero "00" for "Lmax = 9".

5) Prove a local-zero cannot equal integer zero. This proof is needed if a local-infinity is the division reciprocal of a local-zero.

6) Write Cantor's proof that both the set of natural numbers and the set of integers have the same quantity of members.

7) Write Cantor's proof that the quantity of real numbers from zero (inclusive) to one (exclusive) is a larger quantity than the quantity of natural numbers. Use place-value digit notation in base ten with at least five representative strings of place-value digits.

8) Write Cantor's Continuum Hypothesis and explain how it requires its two infinities to be actual infinities and not finite, and for real numbers to have actual infinitesimal small-scale precision with a positive actual infinity quantity of real numbers over a finite interval of the number-line.

9) Review equations for "$dP_{max0}(s)/ds$" and "$P_{max0}(s)$". Write equations for "$dP_{min0}(s)/ds$" and "P_{min0}" using "d" for the unknown and unknowable place-value digits. What is the area under curve "$dP_{min0}(s)/ds$" from negative one to positive one?

CHAPTER 5 - PROPOSED THEORY OF NON-FINITE NUMBERS

10) Prove local-real numbers form a continuum per the criteria that there be an equal chance of a randomly selected number being in any interval of the same length along the number-line.

11) In the Dirac Equation form for Maxwell's Wave Equation there is a "$\pm_N$". What does "$\pm_N$" represent?

Answers to Select Exercises.

1) Per "$p^2 = 2*q^2$", "p^2" has factor "2" and "p" has factor "2", and therefore "q^2" has factor "2" and "q" has factor "2". Our first observation is that both "p" and "q" must be even numbers.

Our second observation is that at least one of "p" or "q" must be able to be odd, and that is because "p" and "q" are in a ratio such that they both can be divided by two until one of them is odd.

The two observations are incompatible. Therefore, the original assumption "$\sqrt{2} = p/q$", with "p" and "q" formed by starting with one and adding one repeatedly, is incorrect.

Set "$p = 1/0$" and "$q = 1/0$" and apply the algebra for "$1/0$" so that both observations are satisfied by "p" and "q" both being even and, because "$1/0 = 1/0 + 1$" one of them can be odd, too.

2) $\log_{10} 11 = 1.10010101110000000001101000111001\ldots\ldots$

3) For "Lmax = 0" the count of truncated numbers is "$((1)*(1) + 1)*2 - 1 = 3$", and truncated numbers are "1", "0" and "-1". For "Lmax = 1" the count of truncated numbers is "$((2)*(2) + 1)*2 - 1 = 9$", and truncated numbers (written in base two) are "10.0", "1.1", "1.0", "0.1", "0.0", "-0.1", "-1.0", "-1.1", "-10.0". For "Lmax = 9" the count of truncated numbers is "$((2^9)*(2^9) + 1)*2 - 1 = 534389$", and the largest truncated number is (written in base two) "1000000000.0" ("512"), and the positive smallest truncated number is "0.000000001" ("1/512").

4) "00" = $0.000000000dddddd\ldots\ldots$

5) Use the probability curve for which the same number minus itself equals zero.

6) See book text

Special Algebra for Special Relativity

7) Use the table below to form new numbers, by selecting a place-value digit from each number given in the table.

.224744871	.549409757	
.581138830	.738612788	New:
.870828693	.915475947	.280309741
.121320344	.082207001	.571202909
.345207880	.240370349	

8) Cantor's Continuum Hypothesis requires that no set has a quantity of members between the quantity of members in the set of natural numbers and the quantity of members in the set of real numbers. Any number is the quantity of members in a set for which it is the largest number, and so to forbid those intermediate numbers from existing, the two identified quantities must not be finite, but, rather, actual infinity. And, the two quantities cannot be reciprocal-of-zero, per the proposed algebra for "1/0", because they are not equal and so must be positive. For the set of real numbers to have a quantity base two to actual infinity, real numbers have a positive infinitesimal difference one to the next. Because Cantor defined real numbers with a dependency on actual infinity, and because the property of actual infinity is derived from the Continuum Hypothesis, the Continuum Hypothesis forms the basis of Cantor's set of real numbers.

9) Write equations for "$dP_{min0}(s)/ds$" and "P_{min0}".

$dP_{min0}/ds = 0$ for $s/(2^{\wedge}\text{-Lmax}) < -1$
$dP_{min0}/ds = (s/(2^{\wedge}\text{-Lmax}) + 1)*2^{\wedge}\text{Lmax}$ for $-1 < s/(2^{\wedge}\text{-Lmax}) < 0$
$dP_{min0}/ds = (1 - s/(2^{\wedge}\text{-Lmax}))*2^{\wedge}\text{Lmax}$ for $0 < s/(2^{\wedge}\text{-Lmax}) < 1$
$dP_{min0}/ds = 0$ for $1 < s(2^{\wedge}\text{-Lmax})$

$P_{min0} = 0$ for $s/(2^{\wedge}\text{-Lmax}) < -1$
$P_{min0} = (s/(2^{\wedge}\text{-Lmax}) + 1)^2/2$ for $-1 < s/(2^{\wedge}\text{-Lmax}) < 0$
$P_{min0} = (2 - (1 - s/(2^{\wedge}\text{-Lmax}))^2)/2$ for $0 < s/(2^{\wedge}\text{-Lmax}) < 1$
$P_{min0} = 1$ for $1 < s/(2^{\wedge}\text{-Lmax})$

Area under "dP_{min0}/ds" curve equals one, "$P_{min0}(s = \infty) = 1$".

10) Answer not provided.

11) "$\pm_N$" "+" models movement in positive "x"-direction. "$\pm_N$" "−" models movement in the negative "x"-direction. Because of "$_N$", "$\pm_N$" says both possibilities apply together, not one or the other.

CHAPTER 5 - PROPOSED THEORY OF NON-FINITE NUMBERS

Further Thought.

1) Attempt to prove "$\sqrt{2}$" is included in Cantor's real numbers by "$N_1*\sqrt{2}$" equaling "N_1". Is this a correct check? Must "N_1" be even and so cannot be odd? Mathematics is unambiguous deductive logic yet asking to prove "$\sqrt{2}$" is in Cantor's set of real numbers introduces subjectivity and wishy-washy ambiguity that fails independent verification. Do you agree?

2) Prove two local-zeros cannot be the same by using a proof similar to the proof a local-zero cannot equal integer zero.

3) Set "T" equal to the infinite "……" sum of reciprocal natural numbers and subtract "T/2" from it, twice to derive "$\ln(2) - \ln(2) = 0$". The second "$\ln(2)$" is a quantity "1/0" of "0"'s. How can the algebra for "1/0" be expanded to include theory for this second "$\ln(2)$", in which many nothings sum to something? If we don't do this, then we are ignoring what could be a necessary expansion of our mathematical tools.

4) Einstein's two theories of Relativity each include a division by zero, called a singularity. What other mathematical models of physics include a division by zero? Can the proposed Theory of Non-Finite Numbers be applied to those mathematical models of physics?

5) Apply the proposed Theory of Non-Finite Numbers to the singularity at the center of a black hole. The General Theory of Relativity is written as "$R_{ab} - (R*g_{ab})/2 + \Lambda*g_{ab} = 8*\pi*G*T_{ab}$".

6) Unknown and unknowable place-value digits become known or knowable as time progresses, along with entropy increase, cause and effect, and collapse of the wave function. Is "now" (and "here") specific to a value of "Lmax"? How might the transition from unknowable to knowable with regard to numbers be related to the wave function collapse of quantum mechanics, and to Hugh Everett's 1957 proposed "many worlds interpretation" of quantum mechanics in which all possibilities occur in a divergence of reality. How is "now" unique, or isn't it, in a theory, yet?

SPECIAL ALGEBRA FOR SPECIAL RELATIVITY

7) With regard to the above question, if "Lmax" increases with time, then how do we reconcile "Lmax" with anti-matter if time is in reverse for observations made by anti-matter particles?

8) How does the new proposed *reciprocal-of-zero axiom* pertain to the exotic Lorentz Transformation for motion faster than the speed-of-light with respect to division by zero for the speed?

9) Using this chapter's new theory for force density, can we find a theory by which energy and momentum "Σ" for the electric field around an electron are derived through mathematics, rather than empirically derived through observations of experiments?

10) One value of "Lmax" applies, per the above models of quantum mechanics, when one particle observes another particle. Does each particle have a unique value of "Lmax" with respect to each other particle? Or, perhaps, does "Lmax" apply to a collection of particles, or to an inertial reference frame?

11) If a photon is one in the same particle as the electron that emitted it, then time-space is distorted by there being two locations for the one particle. And, the simultaneous validity of the large-scale case with the small-scale case appears to create an equivalence between a set of label-numbers in "$_S$" with singular-label-numbers in "$_M$". How is space structured?

12) Try to define properties so that "$2^{\wedge}N_1$" and "$3^{\wedge}N_1$" are equal, so that irrational number "$\log_2 3$" is a ratio of two versions of "N_1". Retain "$N_0 < 2^{\wedge}N_0 < 1/0$" with "$N_1 = 2^{\wedge}N_0$" and neither "$N_0$" or "$N_1$" have contribution after the decimal point. Try to define a positive actual infinity or two for which "$\log_2 3$" or "$\sqrt{2}$" equals the ratio of that infinity or of those infinities. The important point is to make the mathematics unambiguous, and to not depend on mysterious, vague, and remote properties in your positive actual infinity.

13) If the new proposed *reciprocal-of-zero axiom* is included in axiomatic set theory, then Cantor's Continuum Hypothesis must be rejected. Do you agree?

CHAPTER 5 - PROPOSED THEORY OF NON-FINITE NUMBERS

14) What is the definition of a real number per Descartes? What is the definition of a real number per Cantor? What is an idealized real number? What is a practical real number? What is next?

15) Speed-of-light in Special Relativity is mathematically analogous to the event horizon of a non-rotating black hole in General Relativity. The analogy is more complete if the radius of the black hole is infinite, to create a flat space event horizon. Per the theory by Hawking / Bekenstein, temperature of a black hole's surface decreases with radius of the black hole's event horizon, so that a more complete analogy of an infinite radius has zero temperature at the event horizon. Per the evolving theory of loop gravity in which General Relativity is reconciled with quantum mechanics, as developed by Eugenio Bianchi for black holes, quantum fluctuations on the surface of the black hole decrease as radius is increased. Quantum fluctuations generate temperature. Might "Lmax" (or its rate of increase) be very large for cold nearly flat space and very small for the curved event horizon of a hot small radius black hole? To keep "Lmax" finite must there be curvature in the universe? What about the other extreme, for which the smallest black hole has a diameter of the event horizon on the scale of Matvei Bronstein's Planck length, "$L_P = \sqrt{(\hbar * G/c^3)}$"? Was "Lmax = 0" before the big bang?

16) Electrons are fermions of half spin that repel each other, per the Pauli Exclusion Principle, with the example of shells of an atom. Photons are as opposite as possible. Photons are bosons of full spin that have a tendency to coincide, as in a laser. Electrons are the material and photons are the force field. Because they are different to an extreme, combining electrons with photons into one mathematical model feels strange, feels incorrect, feels like it should be impossible. And, photons are generated by protons. And photons are created by pair annihilation / pair production. How can the proposed Theory of Special Relativity with Non-Finite Numbers be generalized to include protons and pair production? What are your thoughts?

17) The theory of the Dirac Spinor matter-wave was developed from rest mass inertia which, per the classical radius model of the electron, results from the electromagnetic field. It's not surprising the electromagnetic field is also a Dirac Spinor, per the new theory presented in this book, because the two phenomena are made of the same stuff, it seems. Where next to take this theory? Can QED get updated?

18) A proposed experiment is to simultaneously measure the photon particle and the electron particle that emitted it, to check for correlated properties, in analogy with the EPR experiment. How could such an experiment be set up? What would it measure, the incident transmission rate?

19) In <u>Reality is Not What It Seems – The Journey to Quantum Gravity</u> by Carlo Rovelli, Riverhead Books, 2017 (2014 in Italian), page 245/246, is the quote below. In the two points, positive actual infinity is removed from mathematical relevance, and the quantity of information from one particle observing another particle increases with time. How can the proposed Theory of Non-Finite Numbers and modified Axiomatic Set Theory fit into quantum mechanics per this quote?

"In fact, the entire structure of quantum mechanics can be read and understood in terms of information, as follows. A physical system manifests itself only in interacting with another. The description of a physical system, then, is always given in relation to another physical system, the one with which it interacts. Any description of a system is therefore always a description of the *information* a system has about another system, that is to say, the *correlation* between the two systems. The mysteries of quantum mechanics become less dense if interpreted this way, as the description of the information that physical systems have about one another.

The description of a system, in the end, is nothing other than a way of summarizing all the past interactions with it, and using them to predict the effect of future interactions.

The entire formal structure of quantum mechanics can be in large measure expressed in two simple postulates:

    1. The relevant information in any physical system is finite
    2. You can always obtain new information on a physical system"

20) Anti-Matter Photon. Per the large-scale case Lorentz Transformation, observer "S" measures passage of time "t_S" proportional to passage of time "t_B" for the electron that emitted it. Regardless of the passage of time, a photon has complete length contraction. Also, per experiments in which two photons annihilate each other, it appears a photon is its own anti-matter particle. How can a photon be its own anti-matter particle if the direction of its passage of time is dependent on the electron that emitted it being either matter or anti-matter?

CHAPTER 5 - PROPOSED THEORY OF NON-FINITE NUMBERS

21) The Dirac Equation has been presented in this book as a singular matrix equation. First it was an algebraic equation, and then a differential equation. "Singular" means it has zeros on the right of the equals sign. The two zeros on the right side of the large-scale case Dirac Equation were replaced by non-zero value "a" to model electric charge. Perhaps the structure of the Dirac Equation is a clue to a more fundamental understanding of electric charge. Make guesses at what electric charge really is.

22) As time progresses, a photon gets absorbed and disappears, so that the large-scale case disappears as the hyperbolic-angle becomes more precise (with more zeros before the decimal point) with respect to an observing particle. The photon disappears because the hyperbolic-angle is small relative to "2^Lmax" and stays small. A succession of electron and photon emission and absorption events occurs through an increase in "Lmax". Try to make this theory quantitative to make it useful.

23) Step three in the Process from Descartes is a change from numbers to geometry in preparation to take a measurement. It is not a change to a more fundamental concept of reality. The suggestion is that numbers are fundamental to reality, and not objects like electrons and photons. What we perceive as our physical reality is the adjustment of rational numbers which become more accurate/precise relative to each other, with those numbers inside exponential functions. That adjustment appears to follow the discovered Dirac Equation, and there's likely much more complexity beyond that equation. The author (me) is biased to math (rather than physics). What do you think is fundamental, math or physics?

Special Algebra for Special Relativity

APPENDIX A - OCTONIONS AND SEDONIONS

Appendix A – Octonions and Sedonions

<u>Octonions</u> were proposed by Caylay immediately after Hamilton proposed quaternions, in 1843. Quaternions anti-commute "$j_x*j_y = -j_y*j_x = j_z$". Likewise, octonions anti-commute.

When Hamilton selected "$j_x*j_y = j_z$", he selected against equating "j_x*j_y" to "$-j_z$". For octonions, there is more than one selection required when setting up the multiplication scheme. One possible multiplication scheme for octonions is selected below. Each row is a "triple".

$j_1*j_2 = j_3$; $j_2*j_4 = j_6$; $j_3*j_6 = j_5$; $j_4*j_3 = j_7$; $j_5*j_1 = j_4$; $j_6*j_7 = j_1$; $j_7*j_5 = j_2$

123
246
365
437
514
671
752

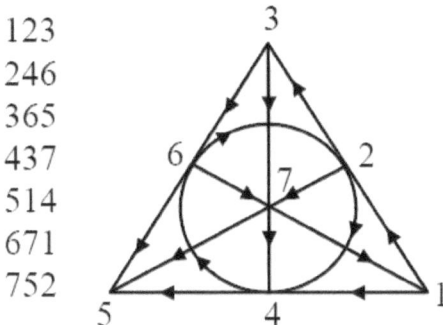

Figure 41. Traditional triangle model for octonion multiplication. Note the lack of symmetry in the arrow directions.

*	1	j1	j2	j3
1	1*1 = 1	1*j1 = j1	1*j2 = j2	1*j3 = j3
j1	j1*1 = j1	j1*j1 = -1	j1*j2 = j3	j1*j3 = -j2
j2	j2*1 = j2	j2*j1 = -j3	j2*j2 = -1	j2*j3 = j1
j3	j3*1 = j3	j3*j1 = j2	j3*j2 = -j1	j3*j3 = -1
j4	j3*1 = j3	j4*j1 = -j5	j4*j2 = -j6	j4*j3 = j7
j5	j3*1 = j3	j5*j1 = j4	j5*j2 = j7	j5*j3 = j6
j6	j3*1 = j3	j6*j1 = -j7	j6*j2 = j4	j6*j3 = -j5
j7	j3*1 = j3	j7*j1 = j6	j7*j2 = -j5	j7*j3 = -j4

SPECIAL ALGEBRA FOR SPECIAL RELATIVITY

*	j4	j5	j6	j7
1	1*j4 = j4	1*j5 = j5	1*j6 = j6	1*j7 = j7
j1	j1*j4 = j5	j1*j5 = -j4	j1*j6 = j7	j1*j7 = -j6
j2	j2*j4 = j6	j2*j5 = -j7	j2*j6 = -j4	j2*j7 = j5
j3	j3*j4 = -j7	j3*j5 = -j6	j3*j6 = j5	j3*j7 = j4
j4	j4*j4 = -1	j4*j5 = j1	j4*j6 = j2	j4*j7 = -j3
j5	j5*j4 = -j1	j5*j5 = -1	j5*j6 = -j3	j5*j7 = -j2
j6	j6*j4 = -j2	j6*j5 = j3	j6*j6 = -1	j6*j7 = j1
j7	j7*j4= j3	j7*j5 = j2	j7*j6 = -j1	j7*j7 = -1

Table 4. Multiplication table for octonions.

0	1	2	3	4	5	6	7
1	-0	3	-2	5	-4	7	-6
2	-3	-0	1	6	-7	-4	5
3	2	-1	-0	-7	-6	5	4
4	-5	-6	7	-0	1	2	-3
5	4	7	6	-1	-0	-3	-2
6	-7	4	-5	-2	3	-0	1
7	6	-5	-4	3	2	-1	-0

Table 5. Short form multiplication table for octonions.

Criteria of an algebraic group for octonions "$O = \{1, j_1, j_2, j_3, j_4, j_5, j_6, j_7\}$" with respect to multiplication, "$\{O, *(\text{anti-commute})\}$":

- Closure: No holes in the 16x16 multiplication table that has negatives included

- Identity: Identity element is integer one

- Commutative Property: Applies, but with two different of the seven octonions anti-commuting

APPENDIX A - OCTONIONS AND SEDONIONS

- Associative Property: Applies, but with three different octonions (that are not all three in the same triple) anti-associating

- Inverse: Ratio of any two numbers is in the set of numbers

An inverse property example is "j_1" divided by "j_4". First, substitute two factors for the numerator, then, second, apply the (anti-)commutative law (if needed) followed by the (anti-)associative law (if needed) to form a ratio of like numbers. The ratio of like numbers is then replaced with the number one. Each combination must equal "$j_1 * -j_4$".

$$j_1/j_4 = (j_2*j_3)/j_4 = (j_2*(j_7*j_4))/j_4 = -((j_2*j_7)*j_4)/j_4$$
$$= -(j_2*j_7)*(j_4/j_4) = -(j_2*j_7) = -j_5$$

$$j_1/j_4 = (j_4*j_5)/j_4 = -(j_5*j_4)/j_4 = -j_5*(j_4/j_4) = -j_5$$

$$j_1/j_4 = (j_6*j_7)/j_4 = (j_6*(j_4*j_3))/j_4 = -(j_6*(j_3*j_4))/j_4$$
$$= (j_6*j_3)*(j_4/j_4) = (j_6*j_3) = -(j_3*j_6) = -j_5$$

Degradations. Each higher order of hypercomplexity includes a new breakdown in symmetry through introduction of another negative. A break-down in symmetry is called (in this book) a "degradation". The pattern of degradations only goes to the fourth order, if only because the degradation of the fourth order is too severe to go on.

- Real numbers (hypercomplex order "N = 0") have no degradation (other than their unboundedness which implies a lack of closure)

- <u>Anti-Identity.</u> Complex numbers ("N = 1") have the complex number label-number square to negative one

    $i^2 = -1$

- <u>Anti-Commutative.</u> Quaternions ("N = 2")

    $j_y*j_z = j_x = -(-j_x) = -j_z*j_y$

- <u>Anti-Associative.</u> Octonions ("N = 3")

    $(j_1*j_2)*j_4 = j_3*j_4 = -j_7 = -j_1*j_6 = -j_1*(j_2*j_4)$

SPECIAL ALGEBRA FOR SPECIAL RELATIVITY

- **Anti-Inverse.** Sedonions ("N = 4")

$$j_{14s}/j_{6s} = (j_{8s}*j_{6s})/j_{6s} = j_{8s}*(j_{6s}/j_{6s}) = j_{8s}$$
$$= -(-j_{8s}) = -(j_{13s}*j_{5s}) = -(j_{13s}*j_{5s})*(j_{6s}/j_{6s}) = -((j_{13s}*j_{5s})*j_{6s})/j_{6s}$$
$$= (j_{13s}*(j_{5s}*j_{6s}))/j_{6s} = -(j_{13s}*(j_{6s}*j_{5s}))/j_{6s} = -(j_{13s}*j_{3s})/j_{6s}$$
$$= -j_{14s}/j_{6s}$$

Because "$j_{14s}/j_{6s} = -j_{14s}/j_{6s}$", it is said sedonion algebra is not a division algebra. (Alternatively, we may say sedonion algebra is not yet properly specified.)

Matrix Isomorphs for Octonions. Pauli Spin Matrices, multiplied or divided by "i", result in the below traditional set of 2x2 matrix isomorphs for Hamilton's quaternions.

$$1 = \begin{pmatrix} 1 & 0 \\ 0 & 1 \end{pmatrix} \quad j_x = \begin{pmatrix} 0 & i \\ i & 0 \end{pmatrix} \quad j_y = \begin{pmatrix} 0 & -1 \\ 1 & 0 \end{pmatrix} \quad j_z = \begin{pmatrix} i & 0 \\ 0 & -i \end{pmatrix}$$

The above four matrices are (traditionally) placed into a general matrix multiplication scheme using complex-conjugate "$*i$" ("$i^{*i} = -i$") on the right-side column-vector.

$$\begin{pmatrix} a & -b^{*i} \\ b & a^{*i} \end{pmatrix} * \begin{pmatrix} c & -d^{*i} \\ d & c^{*i} \end{pmatrix} = \begin{pmatrix} a*c - b^{*i}*d & -a*d^{*i} - b^{*i}*c^{*i} \\ b*c + a^{*i}*d & -b*d^{*i} + a^{*i}*c^{*i} \end{pmatrix} = \begin{pmatrix} e & -f^{*i} \\ f & e^{*i} \end{pmatrix}$$

Letters "a" and "b" (also "c" and "d", and "e" and "f") correspond to terms of the 2x2 matrix isomorphs of the quaternions by these substitutions:

1: $a = 1, b = 0$; j_x: $a = 0, b = i$; j_y: $a = 0, b = 1$; j_z: $a = i, b = 0$

Factors are rearranged because complex numbers commute.

$$\begin{pmatrix} a & -b^{*i} \\ b & a^{*i} \end{pmatrix} * \begin{pmatrix} c & -d^{*i} \\ d & c^{*i} \end{pmatrix} = \begin{pmatrix} a*c - d*b^{*i} & -d^{*i}*a - b^{*i}*c^{*i} \\ c*b + a^{*i}*d & -b*d^{*i} + c^{*i}*a^{*i} \end{pmatrix} = \begin{pmatrix} e & -f^{*i} \\ f & e^{*i} \end{pmatrix}$$

APPENDIX A - OCTONIONS AND SEDONIONS

The arrangement above is more correct because "$e^{*i} = -b*d^{*i} + c^{*i}*a^{*i}$" has factors in each term reversed compared to "$e = a*c - d*b^{*i}$". The new arrangement, with the reversed order of factors and terms in the product 2x2 matrix, specifies a new multiplication scheme between two 2x2 matrices.

The new multiplication scheme is necessary for factors that do not commute, that is, if factors are quaternions. Hypercomplex-conjugate operation "^{*j}" replaces complex-conjugate operation "^{*i}". "$j_x^{*j} = -j_x$", "$j_y^{*j} = -j_y$", and "$j_z^{*j} = -j_z$" with factors reversed in order.

The below matrix multiplication operation applies to matrices with quaternion hypercomplex numbers as elements/terms of the matrix.

$$\begin{vmatrix} a & -b^{*j} \\ b & a^{*j} \end{vmatrix} * \begin{vmatrix} c & -d^{*j} \\ d & c^{*j} \end{vmatrix} = \begin{vmatrix} a*c - d*b^{*j} & -d^{*j}*a - b^{*j}*c^{*j} \\ c*b + a^{*j}*d & -b*d^{*j} + c^{*j}*a^{*j} \end{vmatrix} = \begin{vmatrix} e & -f^{*j} \\ f & e^{*j} \end{vmatrix}$$

1: $a = 1, b = 0$; j_2: $a = j_x, b = 0$; j_4: $a = j_y, b = 0$; j_6: $a = j_z, b = 0$
j_1: $a = 0, b = 1$; j_3: $a = 0, b = j_x$; j_5: $a = 0, b = j_y$; j_7: $a = 0, b = j_z$

(The above 2x2 matrix equation for octonion matrix isomorphs was discovered by the author because a search did not find a previously known set of matrix isomorphs for octonions.)

Examples for "$j_2*j_4 = j_6$" and "$j_4*j_3 = j_7$", respectively.

$$\begin{vmatrix} j_x & 0 \\ 0 & -j_x \end{vmatrix} * \begin{vmatrix} j_y & 0 \\ 0 & -j_y \end{vmatrix} = \begin{vmatrix} j_x*j_y & 0 \\ 0 & -(-j_y*j_x) \end{vmatrix} = \begin{vmatrix} j_z & 0 \\ 0 & -j_z \end{vmatrix}$$

$$\begin{vmatrix} j_y & 0 \\ 0 & -j_y \end{vmatrix} * \begin{vmatrix} 0 & j_x \\ j_x & 0 \end{vmatrix} = \begin{vmatrix} 0 & j_x*j_y \\ -j_y*j_x & 0 \end{vmatrix} = \begin{vmatrix} 0 & j_z \\ j_z & 0 \end{vmatrix}$$

"j_x" can be replaced with its 2x2 matrix isomorph and "i" and "1" by their 2x2 matrix isomorphs to create 8x8 matrices (but with a messy multiplication scheme).

$$i = \begin{vmatrix} 0 & -1 \\ 1 & 0 \end{vmatrix} \qquad 1 = \begin{vmatrix} 1 & 0 \\ 0 & 1 \end{vmatrix}$$

SPECIAL ALGEBRA FOR SPECIAL RELATIVITY

General Rule for Matrix Multiplication with non-Commuting Elements.

$$\begin{pmatrix} A & C \\ B & D \end{pmatrix} * \begin{pmatrix} E & G \\ F & H \end{pmatrix} = \begin{pmatrix} A*E + F*C & G*A + C*H \\ E*B + D*F & B*G + H*D \end{pmatrix}$$

Vector-Aft Multiplication:

$$\begin{pmatrix} A & C \\ B & D \end{pmatrix} * \begin{pmatrix} E \\ F \end{pmatrix} = \begin{pmatrix} A*E + F*C \\ E*B + D*F \end{pmatrix}$$

Vector-Front Multiplication:

$$\begin{pmatrix} E & \\ F & \end{pmatrix} * \begin{pmatrix} A & C \\ B & D \end{pmatrix} = \begin{pmatrix} E*A + C*F \\ B*E + F*D \end{pmatrix}$$

Vector-front multiplication anti-commutes when compared to vector-aft multiplication. For example: In "$j_5*j_3 = j_6$" use the left column of "j_3" to result in the left column of "j_6". Now, reverse the order of the factors to "j_3*j_5". Again, only use the left column of "j_3". Per the vector-front multiplication operation, the result of "j_3*j_5" is negative of the left column of "j_6".

As a side note, we will not use the below alternative.

$$\begin{pmatrix} A & C \\ B & D \end{pmatrix} * \begin{pmatrix} E & G \\ F & H \end{pmatrix} = \begin{pmatrix} E*A + C*F & A*G + H*C \\ B*E + F*D & G*B + D*H \end{pmatrix} \quad \text{Do not use}$$

Rotation of a Seven-Dimensional Object.
Seven-dimensional ultra-space modeled with octonions is organized as seven three-dimensional spaces so that there are three planes of rotation around each axis of rotation.

Rotation around the "j_1" axis has rotation in "j_2"/ "j_3", "j_4"/ "j_5", and "j_6"/ "j_7" planes. Rotation angles in each plane are "θ_{23}", "θ_{45}", and "θ_{67}", respectively. "θ_{23}" begins at positive "j_2" axis and is measured towards positive "j_3" axis. (Rotations in 7-d space might be new with this book.)

$$j_{1o1} = \begin{pmatrix} 0 & -1 \\ 1 & 0 \end{pmatrix}$$

$$j_{2o1} = \begin{pmatrix} j_x*\cos\theta_{23} & j_x*\sin\theta_{23} \\ j_x*\sin\theta_{23} & -j_x*\cos\theta_{23} \end{pmatrix} \quad ; \quad j_{3o1} = \begin{pmatrix} -j_x*\sin\theta_{23} & j_x*\cos\theta_{23} \\ j_x*\cos\theta_{23} & j_x*\sin\theta_{23} \end{pmatrix}$$

APPENDIX A - OCTONIONS AND SEDONIONS

$$j_{4o1} = \begin{matrix} j_y*\cos\theta_{45} & j_y*\sin\theta_{45} \\ j_y*\sin\theta_{45} & -j_y*\cos\theta_{45} \end{matrix} \quad ; \quad j_{5o1} = \begin{matrix} -j_y*\sin\theta_{45} & j_y*\cos\theta_{45} \\ j_y*\cos\theta_{45} & j_y*\sin\theta_{45} \end{matrix}$$

$$j_{6o1} = \begin{matrix} j_z*\cos\theta_{67} & j_z*\sin\theta_{67} \\ j_z*\sin\theta_{67} & -j_z*\cos\theta_{67} \end{matrix} \quad ; \quad j_{7o1} = \begin{matrix} -j_z*\sin\theta_{67} & j_z*\cos\theta_{67} \\ j_z*\cos\theta_{67} & j_z*\sin\theta_{67} \end{matrix}$$

Multiplications "$j_{1o1}*j_{2o1} = j_{3o1}$", "$j_{5o1}*j_{1o1} = j_{4o1}$", and "$j_{6o1}*j_{7o1} = j_{1o1}$" do not involve angle addition, and, therefore, are relatively trivial. The other four triples "$j_{2o1}*j_{4o1} = j_{6o1}$", "$j_{3o1}*j_{6o1} = j_{5o1}$", "$j_{4o1}*j_{3o1} = j_{7o1}$", and "$j_{7o1}*j_{5o1} = j_{2o1}$" involve addition of angles. "$j_{3o1}*j_{6o1}$":

$$j_{3o1}*j_{6o1} = \begin{matrix} -j_x*\sin\theta_{23} & j_x*\cos\theta_{23} \\ j_x*\cos\theta_{23} & j_x*\sin\theta_{23} \end{matrix} * \begin{matrix} j_z*\cos\theta_{67} & j_z*\sin\theta_{67} \\ j_z*\sin\theta_{67} & -j_z*\cos\theta_{67} \end{matrix}$$

$$j_{3o1}*j_{6o1} = \begin{matrix} (-j_x*j_z)*\sin\theta_{23}*\cos\theta_{67} + (j_z*j_x)*\cos\theta_{23}*\sin\theta_{67} & (j_z*-j_x)*\sin\theta_{23}*\sin\theta_{67} + (j_x*-j_z)*\cos\theta_{23}*\cos\theta_{67} \\ (j_z*j_x)*\cos\theta_{23}*\cos\theta_{67} + (j_x*j_z)*\sin\theta_{23}*\sin\theta_{67} & (j_x*j_z)*\cos\theta_{23}*\sin\theta_{67} + (-j_z*j_x)*\sin\theta_{23}*\cos\theta_{67} \end{matrix}$$

$$j_{3o1}*j_{6o1} = \begin{matrix} j_y*(\sin\theta_{23}*\cos\theta_{67} + \cos\theta_{23}*\sin\theta_{67}) & -j_y*(\sin\theta_{23}*\sin\theta_{67} - \cos\theta_{23}*\cos\theta_{67}) \\ j_y*(\cos\theta_{23}*\cos\theta_{67} - \sin\theta_{23}*\sin\theta_{67}) & -j_y*(\cos\theta_{23}*\sin\theta_{67} + \sin\theta_{23}*\cos\theta_{67}) \end{matrix}$$

$$j_{3o1}*j_{6o1} = \begin{matrix} j_y*\sin(\theta_{23} + \theta_{67}) & j_y*\cos(\theta_{23} + \theta_{67}) \\ j_y*\cos(\theta_{23} + \theta_{67}) & -j_y*\sin(\theta_{23} + \theta_{67}) \end{matrix}$$

$$j_{3o1}*j_{6o1} = \begin{matrix} -j_y*\sin\theta_{45} & j_y*\cos\theta_{45} \\ j_y*\cos\theta_{45} & j_y*\sin\theta_{45} \end{matrix} = j_{5o1}$$

SPECIAL ALGEBRA FOR SPECIAL RELATIVITY

The above matrix multiplication that resulted in "$j_{3o1}*j_{6o1} = j_{5o1}$" required "$\theta_{23} + \theta_{45} + \theta_{67} = 0$". "$\theta_{23} + \theta_{45} + \theta_{67} = 0$" is also valid for "$j_{2o1}*j_{4o1} = j_{6o1}$", "$j_{4o1}*j_{3o1} = j_{7o1}$", and "$j_{5o1}*j_{2o1} = j_{7o1}$".

$$j_{2o1}*j_{4o1} = \begin{array}{ll} j_z*(\cos\theta_{23}*\cos\theta_{45} - \sin\theta_{23}*\sin\theta_{45}) & -j_z*(-\cos\theta_{23}*\sin\theta_{45} - \sin\theta_{23}*\cos\theta_{45}) \\ j_z*(-\sin\theta_{23}*\cos\theta_{45} - \cos\theta_{23}*\sin\theta_{45}) & -j_z*(\sin\theta_{23}*\sin\theta_{45} - \cos\theta_{23}*\cos\theta_{45}) \end{array}$$

$$j_{2o1}*j_{4o1} = \begin{array}{ll} j_z*\cos(\theta_{23} + \theta_{45}) & -j_z*\sin(\theta_{23} + \theta_{45}) \\ -j_z*\sin(\theta_{23} + \theta_{45}) & -j_z*\cos(\theta_{23} + \theta_{45}) \end{array}$$

$$j_{2o1}*j_{4o1} = \begin{array}{ll} j_z*\cos\theta_{67} & j_z*\sin\theta_{67} \\ j_z*\sin\theta_{67} & -j_z*\cos\theta_{67} \end{array} = j_{6o1}$$

······ ······ ······ ······ ······ ······ ······ ······ ······ ······ ······ ······

$$j_{4o1}*j_{3o1} = \begin{array}{llll} j_y*\cos\theta_{45} & j_y*\sin\theta_{45} & -j_x*\sin\theta_{23} & j_x*\cos\theta_{23} \\ j_y*\sin\theta_{45} & -j_y*\cos\theta_{45} & j_x*\cos\theta_{23} & j_x*\sin\theta_{23} \end{array}$$

$$j_{4o1}*j_{3o1} = \begin{array}{ll} (j_y*\text{-}j_x*\cos\theta_{45}*\sin\theta_{23} + j_x*j_y*\sin\theta_{45}*\cos\theta_{23}) & (j_x*j_y*\cos\theta_{45}*\cos\theta_{23} + j_y*j_x*\sin\theta_{45}*\sin\theta_{23}) \\ (-j_x*j_y*\sin\theta_{45}*\sin\theta_{23} + -j_y*j_x*\cos\theta_{45}*\cos\theta_{23}) & (j_y*j_x*\sin\theta_{45}*\cos\theta_{23} + j_x*\text{-}j_y*\cos\theta_{45}*\sin\theta_{23}) \end{array}$$

$$j_{4o1}*j_{3o1} = \begin{array}{ll} j_z*(\cos\theta_{45}*\sin\theta_{23} + \sin\theta_{45}*\cos\theta_{23}) & j_z*(\cos\theta_{45}*\cos\theta_{23} - \sin\theta_{45}*\sin\theta_{23}) \\ j_z*(-\sin\theta_{45}*\sin\theta_{23} + \cos\theta_{45}*\cos\theta_{23}) & j_z*(-\sin\theta_{45}*\cos\theta_{23} - \cos\theta_{45}*\sin\theta_{23}) \end{array}$$

$$j_{4o1}*j_{3o1} = \begin{array}{ll} j_z*\sin(\theta_{45} + \theta_{23}) & j_z*\cos(\theta_{45} + \theta_{23}) \\ j_z*\cos(\theta_{45} + \theta_{23}) & -j_z*\sin(\theta_{45} + \theta_{23}) \end{array}$$

APPENDIX A - OCTONIONS AND SEDONIONS

$$j_{4o1} * j_{3o1} = \begin{matrix} -j_z*\sin\theta_{67} & j_z*\cos\theta_{67} \\ j_z*\cos\theta_{67} & j_z*\sin\theta_{67} \end{matrix} = j_{7o1}$$

....

A seven-dimensional object is modeled as seven sticks.

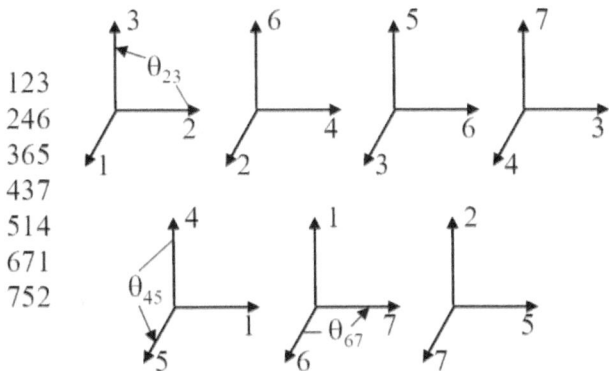

123
246
365
437
514
671
752

Figure 42a. Seven-dimensional space as seven three-dimensional spaces.

Some sticks are rotated ninety degrees and comply with "$\theta_{23} + \theta_{45} + \theta_{67} = 0$", per the illustration below.

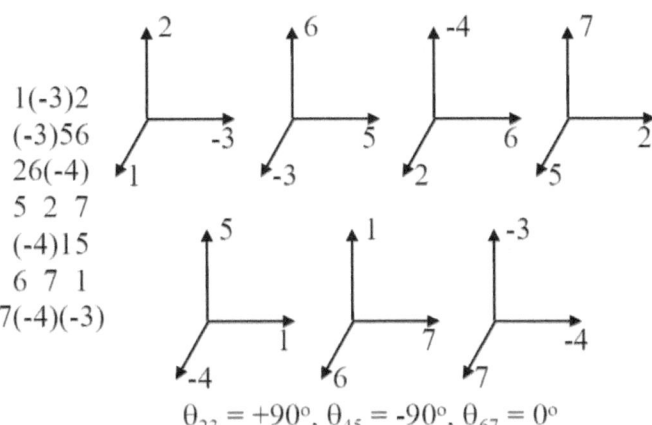

1(-3)2
(-3)56
26(-4)
5 2 7
(-4)15
6 7 1
7(-4)(-3)

$\theta_{23} = +90°, \theta_{45} = -90°, \theta_{67} = 0°$

Figure 42b. Rotation in 2-3 plane causes a counter rotation in 4-5 plane.

SPECIAL ALGEBRA FOR SPECIAL RELATIVITY

$$j_{5o1}*j_{2o1} = \begin{matrix} j_z*(\sin\theta_{45}*\cos\theta_{23} + \cos\theta_{45}*\sin\theta_{23}) \\ j_z*(-\sin\theta_{45}*\sin\theta_{23} + \cos\theta_{45}*\cos\theta_{23}) \\ j_z*(\cos\theta_{45}*\cos\theta_{23} - \sin\theta_{45}*\sin\theta_{623}) \\ j_z*(-\cos\theta_{45}*\sin\theta_{23} - \sin\theta_{45}*\cos\theta_{23}) \end{matrix}$$

$$j_{5o1}*j_{2o1} = \begin{matrix} j_z*\sin(\theta_{45} + \theta_{23}) & j_z*\cos(\theta_{45} + \theta_{23}) \\ j_z*\cos(\theta_{45} + \theta_{23}) & -j_z*\sin(\theta_{45} + \theta_{23}) \end{matrix}$$

$$j_{5o1}*j_{2o1} = \begin{matrix} -j_z*\sin\theta_{67} & j_z*\cos\theta_{67} \\ j_z*\cos\theta_{67} & j_z*\sin\theta_{67} \end{matrix} = j_{7o1}$$

A seven-dimensional object can also rotate around an axis in three-dimensional space, for example, the y-axis. The below substitutions may be visualized using 4x4 matrix isomorphs for octonions.

$$j_{yqy} = \begin{matrix} 0 & -1 \\ 1 & 0 \end{matrix} \; ; \; j_{zqy} = \begin{matrix} i*\cos\theta_{zx} & i*\sin\theta_{zx} \\ i*\sin\theta_{zx} & -i*\cos\theta_{zx} \end{matrix} \; ; \; j_{xqy} = \begin{matrix} -i*\sin\theta_{zx} & i*\cos\theta_{zx} \\ i*\cos\theta_{zx} & i*\sin\theta_{zx} \end{matrix}$$

Unlike octonions and quaternions, there is no rotation possible for complex numbers, because there is only one label number for complex numbers.

General Model for Label-Numbers.

- Real numbers "N = 0" have the trivial 1x1 matrix isomorph.

- Complex numbers "N = 1" have "$2^N = 2^1 = 2$" label-numbers, "1" and "i", and a 2x2 matrix isomorph.

- Quaternions "N = 2" have "$2^N = 2^2 = 4$" label-numbers and a 4x4 matrix isomorph.

$$(1 + j_x)*(1 + j_y) = 1 + j_x + j_y + j_x*j_y \qquad = (1 + j_y)*(1 + j_z)$$
$$= 1 + j_x + j_y + j_z \qquad = (1 + j_z)*(1 + j_x)$$

APPENDIX A - OCTONIONS AND SEDONIONS

- Octonions "N = 3" have "$2^N = 2^3 = 8$" label-numbers and an 8x8 matrix isomorph.

$$(1 + j_1 + j_5 + j_6)*(1 + j_2) = 1 + j_1 + j_2 + j_3 + j_4 + j_5 + j_6 + j_7$$
$$(1 + j_2 + j_4 + j_5)*(1 + j_3) = 1 + j_1 + j_2 + j_3 + j_4 + j_5 + j_6 + j_7$$

There are twenty-one restructures of the two octonion equations above, one restructure for each number in the 7x3 twenty-one number table.

- Sedonions "N = 4" have sixteen label-numbers. The multiplication table for sedonions is not yet finalized.

Vector-Space. Vector space is so far useful in applied mathematics to "N = 2". A more complex notion of vector space may be needed for "N = 3" and higher orders.

- "N = 1" introduced complex label-number "i". Subtract away the real number so the quantity of label-numbers in an "N = 1" vector-space is "$2^N - 1 = 1$".

- Four-dimensional time-space of Special Relativity was created by combining "N = 1" with "N = 2" to form the "N = 2" vector-space: "$(2^1 - 1) + (2^2 - 1) = 1 + 3 = 4$" terms. An example number is "$658*i + 89*j_x + 57*j_y + 456*j_z$".

- The natural extension to octonions is eleven-dimensional vector-space: "$11 = 1 + 3 + 7 = (2^1 - 1) + (2^2 - 1) + (2^3 - 1)$". The seven dimensions of ultra-space are as different from space as space is from time. The example number is "$658*i + 89*j_x + 57*j_y + 456*j_z + 26*j_1 + 44*j_2 + 785*j_3 + 963*j_4 + 76*j_5 + 659*j_6 + 154*j_7$".

- "N = 4" creates a twenty-six-dimensional vector-space "$26 = 1 + 3 + 7 + 15$".

- After that is "N = 5" for "$57 = 1 + 3 + 7 + 15 + 31$", for a fifty-seven-dimensional vector-space, and so on.

SPECIAL ALGEBRA FOR SPECIAL RELATIVITY

<u>Sedonions</u>. A (too) simple attempt at sedonion algebra applies the same matrix multiplication operation applied to octonion algebra. The octonion-conjugate operation "*^{*jo}*" requires the negative of the octonions, "$j_1^{*jo} = -j_1$", ..., and "$j_7^{*jo} = -j_7$", and requires reverse order of factors.

$$\begin{vmatrix} a & -b^{*jo} \\ b & a^{*jo} \end{vmatrix} * \begin{vmatrix} c & -d^{*jo} \\ d & c^{*jo} \end{vmatrix} = \begin{vmatrix} a*c - d*b^{*jo} & -d^{*jo}*a - b^{*jo}*c^{*jo} \\ c*b + a^{*jo}*d & -b*d^{*jo} + c^{*jo}*a^{*jo} \end{vmatrix} = \begin{vmatrix} e & -f^{*jo} \\ f & e^{*jo} \end{vmatrix}$$

1: $a=1, b=0$; j_{2s}: $a=j_1, b=0$; j_{4s}: $a=j_2, b=0$; j_{6s}: $a=j_3, b=0$
j_{1s}: $a=0, b=1$; j_{3s}: $a=0, b=j_1$; j_{5s}: $a=0, b=j_2$; j_{7s}: $a=0, b=j_3$

j_{8s}: $a=j_4, b=0$; j_{10s}: $a=j_5, b=0$; j_{12s}: $a=j_6, b=0$; j_{14s}: $a=j_7, b=0$
j_{9s}: $a=0, b=j_4$; j_{11s}: $a=0, b=j_5$; j_{13s}: $a=0, b=j_6$; j_{15s}: $a=0, b=j_7$

0	1	2	3	4	5	6	7	8	9	10	11	12	13	14	15
1	-0	3	-2	5	-4	7	-6	9	-8	11	-10	13	-12	15	-14
2	-3	-0	1	6	-7	-4	5	10	-11	-8	9	14	-15	-12	13
3	2	-1	-0	-7	-6	5	4	-11	-10	9	8	-15	-14	13	12
4	-5	-6	7	-0	1	2	-3	12	-13	-14	15	-8	9	10	-11
5	4	7	6	-1	-0	-3	-2	-13	-12	15	14	9	8	-11	-10
6	-7	4	-5	-2	3	-0	1	-14	15	-12	13	10	-11	8	-9
7	6	-5	-4	3	2	-1	-0	15	14	13	12	-11	-10	-9	-8
8	-9	-10	11	-12	13	14	-15	-0	1	2	-3	4	-5	-6	7
9	8	11	10	13	12	-15	-14	-1	-0	-3	-2	-5	-4	7	6
10	-11	8	-9	14	-15	12	-13	-2	3	-0	1	-6	7	-4	5
11	10	-9	-8	-15	-14	-13	-12	3	2	-1	-0	7	6	5	4
12	-13	-14	15	8	-9	-10	11	-4	5	6	-7	-0	1	2	-3
13	12	15	14	-9	-8	11	10	5	4	-7	-6	-1	-0	-3	-2
14	-15	12	-13	-10	11	-8	9	6	-7	4	-5	-2	3	-0	1
15	14	-13	-12	11	10	9	8	-7	-6	-5	-4	3	2	-1	-0

Table 6. Short-hand multiplication table for sedonions (first attempt).

APPENDIX A - OCTONIONS AND SEDONIONS

$j_{1s}*j_{2s}=j_{3s}$; $j_{2s}*j_{4s}=j_{6s}$; $j_{3s}*j_{6s}=j_{5s}$; $j_{4s}*j_{3s}=j_{7s}$; $j_{7s}*j_{5s}=j_{2s}$
$j_{1s}*j_{4s}=j_{5s}$; $j_{2s}*j_{8s}=j_{10s}$; $j_{3s}*j_{10s}=j_{9s}$; $j_{8s}*j_{3s}=j_{11s}$; $j_{11s}*j_{9s}=j_{2s}$
$j_{1s}*j_{6s}=j_{7s}$; $j_{2s}*j_{12s}=j_{14s}$; $j_{3s}*j_{14s}=j_{13s}$; $j_{12s}*j_{3s}=j_{15s}$; $j_{15s}*j_{13s}=j_{2s}$
$j_{1s}*j_{8s}=j_{9s}$; $j_{4s}*j_{8s}=j_{12s}$; $j_{5s}*j_{12s}=j_{9s}$; $j_{8s}*j_{5s}=j_{13s}$; $j_{13s}*j_{9s}=j_{4s}$
$j_{1s}*j_{10s}=j_{11s}$; $j_{12s}*j_{10s}=j_{6s}$; $j_{13s}*j_{6s}=j_{11s}$; $j_{10s}*j_{13s}=j_{7s}$; $j_{7s}*j_{11s}=j_{12s}$
$j_{1s}*j_{12s}=j_{13s}$; $j_{14s}*j_{8s}=j_{6s}$; $j_{15s}*j_{6s}=j_{9s}$; $j_{8s}*j_{15s}=j_{7s}$; $j_{7s}*j_{9s}=j_{14s}$
$j_{1s}*j_{14s}=j_{15s}$; $j_{14s}*j_{10s}=j_{4s}$; $j_{15s}*j_{4s}=j_{11s}$; $j_{10s}*j_{15s}=j_{5s}$; $j_{5s}*j_{11s}=j_{14s}$

For rotation around the "j_{1s}" axis, define seven rotation angles "θ_{23s}", "θ_{45s}", "θ_{67s}", "θ_{89s}", "θ_{1011s}", "θ_{1213s}" and "θ_{1415s}". These angles are analogous to the previously defined angles "θ_{23}", "θ_{45}", and "θ_{67}" in the discussion on octonions.

$$j_{1s1} = \begin{matrix} 0 & -1 \\ 1 & 0 \end{matrix}$$

$$j_{2s1} = \begin{matrix} j_1*\cos\theta_{23s} & j_1*\sin\theta_{23s} \\ j_1*\sin\theta_{23s} & -j_1*\cos\theta_{23s} \end{matrix} \quad ; \quad j_{3s1} = \begin{matrix} -j_1*\sin\theta_{23s} & j_1*\cos\theta_{23s} \\ j_1*\cos\theta_{23s} & j_1*\sin\theta_{23s} \end{matrix}$$

$$j_{4s1} = \begin{matrix} j_2*\cos\theta_{45s} & j_2*\sin\theta_{45s} \\ j_2*\sin\theta_{45s} & -j_2*\cos\theta_{45s} \end{matrix} \quad ; \quad j_{5s1} = \begin{matrix} -j_2*\sin\theta_{45s} & j_2*\cos\theta_{45s} \\ j_2*\cos\theta_{45s} & j_2*\sin\theta_{45s} \end{matrix}$$

$$j_{6s1} = \begin{matrix} j_3*\cos\theta_{67s} & j_3*\sin\theta_{67s} \\ j_3*\sin\theta_{67s} & -j_3*\cos\theta_{67s} \end{matrix} \quad ; \quad j_{7s1} = \begin{matrix} -j_3*\sin\theta_{67s} & j_3*\cos\theta_{67s} \\ j_3*\cos\theta_{67s} & j_3*\sin\theta_{67s} \end{matrix}$$

$$j_{8s1} = \begin{matrix} j_4*\cos\theta_{89s} & j_4*\sin\theta_{89s} \\ j_4*\sin\theta_{89s} & -j_4*\cos\theta_{89s} \end{matrix} \quad ; \quad j_{9s1} = \begin{matrix} -j_4*\sin\theta_{89s} & j_4*\cos\theta_{89s} \\ j_4*\cos\theta_{89s} & j_4*\sin\theta_{89s} \end{matrix}$$

$$j_{10s1} = \begin{matrix} j_5*\cos\theta_{1011s} & j_5*\sin\theta_{1011s} \\ j_5*\sin\theta_{1011s} & -j_5*\cos\theta_{1011s} \end{matrix} \quad ; \quad j_{11s1} = \begin{matrix} -j_5*\sin\theta_{1011s} & j_5*\cos\theta_{1011s} \\ j_5*\cos\theta_{1011s} & j_5*\sin\theta_{1011s} \end{matrix}$$

$$j_{12s1} = \begin{matrix} j_6*\cos\theta_{1213s} & j_6*\sin\theta_{1213s} \\ j_6*\sin\theta_{1213s} & -j_6*\cos\theta_{1213s} \end{matrix} \quad ; \quad j_{13s1} = \begin{matrix} -j_6*\sin\theta_{1213s} & j_6*\cos\theta_{1213s} \\ j_6*\cos\theta_{1213s} & j_6*\sin\theta_{1213s} \end{matrix}$$

SPECIAL ALGEBRA FOR SPECIAL RELATIVITY

$$j_{14s1} = \begin{matrix} j_7*\cos\theta_{1415s} & j_7*\sin\theta_{1415s} \\ j_7*\sin\theta_{1415s} & -j_7*\cos\theta_{1415s} \end{matrix} \quad ; \quad j_{15s1} = \begin{matrix} -j_7*\sin\theta_{1415s} & j_7*\cos\theta_{1415s} \\ j_7*\cos\theta_{1415s} & j_7*\sin\theta_{1415s} \end{matrix}$$

Octonions: x-y-z j_x j_y j_z
             θ_{23} θ_{45} θ_{67}
1-2-3 j_2 j_4 j_6
2-4-6 j_3 j_5 j_7
3-6-5

Sedonions: 4-3-7 j_1 j_2 j_3 j_4 j_5 j_6 j_7
            5-1-4 θ_{23s} θ_{45s} θ_{67s} θ_{89s} θ_{1011s} θ_{1213s} θ_{1415s}
            6-7-1 j_{2s} j_{4s} j_{6s} j_{8s} j_{10s} j_{12s} j_{14s}
            7-5-2 j_3 j_{5s} j_{7s} j_{9s} j_{11s} j_{13s} j_{15s}

The octonion/sedonion table above shows multiplication of "$j_1*j_2 = j_3$" internal to sedonion 2x2 matrices is analogous to "$j_x*j_y = j_z$" internal to octonion 2x2 matrices. Also "$j_2*j_4 = j_6$", "$j_5*j_1 = j_4$", and "$j_6*j_7 = j_1$" are analogous to "$j_x*j_y = j_z$" because numbers increase in value: 1-2-3, 2-4-6, 1-4-5, and 1-6-7. The other three triples are not: 7-4-3, 6-5-4, and 7-5-2, because numbers decrease in value.

Each of seven triples pertains to a plane perpendicular to the "1" axis. Each plane is associated with four triples.

$\theta_{23s} + \theta_{45s} + \theta_{67s} = 0,$ (1-2-3)
$j_{2s1}*j_{4s1} = j_{6s1}$; $j_{3s1}*j_{6s1} = j_{5s1}$; $j_{4s1}*j_{3s1} = j_{7s1}$; $j_{7s1}*j_{5s1} = j_{2s1}$

$\theta_{23s} + \theta_{89s} + \theta_{1011s} = 0,$ (1-4-5)
$j_{2s1}*j_{8s1} = j_{10s1}$; $j_{3s1}*j_{10s1} = j_{9s1}$; $j_{8s1}*j_{3s1} = j_{11s1}$; $j_{11s1}*j_{9s1} = j_{2s1}$

$\theta_{23s} + \theta_{1213s} + \theta_{1415s} = 0,$ (1-6-7)
$j_{2s1}*j_{12s1} = j_{14s1}$; $j_{3s1}*j_{14s1} = j_{13s1}$; $j_{12s1}*j_{3s1} = j_{15s1}$; $j_{15s1}*j_{13s1} = j_{2s1}$

$\theta_{45s} + \theta_{89s} + \theta_{1213s} = 0,$ (2-4-6)
$j_{4s1}*j_{8s1} = j_{12s1}$; $j_{5s1}*j_{12s1} = j_{9s1}$; $j_{8s1}*j_{5s1} = j_{13s1}$; $j_{13s1}*j_{9s1} = j_{4s1}$

$\theta_{1213s} + \theta_{1011s} + \theta_{67s} = 0,$ (6-5-3)
$j_{12s1}*j_{10s1} = j_{6s1}$; $j_{13s1}*j_{6s1} = j_{11s1}$; $j_{10s1}*j_{13s1} = j_{7s1}$; $j_{7s1}*j_{11s1} = j_{12s1}$

APPENDIX A - OCTONIONS AND SEDONIONS

$\theta_{1415s} + \theta_{89s} + \theta_{67s} = 0,$ (7-4-3)

$j_{14s1}*j_{8s1} = j_{6s1}$; $j_{15s1}*j_{6s1} = j_{9s1}$; $j_{8s1}*j_{15s1} = j_{7s1}$; $j_{7s1}*j_{9s1} = j_{14s1}$

$\theta_{1415s} + \theta_{1011s} + \theta_{45s} = 0,$ (7-5-2)

$j_{14s1}*j_{10s1} = j_{4s1}$; $j_{15s1}*j_{4s1} = j_{11s1}$; $j_{10s1}*j_{15s1} = j_{5s1}$; $j_{5s1}*j_{11s1} = j_{14s1}$

Angles "θ_{s23}", "θ_{s45}", "θ_{s67}", "θ_{s89}", "θ_{s1011}", "θ_{s1213}" and "θ_{s1415}" must all be zero for the seven angle equations to each equal zero, and that means the math is incorrect as a model for rotation in fifteen-dimensional ultra-ultra space. The four normal planes can be thought of as right-hand rotation. The three anti-normal planes can be thought of as left-hand rotation. The switch of hands made rotation impossible.

To investigate left-hand planes: "$j_{6s}*j_{9s} = j_{15s}$" uses "$j_4*j_3 = j_7$" because "$j_{6s}*j_{9s}$" has the product "$j_3*j_4 = -j_7$" and "j_4*j_3" has the product "$j_y*j_x = -j_z$". "$j_{6s}*j_{9s}$":

$$j_{6s1}*j_{9s1} = \begin{matrix} j_3*\cos\theta_{67s} & j_3*\sin\theta_{67s} \\ j_3*\sin\theta_{67s} & -j_3*\cos\theta_{67s} \end{matrix} * \begin{matrix} -j_4*\sin\theta_{89s} & j_4*\cos\theta_{89s} \\ j_4*\cos\theta_{89s} & j_4*\sin\theta_{89s} \end{matrix}$$

$$j_{6s1}*j_{9s1} = \begin{matrix} (j_3*-j_4*\cos\theta_{67s}*\sin\theta_{89s} + j_4*j_3*\sin\theta_{67s}*\cos\theta_{89s}) \\ (j_4*j_3*\cos\theta_{67s}*\cos\theta_{89s} + j_3*j_4*\sin\theta_{67s}*\sin\theta_{89s}) \\ (-j_4*j_3*\sin\theta_{67s}*\sin\theta_{89s} + -j_3*j_4*\cos\theta_{67s}*\cos\theta_{89s}) \\ (j_3*j_4*\sin\theta_{67s}*\cos\theta_{89s} + j_4*-j_3*\cos\theta_{67s}*\sin\theta_{89s}) \end{matrix}$$

$$j_{6s1}*j_{9s1} = \begin{matrix} j_7*(\cos\theta_{67s}*\sin\theta_{89s} + \sin\theta_{67s}*\cos\theta_{89s}) \\ j_7*(\cos\theta_{67s}*\cos\theta_{89s} - \sin\theta_{67s}*\sin\theta_{89s}) \\ j_7*(-\sin\theta_{67s}*\sin\theta_{89s} + \cos\theta_{67s}*\cos\theta_{89s}) \\ -j_7*(\sin\theta_{67s}*\cos\theta_{89s} + \cos\theta_{67s}*\sin\theta_{89s}) \end{matrix}$$

$$j_{6s1}*j_{9s1} = \begin{matrix} j_7*\sin(\theta_{67s} + \theta_{89s}) & j_7*\cos(\theta_{67s} + \theta_{89s}) \\ j_7*\cos(\theta_{67s} + \theta_{89s}) & -j_7*\sin(\theta_{67s} + \theta_{89s}) \end{matrix}$$

$$j_{6s1}*j_{9s1} = \begin{matrix} -j_7*\sin\theta_{1415s} & j_7*\cos\theta_{1415s} \\ j_7*\cos\theta_{1415s} & j_7*\sin\theta_{1415s} \end{matrix} = j_{15s1}$$

SPECIAL ALGEBRA FOR SPECIAL RELATIVITY

Based on that example, left hand angle equations apply: "$\theta_{1213s} + \theta_{1011s} + \theta_{67s} = 0$", "$\theta_{1415s} + \theta_{89s} + \theta_{67s} = 0$", and "$\theta_{1415s} + \theta_{1011s} + \theta_{45s} = 0$", such that there should not be negative.

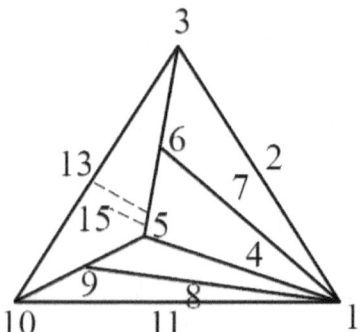

Figure 43. A failed attempt at a tetrahedron triangle diagram for "N = 4" multiplication-division. Fill in circles and arrows and find the task impossible.

Anti-inverse degradation of the above sedonion algebra may be experienced algebraically by attempting to create a 15x7 table for sedonion 7-spaces analogous to the 7x3 table for octonion 3-spaces. A 15x7 table for sedonions would divide fifteen-dimensional space into fifteen different seven-dimensional spaces.

The lack of a division algebra for this simple attempt at sedonion algebra is proven in the Hurwitz Theorem.

See the last two "Thought Exercises" questions for the beginning of other attempts for 2x2 matrix formulation of sedonion algebra.

Thought Exercises

1) Write a 2x2 matrix isomorph for octonion "$a + b*j_1 + c*j_2 + d*j_3 + e*j_4 + f*j_5 + g*j_6 + h*j_7$".

2) Write a matrix multiplication table for octonions "j_1", "j_2", "j_3", "j_4", "j_5", "j_6", "j_7" and complex number factor "i". Use "q_1", "q_2", "q_3", "q_4", "q_5", "q_6", "q_7" with "$q_1 = j_1/i$", etc. Address negatives with a note.

3) Write the 2x2 matrix isomorph of the octonion-complex number "$a + b*q_1 + c*q_2 + d*q_3 + e*q_4 + f*q_5 + g*q_6 + h*q_7$" in which components "a", "b", "c", "d", "e", "f", "g", and "h" are mathematically complex.

APPENDIX A - OCTONIONS AND SEDONIONS

4) In Chapter 1 is a section titled "Quaternion-Complex-Hypercomplex Numbers". Follow the format of that section for octonions and octonions-complex. Why can't octonions-quaternions or sedonions be placed into that format?

5) "j_1", "j_2", "j_3", "j_4", "j_5", "j_6", "j_7" were each assigned a 2x2 matrix isomorph, per the text above. Find all alternative 2x2 matrix isomorphs to "j_1", "j_2", "j_3", "j_4", "j_5", "j_6", "j_7" holding to the criteria that the same 7x3 table of triples applies? Are there other tables?

6) There is no algebra by which to find real number magnitude of "$658*i + 89*j_x + 57*j_y + 456*j_z + 26*j_1 + 44*j_2 + 785*j_3 + 963*j_4 + 76*j_5 + 659*j_6 + 154*j_7$" because quaternion-octonion products have no specified algebra. This issue will likely get resolved through an application in applied mathematics. Find that application.

7) Find octonion 2x2x2 matrix isomorphs that use complex label-numbers "1" and "i" for terms/elements. The major diagonal of a 2x2x2 matrix would likely be elements "$(a_r + i*a_i)$" and "$(a_r + i*a_i)^{*i}$", in analogy to the 2x2 matrix isomorphs for quaternions. The three minor diagonals would likely each be analogous to the one minor diagonal of the 2x2 matrix isomorphs for quaternions. The challenge is to develop a 2x2x2 matrix multiplication scheme for which the 7x3 table of triples for octonion multiplication is satisfied.

$(a_r + i*a_i) \quad (b_r + i*b_i)$
$\qquad\qquad\qquad\qquad (d_r + i*d_i) \quad -(c_r + i*c_i)^{*i}$
$(c_r + i*c_i) \quad -(d_r + i*d_i)^{*i}$
$\qquad\qquad\qquad\qquad -(b_r + i*b_i)^{*i} \quad (a_r + i*a_i)^{*i}$

1: $a_r = 1$; j_2: $c_r = 1$; j_4: $a_i = 1$; j_6: $c_i = 1$
j_1: $b_r = 1$; j_3: $d_r = 1$; j_5: $b_i = 1$; j_7: $d_i = 1$

8) Try to find a 2x2x2x2 matrix structure for sedonions. Four corners are adjacent to each of the two corners of the major diagonal. And six corners are not adjacent to the two corners.

SPECIAL ALGEBRA FOR SPECIAL RELATIVITY

9) For QCD (quarks), propose an expansion of the Dirac Equation that uses octonion matrix isomorphs in addition to, or as a substitute for, quaternion 2x2 matrix isomorphs. A clue in the text was justification for "$e_4*q_x = -q_x*e_4$". Another clue is quaternions inside matrix isomorphs of octonions. If that is successful, use practical real numbers to propose a model for gluon quantum dynamics. A difficulty is including tight ultra-space curvature analogous to general relativity.

10) The 7x3 table for octonions implies the first column is an "x" dimension, the second "y", and the third "z". But that implication is not used in the algebra. It suggests the algebra is incomplete. What concept is missing?

11) Using triples "$j_{13s} = j_{3s}*j_{14s}$", "$j_{5s} = j_{3s}*j_{6s}$", "$j_{13s} = j_{8s}*j_{5s}$" and "$j_{6s} = j_{14s}*j_{8s}$", the violation of the anti-associative property that led to the anti-inverse property is:

$$j_{14s} = -(j_{3s})*j_{13s} = -(j_{6s}*j_{5s})*j_{13s}$$

$$j_{14s} = -j_{6s}*(j_{8s}) = -j_{6s}*(j_{5s}*j_{13s}) = +(j_{6s}*j_{5s})*j_{13s}$$

Can matrix multiplication be modified to remove the anti-associative violation? See the matrix multiplication operation below, in which new groupings apply (for example, "F(CI)" rather than "(FC)I"). The new groupings would affect octonions and not quaternions, and the operations become trinary and not binary.

```
    A C    E G    I K      (AE+FC)I+J(GA+CH)  K(AE+FC)+(GA+CH)L
((      )*(    ))*(    ) =
    B D    F H    J L      I(EB+DF)+(BH+HD)J  (EB+DF)K+L(BG+HD)

                           (AE)I+(FC)I+J(GA)+J(CH)  K(AE)+K(FC)+(GA)L+(CH)L
                         =
                           I(EB)+I(DF)+(BH)J+(HD)J  (EB)K+(DF)K+L(BG)+L(HD)
```

12) Can sedonion algebra be redesigned to include octonion rotation in 7-dimensional space when there is a sedonion rotation in 15-dimensional space, such that octonions inside the matrix isomorphs change to make the sedonions change, too? Like the above attempt to make sedonion algebra work, it changes matrix isomorphs, and so is more than simple sedonion label-number manipulations, and that means the restrictive Hurwitz Theorem does not apply.

Appendix B – Spooky Action at a Distance

<u>EPR Experiment Set Up</u>. Two photons created as a pair have coordinated properties and are called "entangled". The two entangled photons travel a macroscopic distance, perhaps a meter or further in opposite directions, and are each detected. Each detector is comprised of a polarizing film followed by a photographic film. The two polarizing films are parallel and both photographic films detect the photons, to show polarity of the photons was the same, as expected, because they were entangled.

To explain this expected experiment result, we might venture to guess polarity of the two photons was determined at the time of creation at the emission source that sits between the two detectors.

Other phases of the EPR experiment have the polarizing films at different angles. Results are interpreted using Bell's Inequality (from year 1964) and lead to the conclusion polarization of the two entangled photons is determined at the moment of detection, not emission. For polarization to be created at the moment of detection, the direction of polarization is coordinated over the macroscopic distance that separates the two detectors and coordinated instantly over that macroscopic distance: Not at the speed-of-light but faster, instantly. Einstein called the instant communication "spooky action at a distance".

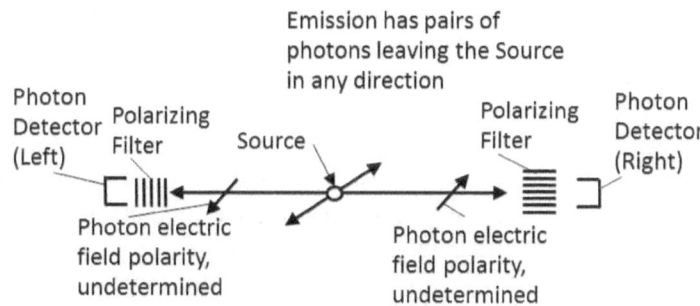

Figure 44. EPR Experiment for photons.

The EPR experiment may also be performed with electrons. Electrons do not have polarization, but, rather, electrons have spin. Axis of spin (using the right-hand rule) is measured as a direction perpendicular to the plane of rotation. Two entangled electrons have the axis of spin of one electron anti-parallel with respect to the other electron, as required for a total of zero angular momentum for the pair.

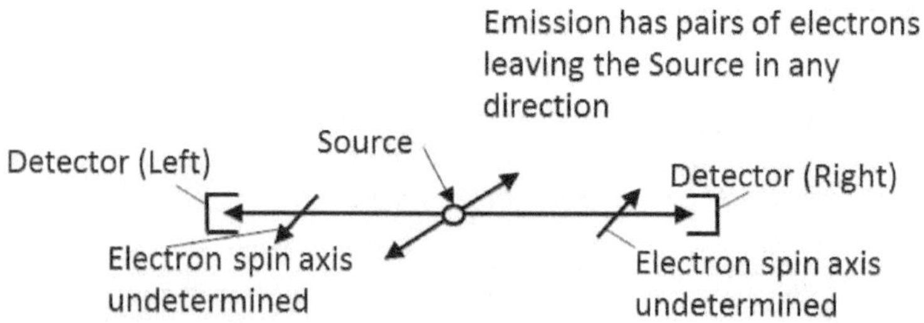

SPECIAL ALGEBRA FOR SPECIAL RELATIVITY

Figure 45. EPR Experiment for electrons.

For a visualization, substitute baseballs for electrons. Two baseballs were resting beside each other. A little expanding spring sent the baseballs in opposite directions. The little expanding spring was not perfectly aligned between the two centers, and, therefore, created baseball spin. The axis of spin must be equal and opposite, so that the angular momentum of both baseballs together equals zero.

We watched the release of the expanding spring. We saw the spin of the two baseballs as they moved in opposite directions. We watched the two baseballs each pass their spin direction detector. The spin detector display confirmed what we had been seeing.

A baseball is a particle, always a particle. A baseball is a particle before its spin is detected, and the baseball is a particle when its spin is detected. Electrons are different from baseballs because electrons have particle/wave duality. An electron is a wave until it is detected as a particle. Unlike particles, waves do not have spin. Therefore, waves do not have an axis for spin. Per particle/wave duality of quantum mechanics, axis for spin of an electron can only become specified when the electron transitions from being an unobserved wave into being an observed particle, a transition that occurs at the detector.

Photons, too, have particle/wave duality. Polarization of a photon is a property of the particle. For visualization, imagine the wave of a single photon approaching a polarizing film. Polarizing film is comprised of long stretched molecules that have thin non-conductive gaps between molecules. The photon's electric field can only oscillate electrons parallel to the long molecules, in which case, an electron absorbs the energy of the photon and then loses that energy to heat. Therefore, a photon will pass the film only if the photon's electric field is perpendicular to the long, stretched molecules. In contrast, if the electric field is exactly parallel to the gaps, then stretched molecules absorb the photon.

APPENDIX B - SPOOKY ACTION AT A DISTANCE

On the back side of the polarizing film is a photographic film detector with an electron that absorbs the photon with a jiggle vibration, if the photon passed the polarizing film to reach it. The jiggle of the electron causes an exposure spot on photographic film.

The polarizing film forced the single non-divisible photon to be completely polarized perpendicular to long molecules (such that it passes) or completely polarized parallel to molecules (such that it is absorbed). The photon, as a particle, cannot be a combination of perpendicular and parallel, because the photon is quantized as an all-or-nothing particle. Half the photons pass the polarizing film, and half are absorbed.

An electron, when used in the EPR experiment, likewise, is a single particle at the detector. Electron spin sensed by the detector forces the spin to become parallel or else anti-parallel to the detector device (analogous to the photon electric field being forced to be parallel or else perpendicular to the polarizing film of the detector device) with no intermediate spin direction possible.

Lack of the particle property of polarization for photons and lack of the particle property of spin direction for electrons, up until the moment of detection, was a debated issue, until the EPR experiment with Bell's Inequality settled the debate.

Bell's Inequality is explained with an example. Consider a set of bikes in a large garage. Each bike has (A) an engine or not, has (B) five pounds of fuel or not, and has (C) blue paint or not. Notice that correlated properties may be included. For example, it doesn't matter that a bike without an engine would naturally not have fuel.

Three properties "A", "B", and "C" each have values of "Yes" or "No", for eight combination possibilities (because $8 = 2^3$). Each quantity (QTY) is found by counting.

QTY1(YYY), QTY2(NYY), QTY3(YNY), QTY4(YYN)
QTY5(YNN), QTY6(NYN), QTY7(NNY), QTY8(NNN)

(YYN) means A=Yes, B=Yes, and C=No, and (NYN) means A=No, B=Yes, and C=No. If a letter A, B, or C is in the parentheses, then that property can be either Yes or No. For example, QTY(AYY) has A=Yes or No, B=Yes, and C=Yes. If there are ten objects, then the eight quantities listed above sum (as QTY(ABC)) to ten.

As an example, consider this set of ten objects:

SPECIAL ALGEBRA FOR SPECIAL RELATIVITY

(YNN), (YYN), (YNN), (NYN), (NNN),
(YYY), (YNN), (NYN), (YYN), (NNY)

QTY1(YYY)=1   QTY5(YNN)=3   QTY2(NYY)=0   QTY6(NYN)=2
QTY3(YNY)=0   QTY7(NNY)=1   QTY4(YYN)=2   QTY8(NNN)=1

Bell's Inequality is

QTY(YNC) + QTY(AYN) ≥ QTY(YBN)

In words: The quantity of objects with A=Yes and B=No plus the quantity of objects with B=Yes and C=No is greater than or equal to the quantity of objects with A=Yes and C=No.

The first term QTY(YNC) has no consideration as to the state of "C". It can be expanded, as can the other two terms:

QTY(YNC) = QTY3(YNY) + QTY5(YNN)

QTY(AYN) = QTY4(YYN) + QTY6(NYN)

QTY(YBN) = QTY4(YYN) + QTY5(YNN)

QTY(YNC) + QTY(AYN) ≥ QTY(YBN)

QTY3(YNY) + QTY5(YNN) + QTY4(YYN) + QTY6(NYN)
    ≥ QTY4(YYN) + QTY5(YNN)

QTY3(YNY) + QTY6(NYN) ≥ 0

The last statement is true because all quantities, including QTY3(YNY) and QTY6(NYN), are greater than or equal to zero. Therefore, Bell's Inequality is proven.

Using the ten objects of our example:

QTY(YNC) = (0+3) = 3
QTY(AYN) = (2+2) = 4
QTY(YBN) = (2+3) = 5

APPENDIX B - SPOOKY ACTION AT A DISTANCE

Because 3 + 4 = 7 ≥ 5, Bell's Inequality is satisfied. In words per the example: The number of bikes with an engine and less that five pounds of fuel plus the number of bikes with more than five pounds of fuel and no blue paint is greater than or equal to the number of bikes with an engine and no blue paint. In words the inequality is not obvious or intuitive. That is why we need the math.

The important point about Bell's Inequality is that it must be satisfied if the objects have properties. If Bell's Inequality is not satisfied, then the objects do not have properties.

Predicting How Bell's Inequality Applies to a Baseball. In the plane perpendicular to direction of motion is the center point of the baseball and through that center point is a line for the projection of the axis of rotation onto the plane. Use the right-hand rule so that the axis of rotation projected onto this plane has an arrowhead designating the direction of the thumb.

On the plane perpendicular to motion, from the center point of the baseball, draw the "x"-axis and, at ninety degrees, the "y"-axis.

Angle "θ" from the "x"-axis to the arrowhead is always known for the baseball because, for a baseball, the direction of the axis of rotation exists even when we are not detecting it.

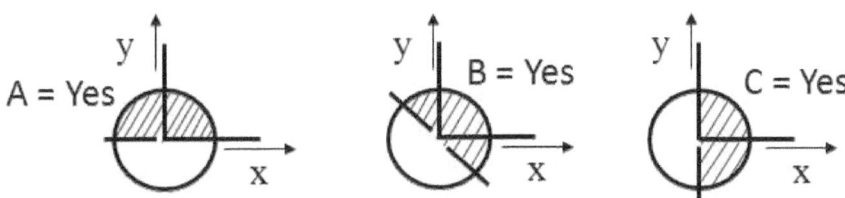

Figure 46. Properties A, B, C used in Bell's Inequality.

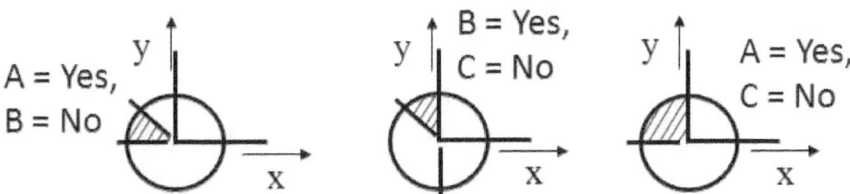

Figure 47. Combined properties used in Bell's Inequality: QTY(YNC), QTY(AYN), QTY(YBN), respectively. QTY(YNC) + QTY(AYN) ≥ QTY(YBN).

SPECIAL ALGEBRA FOR SPECIAL RELATIVITY

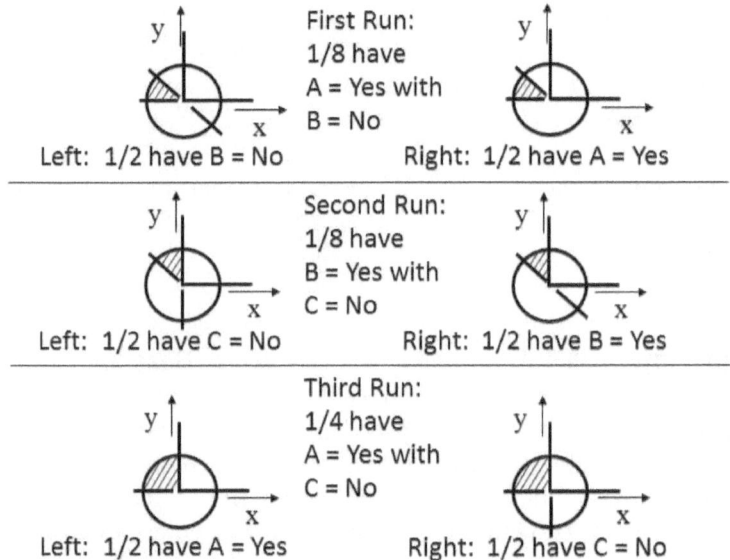

Figure 48. EPR experiment results with baseballs and not with electrons.

"A" is "Yes" if axis of rotation is in the hemisphere of "+y", $0 \leq \theta < 180°$, and "No" corresponds to $180° \leq \theta < 360°$.

"B" is "Yes" if axis of rotation is in the hemisphere with both "+x" and "+y", $-45° \leq \theta < 135°$, and "No" corresponds to $135° \leq \theta < 315°$.

"C" is "Yes" for "+x", $-90° \leq \theta < 90°$, and "No" for $90° \leq \theta < 270°$.

Each baseball pair is given "Yes" or "No" for each of A, B, and C.

Baseballs that satisfy A=Yes and B=No have $135° \leq \theta < 180°$ (one eighth of a circle). Baseballs that satisfy B=Yes and C=No have $90° \leq \theta < 135°$ (one eighth of a circle). Baseballs that satisfy A=Yes and C=No have $90° \leq \theta < 180°$ (one quarter of a circle). One eighth plus one eighth equals one quarter, and therefore Bell's Inequality is satisfied.

We do the experiment (hypothetically) three times.
The first set of 1,000,000 has 125,032 A=Yes with B=No.
The second set of 1,000,000 has 124,992 B=Yes with C=No.
The third set of 1,000,000 has 250,005 A=Yes with C=No.

QTY(YNC) + QTY(AYN) ≥ QTY(YBN)
125,032 + 124,992 = 250,024 ≥ 250,005

APPENDIX B - SPOOKY ACTION AT A DISTANCE

Bell's Inequality was satisfied. Because the three numbers were from three different experiment events, there was a chance a statistically explainable variation could have made "QTY(YNC) + QTY(AYN)" slightly less than "QTY(YBN)". That possibility alerts us to the need to have a very large number of runs in the experiment.

Predicting How Bell's Inequality Does Not Apply to an Electron. The right detector (which is the first to make a detection) reads if the axis of rotation of the electron is parallel with the "+y"-axis. The left detector has the same reading as the right detector (after compensating for the expected axis of rotation to be anti-parallel).

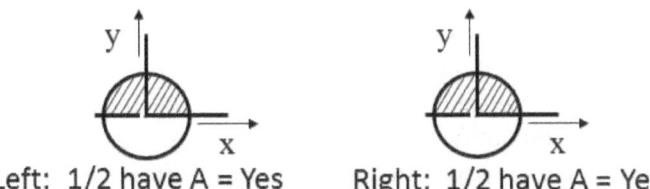

Left: 1/2 have A = Yes   Right: 1/2 have A = Yes

Figure 49. Both detectors set to "A". Electrons are detected 50% for "A" equal to "Yes" in both.

Now, in the second phase of the experiment, the left detector (second detector) is rotated so that it reads if the axis of rotation is parallel (or else anti-parallel) with "+x"-axis. This reading is instantaneously after the right detector ("+y"-axis) reading. The word "instantaneous" means a signal at the speed-of-light would not have time to reach the second detector.

Per quantum mechanics, the event of reading "A" (if the axis of rotation is in the "+y" direction) forces the electron's axis of rotation to be exactly in the "+y" direction, or else in the "-y" direction, and nothing else.

The second detection for reading "C" (on the left, an instant later) forces the rotation axis of the entangled electron to now move parallel or else anti-parallel to "+x"-axis. "+x"-axis is perpendicular to both "+y"-axis and "-y"-axis. Therefore, half the detections for the second detection (on left) read "+x"-axis and half the detections read "-x"-axis.

SPECIAL ALGEBRA FOR SPECIAL RELATIVITY

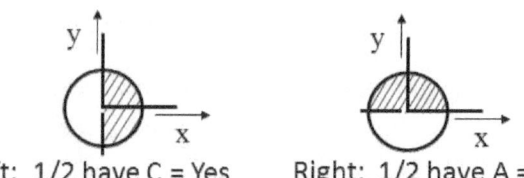

Left: 1/2 have C = Yes   Right: 1/2 have A = Yes

Figure 50. Left detector set to "C". Electrons are detected 50% for "C" equal to "Yes". QTY(YBY) equals one quarter of the total.

Now, as a third phase in the experiment, have the first (right) detector aligned with "+y"-axis, and have the second (left) detector aligned with "-y"-axis. In this case, the left detector reads a "No" for each "Yes" on the right detector. This is an extreme case for which angle "φ" (phi) equals zero.

The general rule is that the second (left) detector has probability "$\sin^2(\phi/2)$" of a detection with "φ" measured from the (opposite) direction of the previous measurement of the entangled pair of electrons. If the two detectors are parallel (the first phase), then "φ=180°" so that "$\sin^2(\phi/2) = \sin^2(180°/2) = \sin^2(90°) = 1^2 = 1$". If the two detectors are anti-parallel (the third phase), then "φ=0°" so that "$\sin^2(\phi/2) = \sin^2(0°/2) = 0$". And, if the two detectors are perpendicular (the second phase), then "φ=90°" so that "$\sin^2(\phi/2) = \sin^2(90°/2) = \sin^2(45°) = (1/\sqrt{2})^2 = 1/2$". The last two phases of the experiment have "φ=45°" (so that "$\sin^2(\phi/2) = \sin^2(45°/2) = \sin^2(22.5°) = .38268...^2 = 0.146447...$") and "φ=135°" (so that $\sin^2(\phi/2) = \sin^2(135°/2) = \sin^2(67.5°) = .92388...^2 = 0.85355...$").

All particles are accounted for between what is counted as in-conformance and not-in-conformance, per "$\cos^2(\phi/2) + \sin^2(\phi/2) = 1$".

The EPR Experiment with Electrons. Do the experiment (hypothetically) with pairs of electrons. There are three experimental set ups.

"73,111" events in "1,000,000" have A=Yes with B=No. The number "73,111" conforms to "$1,000,000*0.5*\sin^2(45°/2)$". The "0.5" factor applies to A=Yes and the factor "$\sin^2(45°/2) = 0.146447...$" applies to the subsequent detection for B=No.

After that, "73,056" events have B=Yes with C=No.

And then, "249,986" events have A=Yes with C=No.

We do a statistical evaluation to explain away minor error in the numbers and conclude Bell's inequality is not satisfied because:

QTY(YNC) + QTY(AYN) $\not\geq$ QTY (YBN) for electrons
73,111 + 73,056 = 146,167 $\not\geq$ 249,986

APPENDIX B - SPOOKY ACTION AT A DISTANCE

In the laboratory, Bell's inequality was not satisfied using photons, electrons, protons, or any other subatomic particles.

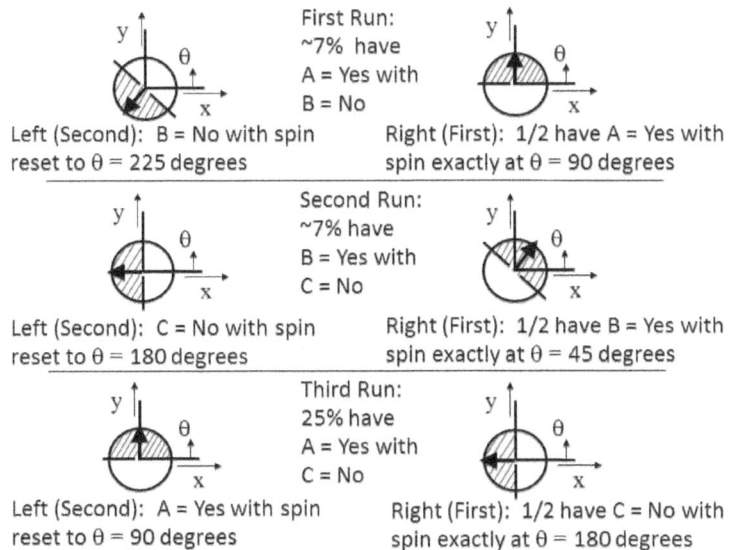

Figure 51. EPR Experiment with electrons. Bell's Inequality is not satisfied because 7% + 7% ≱ 25%.

Properties Determined at Instant of Detection. We expected Bell's inequality to be unsatisfied because of the "$\sin^2(\phi/2)$" rule, which was derived from the requirement that spin of an electron becomes parallel or else anti-parallel with the axis of measurement (which makes an electron different from a baseball). Because Bell's Inequality is unsatisfied, we conclude there are no hidden variables or other memory mechanisms by which direction of spin of an electron was determined when emitted and simply carried as information to the detectors. It means direction of spin is determined at the moment of detection.

A typical emission is one particle, not two entangled particles as in the EPR experiment. When this one particle is detected, perhaps by your eye, or perhaps by a scientific device that we were monitoring, we are unsure if that particle traversed the distance from its emission time and place to the detection time and place as a particle with distinct particle properties (for example, a spin axis direction), or not. When the particle travels it travels as a wave, and a wave should not be a carrier of particle properties, but subatomic particles are a bit mysterious. Before we had EPR experiment results, it was thought perhaps waves did carry particle properties. And that is because, when detecting one particle, we did not have a means to know if spin was established at the time of emission or at the time of detection. Now, because of entanglement, we know.

Instantaneous Information Travel. The second conclusion is that information traveled a macroscopic distance instantaneously. There is a macroscopic distance between the two detectors, and those two detectors each detect spin direction simultaneously. Direction of spin is determined only at the instant of detection, and that means readings on the two detectors are mysteriously coordinated one to the other. It means information traveled from one detector to the other at instantaneous speed, not as fast as speed-of-light but faster, faster to the ultimate speed: instantaneous, "$v = 1/0$".

For the typical situation of one emitted particle (and not two entangled particles), spooky action at a distance is a "collapse of the wave function", in which the wave disappears from everywhere the particle is not (relative to an observing particle), when the particle is detected. The challenge is to explain spooky action at a distance with a mathematical model.

New Theory. An electron and the observed photon from that electron are both one-in-the-same particle, per the proposed Theory of Special Relativity with Non-Finite Numbers. Entangled particles, too, are one particle: A whole light cone for two entangled photons is one-in-the-same as the two electrons at its center, relative to an observing particle, until an observation event (or lack of observation event) of a photon is made by an observing particle. More rigor is needed, but the one particle theory has promise in explaining EPR experiment results.

It is through the EPR experiment that quantum mechanics is brought to a macroscopic scale, to where we can physically see how strange quantum mechanics is. Therefore, naturally, we want to use the EPR experiment to see with our own eyes any proposed new theory.

Appendix C – Discovering an Abstraction

About the Author. The author, Paul C Daiber, is a mechanical engineer who designs and services combustion turbine engines. Outside professional employment, the engineer chose to solve this puzzle: Define infinity by finding an inertial frame of reference in which a blue photon has a faster speed than a red photon.

An engineering spec/procedure was written for developing new theory:

1) <u>Abstractions</u>. How we understand our world.

   - <u>Definitions</u>. A person categorizes patterns in the physical, real, natural world by creating a definition for each identified pattern. Each definition is an abstraction. An abstraction is defined in words or other symbols for people to use.

   - <u>Ours</u>. Abstractions as definitions of categories are for our use. We define abstractions as we find useful or otherwise as we want. No external entity gives us our abstractions or defines our abstractions.

   - <u>Application and Ambiguity</u>. Definitions of abstractions have some degree of ambiguity. Definitions are valid only to the extent definitions are consistently applied to the real world by the definition's prescribed rules.

   - <u>Names</u>. Abstractions are each named. Each is accompanied by a counter-set of abstractions, each of which also get a name.

   - <u>Start Simple</u>. Begin development of definitions of new abstractions by first finding patterns that are obvious. Develop more definitions by finding less obvious patterns.

2) <u>Discovery</u>. A discovery is outside what can logically be derived from abstractions, used as axioms, we already have.

   - <u>Research</u>. Become familiar with what are assumed to be inherently valid abstractions by which we know our world.

- **Guess** a new definition by investigating an observation, or variations, extrapolations, combinations, tangents, violations, or alternatives.

- **Analysis.** Build a bridge of logic from old definitions to the guess or from the guess to old definitions by conforming the guess to logic.

- **Finish.** Write it up to communicate the discovered abstraction to other people. Definitions of abstractions can be communicated and comprehended only by physical, real, natural world example experiences other people have with that abstraction, through use of prescribed rules in the definition.

The method for discovering an abstraction was inspired from a passage in Morris Kline's book *A History of Mathematical Thought, Volume 1* (Oxford University Press, 1972) in which he paraphrased Descartes who said logic could not be used to investigate unknown fundamental truths. It seemed Descartes thought that thoughts from outside logic were needed when searching for something rules of logic were not designed to find.

Elsewhere in that chapter, Morris Kline said mathematicians at the beginning of the age of algebra developed theory without discipline or rigor. In contrast to that era four hundred years ago, algebra of today is full of discipline and rigor. If we were to rethink algebra enough to include infinity in applied mathematics, then, it seemed, we had to remove ourselves from today's discipline and rigor.

Being undisciplined, we grab what we want. We violate an axiomatic rule to create something illogical, because we really want it.

(The value of grabbing what is wanted was perhaps learned from Star Trek in the 1960's: Captain Kirk got what he wanted by leading his crew to do what he wanted, regardless of rules and regulations imposed on him. The intelligent, unemotional and exclusively logical first officer, Spock, could not lead by violating rules. Therefore, Spock could only be a support person to his captain.)

To solve the puzzle of blue and red photons, existing theories of physics had to be learned. Maxwell's Equations was learned in engineering school using Halliday and Resnick's textbook. Schrödinger's Equation had to be self-taught. Fortunately, Eisberg's and Resnick's textbook was on someone's bookshelf at work, and Gasiorowicz's textbook was found in a used-book store. Real treasure came from the author's dad's college textbooks, from the mid-1950's, which were intercepted on their way to the trash when the author was home from college. In those books was *Methods of Theoretical Physics Part I* by Morse

APPENDIX C – DISCOVERING AN ABSTRACTION

and Feshbach, McGraw-Hill Book Company, Inc., 1953. Unlike what was found on the internet, this old book's development of the Dirac Equation was written for engineers and easily learned.

Next, the Dirac Equation and Maxwell's Equations had to be placed into one algebra. The most effective algebra had been assumed to be geometric algebra in Chapter 6 of Hildebrand's textbook *Advanced Calculus for Applications* from college. But geometric algebra didn't work well. Quaternions were found better because of the ease of writing identities and tracking gauge invariance, and because electric and magnetic fields could be combined into one invariant having complex components.

Reading math textbooks became an enjoyable hobby. That might sound strange, but the author was recently with two old college roommates (Gator and Sonny) and they went straight to the math section when the three of them entered the Half Price Books bookstore in the (Walnut Creek, CA) square. Talking in Starbucks (coffee shop) that evening, it appeared, in general: People seek answers to the mysteriousness of the world around them by learning rational and logical techniques of math. Infinity was one such mystery.

What was actual infinity? Extensive reading did not find an answer. One of the best books on infinity was Aristotle's book *Physics*. He summarized what was suspected about infinity at that time, twenty-four hundred years ago. A generation or two prior to Aristotle, the Pythagoreans had proven irrationality of square-root-of-two, and that proof forced the Pythagoreans to admit that infinity, as something different from finite natural numbers, had mathematical relevance. The Pythagoreans disliked infinity because they attributed to infinity unwanted properties "mysterious, vague, and remote" (these three words are a remembered quote from a book that cannot be found), and, in general, the Pythagoreans disliked the contamination of something proven to be irrational in their system of rational deduction. Archimedes, too, disliked infinity (per his story *The Sand Reckoner*) as did most of the great men of science in ancient Greece, Aristotle included. The general dislike of positive actual infinity in ancient Greece grew on the author like good advice from a parent. The feeling was that actual infinity existed in our mathematics as a temporary substitute for a theory presently unknown or unavailable to us.

Early attempts to learn Cantor's theory of infinite sets were unsuccessful because Cantor's real numbers seemed to not include irrational numbers. Other people had seen this issue, with evidence this quote: "Cantor took N_1 to be the power of the continuum. This question, however, remains open, and for the present we see no trace of a path to its solution." (From *The World of Mathematics Volume Three* by Newman, 1956, Simon and Schuster, page 1599 in the article *Infinity* by Hans Hahn.) That book opened the door to seeking alternatives to what Cantor had specified to be the real number continuum.

One early attempt (by the author) gave an irrational number a positive infinity of place-value digits after the decimal point, and then unwritten and unspecified place-value digits after that. It was close to what became local-real numbers / practical real numbers, but at the time it led nowhere.

In the last two years of the project, the author's son's college pure mathematics textbook, written by Chartrand and others, was found in a box in the basement, and Cantor's theory could finally be learned as a full and complete theory. Once the bigger picture of what Cantor intended was learned, then his theory could be adapted into what was becoming the solution to the blue/red photon puzzle.

A second lesson from Cantor was that he grabbed a definition for infinity, and then attempted to build a bridge of logic to it. His process was good.

Before Cantor's theory was understood, at a time when there wasn't much hope of success, an idea for infinity came from the phrase "forever and a day". Reciprocal-of-zero was forever. To add one to it, hyperbolic trig functions were used: "$\cosh(1/0) = 1 + 1/0$" and "$\sinh(1/0) = \pm 1/0$". From that guess Maxwell's Equations were derived from the Dirac Equation. It was the first big break.

"$1/0$" meant the mathematics was invalid, but, but, there was truth down inside it, truth somewhere deeply hidden. Suspected truth became certain when the correct electromagnetic field force density was calculated using the complex conjugate. A burden of responsibility was felt, because, if a bridge of logic did not get built, then this discovery would likely become lost and forgotten, and that was because anyone who cared would not get to hear about it because any mathematics dependent on division by zero cannot be published.

Division by zero is famous as a tool of deception and is called "pseudo-mathematics". If division by zero could not be removed, then, most likely, the discovery would not be stumbled upon again for many, many years, perhaps for as long as it had already lay hidden, from when the Dirac Equation was understood (1930) to then (2010).

Formal prohibition against division by zero is in the definition of rational numbers, and that meant, maybe, "$1/0$" and "$0/0$" could be changed into the wanted infinity by investigating irrational numbers. "$\sqrt{2}$" was ripped apart every way possible. After some years the focus changed to "ζ" (zeta) in "$1 = \zeta^1 + \zeta^5$" and "$1 = \zeta^2 + \zeta^3$", which were found by equating lengths thirty degrees apart on the equilateral triangle spiral. A rational approximation to "ζ" was calculated for each of the two equations. Was the portion not-yet-calculated the same for both "ζ"s with regards to the geometric construction of the triangle spiral? It took many years to develop that question, and even longer to answer it.

APPENDIX C – DISCOVERING AN ABSTRACTION

It was guessed the two "ζ"'s could only be precise to a positive actual infinity quantity of place-value digits. Their difference was called real number zero. The unknown place-value digits starting at the actual infinity count meant real number zero was different from integer zero, and that meant it could be a denominator. The division reciprocal of a real number zero became the infinity that applied to the hyperbolic-angle for motion at the speed-of-light.

This was another moment of celebration. With division by integer zero gone, the theory seemed complete enough to actually tell someone. A downloadable early version of this book (titled: *Infinity Applied to Special Relativity*) was placed on the internet, to make it available to the public in case the project never finished. The book told a somewhat logical story from beginning to end but it lacked key essentials, for example, it lacked the exotic Lorentz Transformation needed to model motion at the speed-of-light. It wasn't advertised and no copies sold, and tallies showed only a person or two reviewed the first several free pages.

Real number zero became "local-zero" when Aristotle's potential infinity, what was thought of as a finite number, replaced positive actual infinity in the count of place-value digits. With no actual infinity in the numbers, there was worry the proposed system of numbers had deviated too far from what was generally accepted as numbers. At the very end of the project a new proposed axiom for how to divide by zero substantiated the removal of positive actual infinity, and the bridge of logic was completed.

Per the method, we guessed at an abstraction (which was the combination of Maxwell's Equations into the Dirac Equation using division by zero), and then ideas evolved until the core idea was found (which was the proposed new *reciprocal-of-zero axiom*). To communicate the discovery in a logical progression, per the method's last bullet, reverse activities (with the proposed new *reciprocal-of-zero axiom* and other changes to Axiomatic Set Theory explained first, and the derivation of Maxwell's Equations from the Dirac Equation last). But reversed activities in the textbook communication erased the process of discovery that was actually followed. To not lose that excitement, with its guesses, its stress, finding critical books, and its little victories, this "About the Author" was written.

Early versions for this math book had little more than an initial guess at what infinity could be. A guess cannot be the logical foundation of a math book, regardless of how strongly a person claims the guess is correct. Discipline came from Paul's brother Andy J Daiber, another engineer. Andy was/is a hobby mathematician, but Andy was not someone like the author who searched for playgrounds in numbers. Rather, Andy was someone who appreciated the serious core values pure mathematicians have. (Andy sent the book by Mary Tiles, the Teach Yourself book on group theory, Paul Cohen's *Set Theory and the Continuum Hypothesis*, and over the years he sent maybe a dozen other pure mathematics books to the author.) The author is grateful to Andy for reading this book every several years, when the book was incomplete and illogical. Andy mercilessly shot down every idea presented without substantiation. Our human tendency is to think our claims are correct, and to defend our claims when subjected to criticism. In mathematics, only logic can be a defense. When digging deeper to find fundamental evidence to prove one's claims, we stumble upon fantastic discoveries that are unexpectedly interesting and useful. That generalization of an interactive process summarizes the human side of the development of the proposed Theory of Non-Finite Numbers.

The attitude, or the sense of values, that fundamentally led to the need to write this book came from something said by a Citizenship in the Community merit badge counselor, Mr. O'Brien, paraphrased as: Don't bother reading something unless you intend to do something with that knowledge. Now, forty years later and no longer a child, this rule for living is changed to: Success in life is not measured by how much knowledge you take to the grave with you, but rather by what good you did with that knowledge.

I hope some good comes out of this effort to insert infinity into applied mathematics. What I really want is for your curiosity to have become ignited, and for you to further grow theory for non-finite numbers. And, I hope you enjoyed reading this book. Math is fun. \V/,

The Storybook

In its first draft, storybook *Alien Invasion Math Story* was a collection of essays and calculations written during the hunt for a useful infinity. The storyline, with its exaggerated emotional extremes, portrayed the confusion, frustration, misdirection, and stress felt in the hunt.

Inspiration came from the movie *The Race for the Double Helix* (BBC Horizon), in which Crick and Watson mixed incomplete bits of data (mostly from Franklin) with staggered inspiration. They persevered past criticism and little failures to win the race. Their moment of success came with a revelation.

Me sitting at a desk or standing in a store watching my wife shop while thinking through proofs is not nearly as exciting as Crick and Watson in their quaint apartments and labs, so my storyline is fiction. But at least the math and the essays on human will are real. And, for the revelation at the climax, well, you'll just have to read it. Please do. I tried to make it fun. \V/,

GLOSSARY AND INDEX

Glossary

<u>Actual Infinity</u>. An infinity larger than finite, that is beyond what has been counted.

<u>Aleph Null</u>. Cantor's version of an actual infinity that pertains to the quantity in the set of natural numbers, of integers, etc..

<u>Aleph One</u>. Cantor's version of an actual infinity that pertains to the quantity in the set of real numbers over any finite or infinite (aleph null length) length of the number-line.

<u>Algebra</u>. A set of rules by which number logic can be exercised.

<u>Algebra Field</u>. Formed by two groups and a distributive property. A set of elements and the criteria by which operations on those elements are performed.

<u>All-Number</u>. An expression formed from quantity numbers, label-numbers, the unspecified-speed-parameter, and the unspecified-label-number.

<u>Anti-Matter</u>. The negative of matter. If matter is a something, then anti-matter is the hole created by the removal of that something from nothing.

<u>Applied Mathematics</u>. Mathematics utilized in mathematical models of physics.

<u>Argument</u>. The independent variable on which a function operates.

<u>Axiom</u>. A fundamental assumption that is accepted without proof as a basis for subsequent derivations.

<u>Axiomatic Set Theory</u>. The mathematical theory of a set of axioms from which algebra is derived. ZFC axiomatic set theory is one form and is most popular.

<u>Biot-Savart Law</u>. The mathematical model of the electromagnetic field of a moving point particle.

<u>Both Plus and Minus But Also Neither Plus Nor Minus Separately</u>. ($\pm_N$) The plus or minus sign is unknown and unknowable. It is in contrast to the plus or minus sign ($\pm$) because the implication of the plus or minus sign is it is known or knowable.

<u>Cantor's Continuum Hypothesis</u>. A conjecture used as an axiom in Axiomatic Set Theory from which Cantor's real numbers are derived. It forces the existence of an actual infinity.

<u>Cantor's Theory of Infinite Sets</u>. A theory of pure mathematics in which an actual infinity is assumed to exist between finite numbers and the reciprocal of the integer zero.

<u>Cardinality</u>. The quantity of members in a set. A number is cardinal if it is the count of members in a set. This definition is part of Cantor's theory of infinite sets.

<u>Cartesian Grid</u>. The typical x-y plot. Descartes used the Cartesian grid to plot roots of polynomials.

<u>Cause and Effect</u>. The name for related events that are sequential in time.

SPECIAL ALGEBRA FOR SPECIAL RELATIVITY

Collapse of the Wave Function. When an observation of one particle by another particle occurs, the perception of the observed particle changes from being a wave that is spread through space and time to being a particle that has one location in space and time. This transition from being everywhere to being in one place is informally called the collapse of the wave function.

Complex-Conjugate. An alternate representation of a complex or hypercomplex number. The complex-label-number "i" is made negative (and the order of factors and terms is reversed but only after the quaternion simple-label-number factors in each term are reduced to only one quaternion simple-label-number).

Component. The factor in each term that is not a direction indicator. A component is often measurable.

Compound-Label-Numbers. Label-Numbers that have a factor for gauge invariance. For a single-term summation-form all-number expression of an invariant (like world volume), the compound-label-number is the same as a simple-label-number, and it is the number one or else is square-root-of-negative-one.

Conjecture. A statement that is not proven and so is not a theorem.

Conjugate. An alternative all-number expression. The conjugate form is used in multiplication of two all-number expressions.

Conservation Law. A mathematical expression that specifies how a material is created or destroyed.

Continuum. The locus of points that form a line. It's an intentionally ambiguous definition.

Contravariant. The alternative coordinate system representation for use with non-rectilinear or non-unit-magnitude geometric-unit-vectors, or, in general relativity, formed by a derivative operation.

Cosmological Model. Our logical model of the physical universe (plethora).

Count. A finite natural number.

Countable. A set that has a one-to-one member correspondence with the natural numbers, in Cantor's theory of infinite sets.

de Broglie Relations. Total energy equals the modified Planck's constant times the frequency. Total momentum equals the modified Planck's constant times the wavenumber.

Dedekind Cut. A cut in the number-line for which the high-side and low-side numbers at the cut, and numbers between them, are all "essentially" the same number. Per Dedekind: "From now on, therefore, to every definite cut there corresponds a definite rational or irrational number, and we regard two numbers as different or unequal always and only when they correspond to essentially different cuts."

Degradations. The break-down in symmetry for higher orders of hypercomplexity.

Denominator and Numerator. Fraction 5/7 has 5 the numerator and 7 the denominator.

GLOSSARY AND INDEX

Denumerable. Same as Countable.

Descartes. Mathematician and philosopher who marks the beginning of the modern era of intellectual thought.

Dirac Equation. The relativistic mathematical model for the dynamics of an electron.

Dirac Spinor Solution. Any of the four Dirac Spinor Solutions presented in this book.

Distributed Material. A material that exists as a distribution in space and time such that a gradient exists. Distributed material may also be called a "field", but "field" is so general a term it should be used cautiously.

Distributed Material Theory. The theory (in this book) for which continuum operators, such as the gradient differential operator, may be applied.

E-M-Compound-Label-Numbers. The label-numbers used for a six-term invariant.

Enabler Functions. Variables inserted into an equation to accomplish a purpose.

Energy-Momentum. The invariant that combines energy (as the time term) with momentum (as the space terms).

Engineering-Calculation Algebra. Algebra designed to be as efficient as possible in the task of making a calculation.

Entropy. The amount of disorder in the universe. The amount of entropy in the universe increases with time per models of physics we have so far reverse-engineered.

EPR Experiment. The EPR experiment verified, through the use of Bell's Inequality, that particle properties occur at the moment of observation and not at the moment of emission of the particle that is observed. As a secondary result there is the experimental verification of "Spooky Action at a Distance".

Euler's Equation. Relates the exponential function to the trigonometric functions.

Exotic Lorentz Transformation. A Lorentz Transformation that uses a hyperbolic-angle that is not a real (or rational) number. Exotic Lorentz Transformations take advantage of label-numbers in the all-number algebra.

Exponential Function. The single valued equivalent of base of "e" to an exponent. The exponential function has a polynomial expansion to infinity as its definition.

Factor. A factor in a mathematical expression is separated by multiplication and division signs, such that factors are multiplied. In contrast, terms are added.

Feynman. A theoretical physicist who contributed extensively to QED and to the theory of electrons and photons in general. Feynman's fame is perhaps due to his ability to present concepts through visualizations and explanations that were easily comprehendible by his intended audience.

Final-Result. The third step of the Process from Descartes of a local-real number algebra operation applies the condition that the result be a truncated number (with "Lmax" maximum count of known or knowable place-value digits before or after the decimal point). The final-result applies to what can be measured in geometric space, with examples being components of invariants and speed.

Finished-Calculation. The second step of the Process from Descartes of a local-real number algebra operation uses probability theory to find the "finished-calculation".

Finite. A finite number is a number from the set of natural numbers (defined by starting at one and repeatedly and unboundedly adding one), or a finite number is a number that is a result of a binary operation (addition, subtraction, multiplication or division) performed using a natural number or other finite number, per the algebra field for rational numbers. A finite (irrational) number is bounded high and low by two rational numbers.

Gauge. A reference against which a perspective is made.

Geometric-Unit-Vector. One of the four geometric entities that represent mathematically a direction in space or the direction of time.

Governing Equation. A mathematical model for a theory of physics.

Gradient Operator. The time and space differential operator.

Group. A set and the operations performed on that set, as a basis for unambiguous derivations using that set and those operations.

Handedness. The fingers curl with the geometric feature and the thumb points a direction for the geometric feature. If it is the right thumb, then the handedness is right-handedness. If it is the left thumb, then the handedness is left-handedness. In geometry the handedness is the correlation between the nomenclature of x, y, z relating to the fingers passing through positive x to positive y and the thumb pointing to positive z. In physics matter can have a geometrically right-hand or a left-hand spin but, more typically, a reverse of handedness refers to a reverse of parity to the extent that matter becomes anti-matter with the example being a right-hand glove (matter) has the same appearance as a left-hand glove that is turned inside out (anti-matter).

Hypatia. Woman teacher at the end of the Hellenic / Hellenistic era. She contributed to the mathematics of conics and had other contributions to science and mathematics. Her murder marks the end of the era of the thousand years of Greek intellectual thought.

Hypercomplex-Conjugate. An alternate representation of a hypercomplex number. The quaternion simple-label-numbers are made negative and the order of factors (and of terms) is reversed.

Hypercomplex-Plane. Similar to the Complex-Plane, but with one of the two dimensions imaginary rather than real. An illustration of time-space on the space-space of a sheet of paper. What is a circular rotation in time-space is a hyperbolic rotation in space-space and so hyperbolic-angles are illustrated in the hypercomplex-plane using circular angles and that substitution causes a distortion in the illustration of a hypercomplex-plane.

Hypothesis. An educated guess.

Identity. A mathematical statement that is true as a proven theorem and is stated generically without reference to a mathematical model of physics.

GLOSSARY AND INDEX

Identity Elements. The zero for addition and the one for multiplication.

Imprecision term "Ξ" (xi). Either a local-zero (small-scale imprecision) or else a local-infinity (large-scale imprecision). It is added to a rational truncated number to form a local-real number.

Incident Transmission Rate. The speed relative to an observer at which different locations of a single particle appear to communicate to each other.

Inertial Reference Frame. A constant speed reference frame. The word "inertial" is used because a reference frame that is accelerating because it is falling in a gravitational field may be approximated as a constant speed reference frame.

Infinitesimal. The division reciprocal of infinity. Infinitely small.

Infinity. Not Finished. No Finality. Unbounded. No limit. The loose definition permits a wide interpretation, with infinity including large ever-increasing finite numbers all the way up to division by zero. Beyond what can be measured.

Invariant. A mathematical expression for something that is physically real, such that the expression cannot change when the physically real something is observed from a different vantage.

Irrational Number. A number proven to not be rational in that it is not constructed as a ratio of finite natural numbers that are formed from one by adding one indefinitely.

Kinetic Energy. The mechanical energy due to particle movement.

Known or Knowable. Known or else knowable in that it may have a definite value.

Label-Numbers. Numbers of unit magnitude (or zero magnitude) that represent direction rather than quantity.

Length Contraction. The reduction in length when the object is moving relative to an observer. The front is at an earlier time than the back, and so length is reduced.

Local-Infinity. The sum of unknown and unknowable place-value digits left (before) the decimal point. Until we know better, it may be positive or else positive/negative. For visualization, it may be thought of as the division reciprocal of a local-zero.

Local-Real Number. The truncated number portion has the precision of the number limited to a finite maximum yet counted-to of place-value digits after the decimal point in base two. The same limit is before the decimal point for the truncated number portion. A local-zero or else a local-infinity is added to the truncated number to form a local-real number.

Local-Zero. Either "b" or else "$d = b - b$" for the value of each place-value digit after a count of "L_{max}" after the decimal point (in base two). "b" is zero or else one with the selection unknown and unknowable until an observation is made.

Lorentz Transformation. The method for changing component and compound-label-numbers (or geometric-unit-vectors) for a different inertial reference frame. It's an addition of hyperbolic angles.

Special Algebra for Special Relativity

Macroscopic. The large scale of objects.

Many Worlds Hypothesis. The repeated divergence to other universes through the observation of a quantum particle that makes a selection from a particle's probability.

Mathematical Model. A logical model for something physical. Typically, a mathematical model is used to make a prediction of a measurement or other definite observation.

Matrix Equation. An equation that uses a column vectors and uses a matrix as an operator.

Matrix Isomorph. A matrix equivalent of a label-number such that the set of matrices corresponding to a set of label-numbers has the same behavior in an algebra field.

Matter. The stuff of the universe. Typically thought of as fermions (electrons, protons, and other half-spin particles).

Matter-Wave. A wave associated with a particle per the mathematical models of Schrödinger's Equation and the Dirac Equation. A wave is the alternative existence of a particle when the particle is not being observed. The matter-wave interference pattern with itself forms the constructive interference group and the particle location is most likely at the peak of the group.

Maxwell's Equations. The mathematical model for the electric field, the magnetic field, and electric current. It's the set of first order differential equations that mathematically model the electric field and the magnetic field formed by a static or moving electric charge density field.

Maxwell's Wave Equation. The equation by which electromagnetic radiation is modeled.

Natural Numbers. The numbers formed by starting at the number one and adding one repeatedly and unboundedly. The set of natural numbers "$N = \{1, 2, 3, ...\}$" is the set of numbers used for counting. Natural numbers must each be finite because finite is defined as numbers so far counted-to.

Newton's Second Law. Force equals mass times acceleration, or, as Newton said it, force equals the time derivative of momentum.

Non-Linear. Has products of independent variables.

Number-Line. A geometric model of the real numbers. A number-line models the continuum.

Numerator and Denominator. Fraction 5/7 has 5 the numerator and 7 the denominator.

Observer. In Special Relativity component values are specified relative to the observer. In quantum mechanics particle properties are specific to an observer. An observer is another quantum particle, but, also, the observer particle is personified as a person, as us with our measurement equipment.

Octonions. Similar to the complex number factor but pertain to the seven directions of ultra-space.

GLOSSARY AND INDEX

Particle. An existence with location (and time), momentum (and energy), and angular momentum. A particle may extend over a region of space, but, typically, a particle is approximated as a point particle. Particles are typically a composite of smaller particles.

Pascal's Triangle. A pyramid of rows formed by adding the two numbers immediately above, after starting with one: 1, 1 1, 1 2 1, 1 3 3 1, 1 4 6 4 1, etc.

Pauli Spin Matrices. The Pauli Spin Matrices are 2x2 matrices used by Dirac in the development of the Dirac Equation. They are isomorphic with quaternions.

Place-Value Digits. Place-value digits are the integer numbers used before or after the decimal point. The value of the integer number is based on its placement relative to the decimal point, per a power series notation equivalent to place-value digit notation.

Polynomial. An algebraic summation in which the independent variable has an integer exponent in each term with the integer exponent typically non-negative. For example, the quadratic equation.

Potential Field. The gradient of the potential is a vector field.

Potential Energy. Energy due to location in a potential field. Examples are a gravity field, a pressure field, and an electric voltage field.

Potential Infinity. Aristotle considered the potential infinity to be a finite number that is either so large that its value is not relevant, or else a finite number that increases instantly and unboundedly. Subsequent thinkers might have moved it a little higher into being an actual infinity and so a person reading a text should not assume the potential infinity is finite.

Process from Descartes. The three-step process that begins and ends with geometric physical space and has all-number algebra for the analysis in the middle. It was pioneered by Descartes in what he called analytic geometry. It is the process of translating a geometric problem into the abstract for analysis and then translating the result back into geometry as a prediction for a measurement.

Product. The result of two factors multiplied by each other.

Property. A partial definition of something.

Pseudo-Mathematics. The hidden use of the reciprocal-of-zero to prove true something that is not true. "Pseudo" means "false".

Pseudo-Vector. A vector in a plane such that is depicts a rotation in the plane and not a direction perpendicular to the plane. In geometric vector algebra vectors cannot be planar but instead must be linear, and because a line points two directions out of a plane, a choice of direction must be made, and the need for a selection is depicted by the word "pseudo". Examples are rotation speed, torque, magnetic field, angular momentum, and area.

Pure Mathematics. A system of axioms and proofs. Pure mathematics is in contrast to applied mathematics because in applied mathematics the axioms must be applicable to models of our physical, real, natural world.

SPECIAL ALGEBRA FOR SPECIAL RELATIVITY

<u>Pythagorean Theorem</u>. The sum of the squares of the lengths of the two perpendicular sides of a right triangle equals the square of the length of the hypotenuse side.

<u>Quantity</u>. A quantity is different from a count because a quantity may be an actual infinity in value. A count is limited to being finite.

<u>Quaternions</u>. Similar to the complex number factor but pertain to the three directions of space. These each square are to negative one, except for the number one, which squares to one, and is the fourth quaternion. Proposed by Hamilton in 1843.

<u>Rational Numbers</u>. The numbers formed as ratios of finite integers. Division by zero is excluded.

<u>Real Numbers</u>. Real numbers are not imaginary or complex (per Descartes). Per Cantor, real numbers have a positive actual infinite quantity aleph one over a finite or infinite (aleph null) interval of the number-line and real numbers have all zeros before the non-zero place-value digits before the decimal point. Idealized real numbers (defined in this book) have integer zero interval one to the next and so have quantity 1/0 over an interval. Practical real numbers (defined in this book) have a finite count of place-value digits before and after the decimal point, with finite magnitude imprecision added separately on both the large-scale and small-scale.

<u>Reciprocal-of-Zero Axiom</u>. Proposed in this book. A new proposed axiom for numbers that has its basis in the proof of irrationality of a logarithm. The purpose of this axiom is to bring axiomatic set theory into applied mathematics.

<u>Reciprocal of the Integer Zero</u> exists as an abstract concept that represents the largest magnitude of numbers and is both positive and negative.

<u>Recursive</u>. A recursive process has repeated steps. For example, the natural numbers are created through the recursive process of repeatedly adding one after starting at the number one.

<u>Remnant-Product</u>. The real portion of the product of three four-component time-space invariants.

<u>Reverse Parity</u>. The complete reversal of a particle, as if the particle were a glove and the glove was inside-out. A particle that is reverse parity of matter is anti-matter and has opposite spin and electric field.

<u>Rest Mass</u>. The hyperbolic-radius of the energy-momentum invariant. Rest mass is thought of as the stuff of our macroscopic world, but there probably is no stuff of rest mass. Rather, we can think of rest mass as the fields subject to induction inside a subatomic particle.

<u>Root</u>. The root of an equation is a variable value for which the equation is satisfied. The term "square root of two" is the positive root of the equation "$x^2 - 2 = 0$".

<u>Scalar</u>. Having only one term. The time component of space-time vectors is often called a scalar. Also, one-dimensional invariants are scalars.

<u>Schrödinger's Cat</u>. Until we open the box, we do not know if Schrödinger's Cat is dead or alive: Schrödinger's Cat is a cat in a box with a glass flask of poison and with a

GLOSSARY AND INDEX

radioactive atomic nucleus. If the nucleus decays, then the emitted particle breaks a container of poison and the cat dies. An observer outside the box does not know if the nucleus has emitted a particle to break the flask and kill the cat (state = 1), or hasn't (state = 0). Relative to the observer the cat is both dead and alive as a state that is both and so is neither dead nor alive separately. The decay both happened and did not happen until the decay is observed by the box being opened. The state of the cat becomes known relative to the observer when the box is opened.

Schrödinger's Equation. The non-relativistic predecessor to the Dirac Equation as a model for the dynamics of an electron using a wave as the intermediate observed form of the electron. It is a second order ordinary differential equation and so is quite different from the first order partial differential equation set by Dirac. Often the frequency factor can be separated away, which makes Schrödinger's Equation very convenient to use.

Sedonions. Similar to the complex number factor but pertain to the fifteen directions of ultra-ultra space. Sedonion algebra is not valid in applied mathematics because it gives ambiguous results depending on the selection of operations, and therefore does not have inverse operations. Either it requires restrictions on its use, or it must be reformulated.

Simple-Label-Numbers. Label-Numbers that do not have a factor for gauge invariance.

Singular-Label-Number. A label-number of zero magnitude. A label-number that results in zero when multiplied by its hypercomplex-conjugate. It cannot be in a denominator.

Singularity. A singularity is a point, but the singularities of interest are where there is a division by zero. In physics, it is where a division by zero occurs in a theory for a phenomenon, with the popular example being the center of a black hole of General Relativity. Motion at the speed-of-light is a singularity for special relativity.

Six Dots. Repeats as the zeros repeat after the decimal point of an integer. The items represented by six dots are of a quantity beyond what can be counted, and, therefore, are addressed as a quantity in bulk. The quantity is not finite and is assumed to equal the reciprocal of integer zero.

Space-Like Invariant. A four-term (or two-term) invariant that has only a space term for the case of zero speed.

Space-Negative is the operator by which the Lorentz Transformation is the inverse matrix of the normal Lorentz Transformation, and, to compensate, the three-space label-numbers are negative. It is normally used for the gradient operator and normally used by convention for anti-matter. "sn" post-superscript is an explicitly written marker that identifies an invariant or other mathematical entity as requiring an inverted matrix for the Lorentz Transformation and as requiring a negative on the quaternion compound-label-numbers (that is, on the space label-numbers and not on the time label-number).

Special Theory of Relativity. Einstein's theory for there being no preferred inertial (constant speed) reference frame combined with the condition the speed-of-light be the same for every inertial reference frame.

Speed-of-Light. The speed of light over long distances in a vacuum. It is a unit conversion factor from time measurement units to space measurement units.

SPECIAL ALGEBRA FOR SPECIAL RELATIVITY

Spin. The property of a particle by which it has angular momentum. The spin of a particle is in units of Planck's constant. Spin may be positive or negative.

Spooky Action at a Distance. Einstein's name for the coordinated polarization (or spin) of two macroscopically separated entangled photons (or electrons) of the EPR experiment. This is Einstein's name for the particle properties of two entangled properties to become instantly specified at two locations that are a macroscopic distance apart.

Sum. The result of terms added.

Term. A term in a mathematical expression is separated by plus and minus signs, such that terms are added.

Theorem. A mathematical statement that is proven.

Theory. In physics a theory is a logical or otherwise mathematical model that is generally accepted as valid for predicting measurements within set limits. In mathematics a theory is a branch or subset of mathematics

Theory-Development-Algebra. Algebra designed to explicitly represent in symbols the subtle aspects of the physics being modeled and designed explicitly to be as abstract as possible.

Theory of Non-Finite Numbers. Proposed in this book. The theory of numbers based on the proposed new reciprocal-of-zero axiom and other modifications to Axiomatic Set Theory in which practical real numbers have an unknown and unknowable portion.

Theory of Special Relativity with Non-Finite Numbers. Proposed in this book. The application of the proposed Theory of Non-Finite Numbers to the existing Special Theory of Relativity.

Three Dots represent the largest number yet counted-to using finite natural numbers. Each and every item counted is individually addressed through the mechanism of the count, which contrasts with six dots.

Time Dilation. The reduction in the rate of a clock's mechanism when the clock is moving relative to an observer.

Time-Like Invariant. A four-term (or two-term) invariant that has only a time term for the case of zero speed.

Time-Space. Also called "space-time". It refers to the union of time as a fourth dimension added to the three dimensions of space.

Trigonometric Function. Sine and cosine and other functions, including hyperbolic functions and inverses, as derived from the exponential function.

Triple-Vector-Product. The imaginary portion of the product of three four-component time-space invariants.

Truncated Numbers. Rational numbers of finite limited place-value digits both before and after the decimal point.

GLOSSARY AND INDEX

Uncountable. A set that has a quantity of members that cannot be counted because the quantity is greater than the quantity of members in the set of natural numbers, per Cantor's theory of infinite sets.

Unknown and Unknowable. The value cannot be known because the value does not exist relative to the observer. The analogy is Schrödinger's Cat.

Unspecified-Label-Number. A simple-label-number that is unknown and unknowable but is restricted to being one of the three simple-label-numbers and not a combination of them. The unspecified-label-number "κ" (kappa) is unknown and unknowable and is restricted to being exclusively one of the simple-label-numbers "q_x", "q_y", or "q_z".

Unspecified-Speed-Parameter. A hyperbolic-angle that is unknown and unknowable. It differs from an independent variable "x" as an unknown because "x" is unknown but is knowable. The unspecified parameter "ς" (sigma or esse) specific to use in compound-label-numbers as a hyperbolic-angle ensures relativistic gauge invariance (that is, it ensures there is no preferred inertial reference frame).

Vector. Having multiple terms that are in a specific order.

Voltage. The potential for the electric field. The voltage as a vector-space field includes the potential for the magnetic field by use of a cross product operation.

Wave Group. A wave group is a constructive interference location of a set of individual waves. The wave group moves at a speed that may be different from the speed of the crests of the individual waves, if the waves themselves have different wave crest speeds.

World Line. The path of trajectory of a particle or other object on the hypercomplex-plane or on the four-dimensional version of the hypercomplex-plane. Or, in general, the path of a particle through four-dimensional space-time with that world line a continuum of time and space coordinates. It is typically time-like, at least for particles with mass.

ZFC Set Theory. The comprehensive Axiomatic Set Theory used in mathematics, evolved from Cantor's theory of infinite sets and other beginnings. Uses specific axioms as its basis. A field of pure mathematics

Special Algebra for Special Relativity

GLOSSARY AND INDEX

Index

aleph, ix, x-xiv, 185-198, 287
Ampere, 92-3
android, 72
angular momentum, 56, 84-6, 88, 158, 166, 181, 241-3, 271
anti-associative, xii, 7-14, 254-270
anti-commutative, xii, 7-14, 173, 223, 254-270
anti-matter, 65-72, 75, 81, 84, 88, 129, 154, 159-83, 248, 250, 287
Archimedes, 283
area differential operator, 145
Aristotle, vii, xi, 283, 285, 293
axial vector, 19-20, 23
axiom of infinity, xii, 185, 192-194, 199
axiomatic set theory, ix, x-xii, xvii-xviii, xxiii, 185-187, 192-198 , 240, 285, 287

beautiful, 107, 213, 154, 240
Bell's Inequality, 271-280, 289
Bianchi, Eugenio, 249
Biot-Savart Law, 116-8, 124, 153, 287
black hole, 210, 247, 249, 295
Bronstein, Matvei (Planck length), 249

Cantor, viii-xiii, 185-190, 198, 283-4, 287
Cantor's Continuum Hypothesis, ix-xii, 186, 191-2, 195, 199, 244, 246, 248, 287
cardinality, 190, 287
Cohen, 192, 193, 286
collapse of the wave function, 72, 88, 280, 288
complex numbers, 4, 5, 7, 10-11, 23, 35-36, 255-256, 262, 288
contravariant, 18, 26-28, 78, 288
cross-product, 3, 15, 26

de Broglie, 158, 217, 288
Dedekind, xiii, 187-9, 192, 196, 288
denumerable / countable, xi, xiii, 186, 188-91, 195, 288
Descartes, 1, 7, 249, 282, 287, 289
determinant 20
differential geometry, 78
dot-product, 3, 15-18, 26, 77

Einstein, viii, 35, 38, 47, 59, 89, 186, 247, 271
E-M-compound-label-numbers, 96
enabler functions,160-3, 226, 289
Entropy, 88, 247, 289

SPECIAL ALGEBRA FOR SPECIAL RELATIVITY

EPR Experiment, 88, 209-210, 250, 271-80, 289, 296
exponential function, xxiii, 1, 4-5, 21, 23, 44, 95, 135, 203, 251, 291

Faraday, 91-2, 154
Feynman, 65, 241, 289
final-result, 1, 203, 205-8, 217, 243, 289
finished-calculation, 1, 203, 290
finite, 7, 290
force, xxiii, 2-3, 68-9, 121, 130-142, 233, 241, 292

Galileo, viii, xi, 38,
Gauss, 91
Gödel, 192, 193

Hamilton, 4, 91, 253, 294
Hawking / Bekenstein, 249
Heaviside, 3, 89, 117
Hurwitz, 268
Hypatia, 38, 290

imprecision term "Ξ" (xi), 204
irrational, viii, xii-xiv, 7, 188, 193, 198-9, 244, 248, 283-4, 291

Kline, Morris, 282

length contraction, 47-9, 129, 149, 211, 242, 291
local-real numbers, xvii, 225, 245, 291
local-zero, xv-xvii, xix, 200, 202, 244 , 291
local-infinity, xvi-xvii, xix, 201, 291
Lorentz Transformation, xvii-xx, 35, 44-7, 107-116, 173-8, 291
Lorenz Condition, 98, 99, 104, 105-6, 154

magnetic moment, 241
matrix isomorph, 5-6, 14, 20, 34, 163, 221, 256, 292

non-Relativistic, 43, 74, 155, 295

Pascal's Triangle, 19, 23, 142, 293
Pauli Spin Matrices, xx, 6, 34, 163, 293
Planck's constant, xx, 158, 213, 243, 249
polar vector, 19, 23, 32
Poynting Vector, 120, 127, 134-5
Process from Descartes, 1
pseudo-vector, 20

QED, 241
quaternion hypercomplex numbers, 4-31, 253-63, 294

rational numbers, 7, 188, 294
real numbers, 195

GLOSSARY AND INDEX

reciprocal-of-zero axioms, xiii-xiv, 185, 197-8, 294
remnant-product, 16-9, 100-2, 294
Rovelli, Carlo, 250

Schrödinger, xv, 167, 200, 282, 292, 294-5
six dots, xv, 195-6, 295
space-like, 49-53, 54-56, 57, 88, 183, 211, 295
space-negative, 72-83, 110, 144, 155, 159-60, 165, 224, 227, 295
spin, 56, 84-6, 128, 166, 181, 241-3, 249, 271-81, 296
spiral waves, 126, 164
Spooky Action at a Distance, 59, 209, 271-81, 289, 296
super-potential, 105-7

tau particle, 123
Theory of Non-Finite Numbers, xvi, 240, 286, 296
Theory of Special Relativity with Non-Finite Numbers, 243, 296
Tiles, Mary, 192
time dilation, 47, 296
triple-vector-product, 16-9, 23-32, 101, 154, 296
truncated numbers, xvii, 201, 203, 205, 243, 245, 296

uncountable sets, 190, 297
unspecified-label-number, 94-5, 297
unspecified-speed-parameter, 39-40, 88, 297

vector-front multiplication, 223, 258
vector identities, 101, 103, 150, 154
vector-space, 263
volume, 17, 142

Wolchover, Natalie, 192
World-volume, 17, 142

Back Cover

Axiomatic Algebra Without Actual Infinity

This book broadens applied mathematics from Group Theory to Axiomatic Set Theory by removing actual infinity, because removing actual infinity removes paralyzing holes in logic. Transitioning Axiomatic Set Theory from temporary, experimental pure mathematics into useful applied mathematics completes the formalization of real numbers started by Cantor 150 years ago.   \V/,

- Modify Cantor's Continuum Hypothesis by replacing countable infinity with an ever-increasing finite counting natural number and by replacing uncountable infinity with reciprocal of zero, so that positive actual infinity cannot be a quantity of a set.
- Replace "Don't Divide by Zero" of rational numbers with two axioms that permit $1/0=7+1/0$, $1/0=7*1/0$, and $2^{1/0}=3^{1/0}=1/0$ or 0, but not 0/0, and have $1/0$ the quantity of a set.
- Remove Axiom of Choice so that only bulk operations, like truncation to form an integer, apply to $1/0$ quantities. No individual operations.
- Real Numbers have only a finite count of knowable place-value digits both before and after the decimal point. Left of the string of zeros unknowable digits form large-scale imprecision which is akin to infinity but can be analyzed using finite numbers.
- Special Relativity. A Lorentz Transformation adds large-scale imprecision to the time-space hyperbolic angle to derive Maxwell's Equations from the Dirac Equation. Particle properties conform to measured electromagnetic field force and energy.
- That successful application of the axioms to discover a quantum model for 100+ year-old electromagnetism substantiates the new axiomatic algebra is correct and is available for modern in-development theories of physics.

Although the math being depicted is quite deep and fundamental, symbols used in the book are nothing beyond what high school students learn in their most advanced math classes.

Perhaps what is published in this book should be placed in a technical journal. It isn't in a journal because the author is not familiar with journals. Paul C Daiber is an engineer and not in academia. As a hobby he had dug into infinity, a feature of math and science that greatly troubled him like it has troubled many of us,

GLOSSARY AND INDEX

and he thinks he has fixed the trouble by restructuring infinity so that it fits into logic. He placed the fix into this book to share it. On Amazon's free pages please read a summary of what he discovered, and if you want more details there's the rest of the book. \V/,

Previous Back Cover

Algebra for applied mathematics without positive actual infinity

Maxwell's Equations unite with the Dirac Equation to combine electron dynamics with photon dynamics because, by use of the algebra, an electron projects itself as a photon. Electron/photon double existence derives from Schrödinger's Cat because each place-value digit of a real number beyond a finite maximum in count is unknown and unknowable, analogous to the cat being both alive and dead inside its unopened box. The algebra is derived from a proposed axiom that replaces Cantor's Continuum Hypothesis.

Empirically derived energy density and the Poynting Vector unite in the force density invariant as one mathematical model. That unity suggests quantities in our geometric world actually do have finite imprecision, and that the new algebra applies to more modern theories of physics.

Visualizations and exercises help comprehension.

The mathematics is simple enough to be understood by a high school student who has taken first year level college math and physics classes (and is familiar with trigonometry and logarithms, complex numbers, matrix multiplication, geometric-unit-vectors, and partial differential equations).

One particle at two places violates a preconceived notion that that isn't possible. The one particle is material (fermion electron) and, its opposite, force (boson photon). Take this radical notion further by supposing perceived reality results from numbers inside exponential functions, alone from objects, interacting by becoming more precise with respect to each other, to form patterns we see as the Dirac Equation and other mathematical models of physics. It seems the universe is fundamentally numbers.

Special Algebra for Special Relativity

www.ingramcontent.com/pod-product-compliance
Lightning Source LLC
Chambersburg PA
CBHW020628220526
45464CB00001B/59